AF784128

Future of Business and Finance

The Future of Business and Finance book series features professional works aimed at defining, analyzing, and charting the future trends in these fields. The focus is mainly on strategic directions, technological advances, challenges and solutions which may affect the way we do business tomorrow, including the future of sustainability and governance practices. Mainly written by practitioners, consultants and academic thinkers, the books are intended to spark and inform further discussions and developments.

Peter Wollmann • Doris Pemler
Mersida Ndrevataj
Editors

The Sustainable Organization – Volume 3

Academic, Cross-Sectoral, and Industrial Pathways to Achieving the UN SDGs

Springer

Editors
Peter Wollmann
Bonn, Germany

Mersida Ndrevataj
IUAV University of Venice
Venice, Italy

Doris Pemler
Ciudad Condal GmbH
Zürich, Switzerland

ISSN 2662-2467 ISSN 2662-2475 (electronic)
Future of Business and Finance
ISBN 978-3-032-28948-3 ISBN 978-3-032-28949-0 (eBook)
https://doi.org/10.1007/978-3-032-28949-0

© The Editor(s) (if applicable) and The Author(s), under exclusive license to Springer Nature Switzerland AG 2026

This work is subject to copyright. All rights are solely and exclusively licensed by the Publisher, whether the whole or part of the material is concerned, specifically the rights of translation, reprinting, reuse of illustrations, recitation, broadcasting, reproduction on microfilms or in any other physical way, and transmission or information storage and retrieval, electronic adaptation, computer software, or by similar or dissimilar methodology now known or hereafter developed. Unless otherwise expressly agreed by the Publisher in writing, the Publisher does not authorise use or reproduction of all or any part of this work in any manner for the purpose of training, fine-tuning, evaluating, or developing artificial intelligence models, technologies, or systems. In accordance with Article 4(3) of the EU Directive 2019/790 on copyright and related rights in the Digital Single Market, the Publisher expressly reserves all rights with respect to this work under the text and data mining exception of the directive.
The use of general descriptive names, registered names, trademarks, service marks, etc. in this publication does not imply, even in the absence of a specific statement, that such names are exempt from the relevant protective laws and regulations and therefore free for general use.
The publisher, the authors and the editors are safe to assume that the advice and information in this book are believed to be true and accurate at the date of publication. Neither the publisher nor the authors or the editors give a warranty, expressed or implied, with respect to the material contained herein or for any errors or omissions that may have been made. The publisher remains neutral with regard to jurisdictional claims in published maps and institutional affiliations.

This Springer imprint is published by the registered company Springer Nature Switzerland AG
The registered company address is: Gewerbestrasse 11, 6330 Cham, Switzerland

If disposing of this product, please recycle the paper.

Contents

About the Editors

Peter Wollmann has been a responsible manager, initiator, mentor and facilitator in large, predominantly global transformations and strategic developments for almost 40 years in large organizations and since 2017 as independent consultant. He is also a member of an advisory board of a R&D oriented company. Peter holds a degree in mathematics and physics from the University of Bonn. He is the author and publisher of numerous books and articles on strategy, leadership, organization design and project and project portfolio management.

Doris Pemler is an entrepreneur, artist and researcher based in Zurich. She received her M.A. in Egyptology, Sudan Archaeology and Archaeology of the Ancient Near East from Humboldt University zu Berlin and FU Berlin. Her research focuses on iconography, gender and intercultural communication in Ancient Egypt and Sudan. As a photographer she contributed to several exhibitions in Germany and Switzerland. As CEO of Ciudad Condal GmbH she leads an aspiring company focused on import and distribution of high-end sustainable products in Switzerland

Mersida Ndrevataj is an urban scholar with a background in architecture, urban planning, and environmental psychology. She holds a PhD in Urban Planning and Public Policy from IUAV University of Venice, where her research focused on the implementation of public policies in cultural heritage management. She is currently a lecturer at IUAV University of Venice and collaborates with several COST Action research networks focused on urban studies. Her work brings together academic research and field-based inquiry by adopting audiovisual and sensory approaches to investigate urban spaces, everyday socio-spatial practices, and lived experiences.

"Experience is one thing you can't get for nothing"
—Oscar Wilde

Part I

Introduction

1 The Framing of the Book: Why a Second Volume Is Needed and How It Reflects Current Developments of Sustainability and UN SDG Coverage Activities

Peter Wollmann, Doris Pemler, and Mersida Ndrevataj

Abstract

It is important to know that the second and the third volume of The Sustainable Organization were created at the same time and in close coordination by a joint author community and with a joint general framing which means that the two books complement each other and refer to each other. In this first chapter of the third volume, the general direction and framing of the book will be explained, its close thematic connection to the second volume, its necessity after the publication of the first and the second volume, its purpose and its special focus stressed and further considerations, additional and complementary to the second volume, on the current state of the broad-based Movement represented by the nearly infinite Activity Landscape to reach Sustainability and UN SDG Coverage recapitulated and continued. To guarantee the same framing for the third volume as it was presented for the second volume, the facts on perception traps for judgment intensively stressed in the second volume are taken up again and the book's structure is explained. For this, the general linkage between the first, second, and third volume is described which includes the continuation of using the same terminology, the same underlying thinking styles represented by the used joint "language" for the explorations and evaluations (the so-called Three-Pillar Model, see: https://www.peterwollmann.com/three-pillars), the same categoriza-

P. Wollmann (✉)
Bonn, Germany
e-mail: pw@peterwollmann.com

D. Pemler
Ciudad Condal GmbH, Zürich, Switzerland
e-mail: doris@ciudad-condal.ch

M. Ndrevataj
IUAV University of Venice, Venice, Italy
e-mail: mndrevataj@iuav.it

© The Author(s), under exclusive license to Springer Nature Switzerland AG 2026

P. Wollmann et al. (eds.), *The Sustainable Organization – Volume 3*, Future of Business and Finance, https://doi.org/10.1007/978-3-032-28949-0_1

tion of the activities and actors of the Movement, the activity landscape, as in the previous two books, which makes a brief recap in this chapter necessary. In the context of the recap, the fundamental connection between the term "Sustainability" from a UN point of view and the UN SDGs, the Sustainable Development Goals (https://sdgs.un.org/goals), is touched again. To complete the second volume's rough evaluation of the current state of the Movement's nearly infinite landscape of activities and projects to reach sustainability for organizations, regions, countries or for the world as a whole and to cover the UN SDGs, some new significant use cases are discussed and some use cases from the last book recapitulated to achieve a comprehensive picture. The interview of a former high-ranking UN official from the second volume is mentioned again because of its importance as well as the discrepancy between the public perception of the Sustainability Movement and its actual state—and some neurological reasons for the occurring perception biases (already briefly touched in the first two books in chap. 1 of each). As a result, the cautious optimism regarding the Sustainability Movement's (or its Activity Landscape's) unbroken momentum is further substantiated and the stage prepared for the numerous detailed use cases described in the following chapters of the third volume. The book again will not at all strive for a detailed systematic and comprehensive overview but try to give a taste of the dimension of the Movement's Activity Landscape to reach Sustainability and UN SDG Coverage and the reached status. Additionally, an outlook on the upcoming Volume 4 at the end of the decade as well as on further related book projects in the future is provided.

1.1 Introduction: Reasons and Background for a Third Volume[1]

This book is the third volume of "The Sustainable Organization—How Organizations Address the 17 UN SDGs Using the 3-P-Model," published by SpringerNature in June 2025[2] and the successor of the second volume to be published in September 2026.[3] This third volume was created at the same time and in close coordination with the second volume, and the authors of the second and the third volume acted as joint author community. The two books therefore have a joint general framing which means that the two books complement each other and refer to each other. Thus, the third volume is also based on the same methodical setting as the first and second book, which includes the clarification and origin of key terms and concepts like Sustainability, UN SDGs, UN SDG Coverage, and the Three-Pillar Model[4]

[1] Parts of the introduction quoted almost verbatim from the previous book, "The Sustainable Organization" Volume 2, s. https://link.springer.com/book/9783032238016

[2] Wollmann, P., Pemler, D., Ndrevataj, M.: The Sustainable Organization—How Organizations Address the 17 UN SDGs Using the 3-P-Model; Cham 2025, SpringerNature (https://link.springer.com/book/10.1007/978-3-031-89549-4)

[3] https://link.springer.com/book/9783032238016

[4] See: https://www.peterwollmann.com/three-pillars)

(details see below) as the leading thought model. As the previous two books, the third volume understands the nearly infinite Landscape of Sustainability and UN SDG coverage activities with all its acting players and affected stakeholders as a sort of an overwhelming Whole World Transformation Project Portfolio with millions of single activities—we therefore name it the Sustainability and UN SDG Coverage Movement.

As already underlined in Volume 1 and 2, a comprehensive and detailed overview of this complex—partly top-down driven,[5] partly bottom-up driven[6]—movement is impossible in the scope of three or even more books; therefore, we continue following our concept and goal from the first two books to provide for the reader solid meta-level ideas (of parts) of the Movement's activity landscape, its complexity and its former and current momentum by describing some few fundamental high-level considerations and some interesting and insightful activities of key players including their related individual case studies—without claiming in any way to be exhaustive. As stated in the second volume, we are fully aware that the Movement's situation is contradictory and confusing,[7] not only in terms of public perception[8] (and in media) but also in terms of content. We also know that supposedly good and bad news occur in a mix that is difficult to assess, and that it is demanding to form a uniform, consistent opinion about where the Movement actually stands and what its chances of success are—especially since people tend to overestimate bad news in their perception (more on this later in part 1.2.4). The aim here—as in the first and the second volume—is rather to create a basic, flexible understanding on a meta-level, more from top-down, and to underpin this with some case studies (see 1.2.2) and therefore to generate curiosity and openness for further research and evidence.

In the first two books we documented in the first chapter of each some key messages for our readers, based on our evaluation, knowledge, and insights as of 2024[9]—and it might be interesting in the third volume to challenge these messages again and to check if they are still valid:

- There is a broad movement of activities and actors in place.
- There is a lot more progress already achieved than perceived in the public perception.

[5] For example, as an impact of the UN SDG planning and reporting system

[6] For example, also many grassroots activities (https://en.wikipedia.org/wiki/Grassroots), often according to the grassroots organizing model of Marshall Ganz (https://en.wikipedia.org/wiki/Marshall_Ganz)

[7] As also the situation of the world as a total: in an obviously very confusing state, perhaps at a turning point with many contradictory and conflicting developments occurring and where a "new clear pattern" is not (yet) foreseeable (a typical Terra Incognita Situation).

[8] It has nevertheless to be stated that there is a systematical underestimation of movements in place, see, for example, the discussion of the so-called 3.5% rule for the success of Movements, which nonviolently protest against an authoritarian government (see: https://en.wikipedia.org/wiki/3.5%25_rule), which shows positive and negative use cases. As there are examples supporting the hypothesis, the statement that the impact of the Movement might be underestimated can be regarded as proven.

[9] Cited from chap. 1 in Volume 1, https://link.springer.com/book/10.1007/978-3-031-89549-4 and from chap. 1 in Volume 2, https://link.springer.com/book/9783032238016

- In general, there is a perception bias in the judgment of large transformational movements as people tend to expect linear developments (also as a consequence of neurological imprints).
- The Movement is to a very large degree well rooted in the middle of society and irreversible.
- The Movement will force organizations' (and private persons') journeys through unknown territories which requires the mindset of a so-called Travelling Organization (more details see below).
- The coordination and cooperation across activities, projects, and transformations is a Herculean task and can only be handled in an explorative and agile way.

The second, third, and fourth statements were already challenged in the second volume and their validity confirmed. This result will not be questioned in the third book. On the contrary, we will complete this picture in this book regarding especially some new (political) developments in the first months of 2026 (first comments see sections 1.2.1–1.2.5). We still assume that the other statements are less controversial. We will run the full overall resumé in the last chapter (Conclusions).

Another reason for a second and now for a third volume—already mentioned in volume 2—was that we experienced after the publication of the first volume a significant increase of our networks by new players in this field offering many so-far unknown use cases from different fields and parts of the world, broadening our perspective and focus tremendously.

Overall, we might sum up the following key reasons for a third book (which are similar to the reasons for the second book):

- Overwhelming mass of new material, new exciting use cases and considerations arising after the publication of the first book, complementing the key messages of it—the volume exceeded the capacity for one book so that we created in parallel a second and a third volume.
- Significant number of new authors joining the author community and contributing new insightful use cases and insights.
- Urgent need to check the validity of our statements, outcomes, and insights from the first book.
- New insights on the current intensive controversial discussion on the state and future development of the momentum of the Sustainability and UN SDG Coverage Activity Landscape—is there a serious setback or is the development irreversible and unstoppable long term?
- Some new additional fundamental considerations on the difficulties to evaluate and judge historical (political, societal, economic) disruptions and connected developments/transformations and their underlying neurological dispositions.

Thus, in our third book we will—as well as we did in the second volume—try to find an additional convincing (but always preliminary) answer to the question about the situation of the Sustainability and UN SDG Coverage Activity Landscape from above: is there a serious setback or is the development unstoppable long term. To anticipate: despite all the adversity, the newly gained pieces of the puzzle unexpectedly show—contrary to public perception in the media—a clear and unbroken momentum, a growing gap between what is officially happening at the political level and what is actually taking place in the private and public sectors.

We underlined in the second volume that this is plausible also on the second view as it seems impossible for such a complex, diverse system to be controlled by just a few individual players[10]—even if simplified linear thinking suggests otherwise (more on thinking and perception traps see below under 1.2.4)—a lot of control also comes from the bottom up, from the sheer multitude of activities. We will elaborate on this in more detail later on, but it might already here we stated that interestingly, current developments in the Middle East have led to the decarbonization movement unexpectedly regaining momentum, clear evidence that history does not unfold linearly but dialectically, especially when new developments and constellations arise.

1.2 The Sustainability and UN SDG Coverage Movement's Activity Landscape and Its Momentum

1.2.1 Start with a Fundamental Debate Contribution from Florence Gaub[11]

In the second volume we discussed the widespread impression that we are living in unprecedented historic times—meaning times of great uncertainty and upheaval for us all. Historic upheavals and the developments they trigger with often long-term effects are frequently difficult to assess, particularly with regard to their further course and consequences. We stressed that it lies in human nature that most people find this very unsettling (we will discuss the underlying neurological dispositions in detail in the following sections). In this context, we introduced the following anecdote as food for thought to change the thinking perspective: During a conversation between Henry Kissinger and Zhou Enlai in 1972 in the course of the diplomatic rapprochement between the People's Republic of China and the USA. Henry Kissinger is said to have asked Zhou Enlai about his opinion on the impact of the French Revolution. "Too early to say," Zhou Enlai is said to have replied. Independently if Zhou Enlai really meant the French Revolution of 1789 which had occurred nearly 200 years before or wanted to comment on the more recent days of political disruption in 1968 in Paris and France, he was for sure expressing the long view of history in a very witty and Oscar Wildean way which was soon used as evidence that the Chinese (especially Chinese intellectuals and leaders) took the long view of things, that they were a patient civilization, and that, when they thought about the future, it was hundreds of years distant.[12]

[10] One could also formulate this as "fundamental system logics beats individual power and influence"

[11] Florence Gaub (born 1977) is a Franco-German researcher, security expert, and futurist who focuses on foresight-based policy formation for international relations and security policy. She is director of the research division at the NATO Defense College (Rome). Her contribution was published at the renowned German newspaper Süddeutsche Zeitung on April third, 2026, under the title (translated into English): "The Future Has Always Sucked" (see: www.sz.de/li.3457426)

[12] https://professorbuzzkill.com/2017/08/07/qnq-26-zhou-enlai/

Florence Gaub added some very insightful perspectives to this debate,[13] asking first if the world has ever been in such a bad shape like now in April 2026? And she gives already at the beginning of her contribution the answer that this is not the case as "humanity has always been pessimistic—and that's exactly why it's searching for better solutions." And she stresses that she regards her contribution as a "brief guide to optimism."

First, she returns to her initial question and explains that "just as today, every generation before us was convinced that things had never been as bad as they are now. Today, Peter Thiel proclaims the end of the future; in 1918, the German philosopher Oswald Spengler wrote an entire book titled 'The Decline of the West', in which he claimed that the age of great discoveries and cultural innovations was over. In the late 1940s, the British historian Arnold Toynbee sang the same tune, convinced that Western civilization was showing signs of creative exhaustion and intellectual decline. In the 1960s, Daniel Bell, much like Fukuyama's "End of History" later, proclaimed 'The End of Ideology' and later spoke in a manner very similar to Thiel today—lamenting the end of innovation. The doomsayers are wrong precisely because their fears trigger countermeasures.

The irony: Each of these works was followed by an era that proved exactly the opposite. The decade following Spengler's thesis of decline brought us television and radio, nuclear physics, antibiotics, aviation, synthetic materials, cinema, jazz, the Bauhaus, modernism, democracy, decolonization, and women's suffrage. Toynbee was refuted by space travel, computers, the birth control pill, the discovery of DNA, the beginnings of a global financial order, and the European Union. Bell was followed by the personal computer revolution, the biotechnology boom, and the explosion of the Internet—all driven by precisely that entrepreneurial—all driven by the very entrepreneurial vision he believed was dying out. And in the 15 years since Peter Thiel lamented the end of innovation, we have seen private spaceflight, new vaccines, artificial intelligence, solar and wind energy, and the smartphone. So, have these authors worked themselves up into a panic about the future for no reason? No—they are part of that general cycle of production that is, after all, the future. They are wrong precisely because their fears trigger countermeasures."[14]

We have rarely read contributions to the debate in which the dialectics of history have been presented so clearly and simply. Florence Gaub goes even further: "The future emerges something like this: humans—possibly the only species capable of vividly imagining a time that has never existed before—take cues from the past and present and construct the future as a space full of possibilities, both good and bad. This future can trigger two things: either the desire to make it a reality, or fear, if it appears undesirable. Where one lands on this emotional scale depends less on intellect or knowledge than on personality—and the feeling itself has only limited predictive power. Optimists tend to achieve better results, that is true; but a certain degree of fear—not too much, as that leads to paralysis, known as the Cassandra

[13] Süddeutsche Zeitung, April third, 2026, (see: www.sz.de/li.3457426). Quotes marked were translated into English using DeepL.

[14] ibid.

effect—can spur us to act. And this action then changes the course of the future. Put simply: The reason why the future ultimately turns out to be a better present than today is that we fear it."[15]

Florence Gaub mentions respective examples—history is full of examples of:

- "The grim predictions in Paul Ehrlich's 'The Population Bomb' (1968) foresaw millions of deaths from starvation well into the 1980s. But Ehrlich did not account for either technological or social progress—and both thwarted his scenario. Thanks to the 'Green Revolution', corn production quadrupled, and thanks to rising literacy rates, birth rates began to decline.
- Science fiction films like 'Soylent Green' (1973) and 'Animal House' (1977) depicted a future full of pollution and without an ozone layer, thereby triggering an avalanche of laws that not only saved the ozone layer but have also ensured that the air in most Western major cities is cleaner than it was in the 1850s.
- NATO spent decades imagining and preparing for the worst-case-scenario war with the Soviet Union—which then failed to materialize precisely because of that preparedness.
- And finally, the fear for democracy, a constant feature of the system itself since its inception. Every single decade of the past century was marked by deep concern whenever the toxic triumvirate of social polarization, economic downturn, and constitutional vulnerability came together."[16]

As a conclusion and key insight Florence Gaub stresses: "Every single decade of the past century was marked by deep concern whenever the toxic trio of social polarization, economic downturn, and constitutional vulnerability came together. And although we find ourselves in such a phase once again today, we easily forget that a hundred years ago, democracy was still a minority system fighting against various dictatorships—and yet it prevailed. It has proven not only resilient but also capable of renewal after setbacks, as in Malawi, Poland, or Brazil. None of this is to say that fears about the future are unfounded. They are entirely justified—but our unease about them should not obscure the fact that they serve a purpose. Humans possess the ability to imagine a future so that they can act in the present. This means: These fears are perfectly normal and are an inseparable part of that magnificent tool that is the future. We have no right to be free of fear—we are born with it. Instead of lamenting what could all go wrong, we must act to ensure it doesn't."[17]

This conclusion is closely connected to our oft-cited insight that—primarily for neurological reasons—we tend to favor linear developments as simple patterns, yet in reality (especially in history) these do not occur all that frequently; consequently, we must very consciously adopt different ways of thinking, particularly in complex or unfamiliar situations, in order to assess and act truly effectively and efficiently. The following cases show the validity of our assumption.

[15] ibid.

[16] ibid.

[17] ibid.

1.2.2 Some Unexpected or Often Missed Perceptions of Developments

1.2.2.1 Use Cases from Volume 2 Summarized

In volume 2 we presented some use cases of developments which we did not expect, and which look completely controversial to the public perception of a backlash for Sustainability and UN SDG Coverage.[18] We begin our exploration with a few initial unsystematic and random observations and then examine them systematically in order to arrive at a plausible overall picture.

The first use case presented in volume 2[19] is based on an article from the leading German magazine Der Spiegel. Christian Stöcker[20] wrote 06.04.2025 under the headline "The overlooked revolution—A column by Christian Stöcker"[21] about the new "Global Energy Review 2025" from IEA (International Energy Agency) on worldwide electricity generation and demand. The demand increased in 2024 by 4.3% for different reasons (one is the demand to run AI) and the generation development looks like this:

"In 2024, 80% of the growth in global electricity generation was provided by renewable sources and nuclear power. Together, they contributed 40% of total generation for the first time, with renewables alone supplying 32%. New renewable installations hit record levels for the 22nd consecutive year, with around 700 GW of total renewable capacity added in 2024, nearly 80% of which was solar PV. Generation from solar PV and wind increased by a record 670 TWh, while generation from natural gas rose by 170 TWh and coal by 90 TWh. In the European Union, the share of generation provided by solar PV and wind surpassed the combined share of coal and gas for the first time. In the United States, solar PV and wind's share rose to 16%, overtaking that of coal. In China, solar PV and wind reached nearly 20% of total generation."[22] This development is confirmed and supported by other sources (like from the International Renewable Energy Agency, abbreviated IRENA). Overall, the generation capacity of renewable sources was extended significantly faster than coal and gas and also nuclear sources, probably as a consequence of the unbeatable low price. This finding contradicts the current public discussion which transports another perception so far, even though this perception seems to change as a consequence of impacts from the wars and conflicts in the Middle east in the first months in 2026. This confirms that in the end, energy generation follows economic principles like the sustainable (lower) price.

---

[18] Volume 2, chap. 1, section 1.2.2 (s. https://link.springer.com/book/9783032238016)

[19] quoted below, in part verbatim, from volume 2, Volume 2, chap. 1, section 1.2.2 (s. https://link.springer.com/book/9783032238016)

[20] Professor at HAW University of Applied Sciences Hamburg: Hochschule für Angewandte Wissenschaften Hamburg

[21] https://www.spiegel.de/wissenschaft/mensch/energiemarkt-strom-aus-sonne-und-wind-die-uebersehene-revolution-a-483e4a52-b3ee-492d-9f76-41e573ebcb42

[22] https://www.iea.org/reports/global-energy-review-2025/key-findings

The second case was presented on 12.10.2025 by the same author who summarized under the headline "These facts will change your worldview"[23] an overview of enormous changes which are currently taking place around the world. Because of their importance we present again those for five of his ten fact statements already selected for volume 2[24]:

1. Around the planet, nothing is being expanded as massively in terms of electricity as renewable energies, i.e., electricity from the sun and wind, as well as battery storage. Investment in so-called ElectroTech is about twice as high as in fossil fuels. 92.5% of newly built electricity generation capacity worldwide in 2024 was renewable. In 2023, the figure was 86%, and in 2022 it was already 83%.
2. In the first half of 2025, China added twice as much renewable energy capacity as the rest of the world combined.
3. Since at least 2024, China has been earning tens of billions of dollars more from exports of ElectroTech than the US, the world's largest fossil-fuel exporter, earns from exports of oil and gas. The gap is growing. Between 70% (wind) and 90% (solar, batteries) of global production capacity for ElectroTech is located in China.
4. In China and India, CO_2 emissions from electricity generation are now falling, which is sensational news. This is due to renewable energies.
5. The market for combustion engines has been shrinking since 2017. Just as only renewables and storage are growing in electricity generation, only electric motors are growing in mobility, and at an accelerating rate. This is called disruption: a superior, easier to manufacture, and ultimately cheaper product is displacing the old, expensive competition. This is how the market economy works.

The interesting outcome of this—reduced and focused—overview is that a systematic momentum has clearly been triggered, which is already having a demonstrable impact on national economies and was triggered by price-driven disruption, among other factors. The mentioned facts in the five statements above are publicly available, so it is worth asking why public perception, which often focuses only on narrowly defined areas, is so different. It seems as if there are various barriers to forming an overall picture (and basing this on pure economic logic).

The impression gained from these presented use cases based on the mentioned two analytical articles in the magazine can be easily confirmed many times over when accessing the official information platforms of the IEA and IRENA with their collection of data and facts verified multiple times. There are additionally more pieces of information—especially on the role of innovation—available which might be taken from details in chap. 1 of volume 2.[25] In general, it has to be taken into

[23] https://www.spiegel.de/wissenschaft/mensch/energiewende-diese-fakten-werden-ihr-weltbild-veraendern-a-37b25092-46d0-403e-8f49-d0afea2d9758

[24] Cited from ibid., translated into English by DeepL. Identical to the explanations in Volume 2

[25] See as examples https://www.iea.org/commentaries/how-can-innovation-help-secure-future-battery-markets-and-mineral-supplies, EA Report on Cross-Cutting Technologies & Infrastructure, see: https://www.iea.org/reports/cross-cutting-technologies-infrastructure,

account that today's innovation landscape has shifted. China became the largest single country for energy patenting in 2021, overtaking Japan and the United States. More than 95% of China's energy patenting in 2022, the latest year for which data are available, was related to low-emissions technologies.[26]

A third use case in volume 2 was also based on an article on 04.10.2025 in the leading German magazine Der Spiegel, this time under the headline "Ecological Restructuring—Trump's fight against windmills will fail in the long run. A column by Ullrich Fichtner."[27] Ulrich Fichtner mentions in his article a trade fair in Lyon for environmental technology, energy and waste issues, water and recycling management called Pollutec,[28] which turns out as more exciting than at first expected. At Pollutec, 1500 exhibitors set up their stands in Lyon, sit down in conferences and explain their projects. This might not sound like much compared with the world as a whole, but Ulrich Fichtner regards Pollutec, this one event, only a placeholder for the millions of players involved in the ecological and economic transformation currently taking place in the world's economies: "Assuming that there are around 30,000 notable trade fairs worldwide, which is a plausible estimate. If we further assume that 10 to 15% of these are dedicated to CleanTec[29] topics, then we arrive at around 3000 to 4500 exhibitions per year that focus on environmental and sustainability issues.

If each of these 3000 to 4500 trade fairs had only 1000 exhibitors and visitors, that would already amount to three to four and a half million people. It is likely that the actual number is much higher. So, we are talking about a whole host of experts in ecological transformation and a mass of people who are actively interested in it.

I'm harping on about this a little because it reflects a pattern that has recently emerged concerning our distorted perception of the reality that surrounds us. As we know, modern media behavior is dominated by quick headlines and bad news. Spectacular scandals are appealing, while protracted reform successes are not. The former immediately seem immensely important, while the latter seem insignificant and 'boring.' However, they only become apparent in retrospect, sometimes with great force."[30]

As already presented in volume 2 and because of its importance stressed again in volume 3, we want to mention the references with which Ullrich Fichtner continued introducing to so far unknown but important companies or organizations like "Plastic Vortex"[31] which might actually be able to capture 100% of all plastic waste

[26] https://www.iea.org/topics/energy-innovation

[27] https://www.spiegel.de/ausland/klimapolitik-donald-trumps-kampf-gegen-windmuehlen-wird-langfristig-scheitern-a-db1631f7-bb55-4a4b-9d79-2e48ef6b2766 (quotes from the article were translated into English by DeepL)

[28] https://www.pollutec.com/en-gb.html#/

[29] Abbreviation for „Clean technologies"

[30] https://www.spiegel.de/ausland/klimapolitik-donald-trumps-kampf-gegen-windmuehlen-wird-langfristig-scheitern-a-db1631f7-bb55-4a4b-9d79-2e48ef6b2766 (quotes from the article were translated into English by DeepL)

[31] https://www.plasticvortex.co

from rivers and streams before it reaches the oceans; or AC Biode,[32] with its "Plastalyst," which found a CO2-free, energy-saving, endlessly scalable method for chemically breaking down plastics that are difficult to recycle; or Cactile,[33] a company whose "Cereus" process turns roofs, facades, and garden walls into rainwater reservoirs and water tanks, without sealing any additional surfaces and powered solely by natural gravity, only to show some exciting companies of Ullrich Fichtner's list.

The special takeaways from this case are, on the one hand, the fast development of innovation and game-changing new products and procedures (btw: this time not connected with AI) and, on the other hand, his critics on modern media creating wrong perceptions. It shows again that the world is larger and more diverse than expected, and that one needs the whole picture to solidly judge the situation.

1.2.2.2 A New Use Case

After having digested the fundamental comments of Florence Gaub on the dialectic forces to shape the future (and the historically proven inability to predict the future with any degree of accuracy) and having read the summary of the three use cases already presented in volume 1, it might make sense to introduce a fourth use case: Claus Hecking from the already mentioned magazine The Spiegel explained[34] to which degree renewable energy continues to gain ground in the U.S. "From February 2025 to January 2026, solar capacity increased by more than 33 gigawatts, according to an analysis of data from the U.S. Energy Information Administration (EIA). Solar energy now accounts for more than one-sixth of installed electricity generation capacity in the U.S. In 2025, several major solar farms began operations, for example in Texas, Colorado, and Arizona. And the boom continues. In February 2026, energy giant TotalEnergies announced it would supply massive amounts of solar power to the US tech giant Google; construction of the plants in Texas is set to begin in the second quarter.

Furthermore, large-scale storage systems saw significant growth, increasing by 16 gigawatts. Wind energy grew by 6 gigawatts. The capacity of wind and solar energy now accounts for nearly 30% of total U.S. capacity. Fossil-fuel power plants and nuclear power increased only slightly—by just under one gigawatt. As a result, wind and solar power plants are generating steadily more electricity in the U.S. For example, the contribution of solar energy grew by about 15% in January, while that of all renewable energy sources increased by 11.5% compared to January of the previous year.

Coal and gas power plants, on the other hand, generated less electricity."

This very real development is again contractionary to usual public perception and communication. The reason for this development by the way is not primarily

[32] https://acbiode.com

[33] https://www.cactile.fr

[34] https://www.spiegel.de/wissenschaft/usa-donald-trump-will-windkraft-und-solarenergie-abwuergen-doch-es-passiert-das-gegenteil-a-37a79ff4-2037-4d7e-9b79-ef8f32b101f1. Title of the analysis translated: "Trump wants to halt wind and solar energy—but they just keep booming"

ethical considerations or an ethic of conviction but the fact that energy produced from renewable sources is simply significantly cheaper than that from fossil fuels. And as in these days batteries for storing electricity are becoming increasingly affordable, the economic logic is overwhelming. Additionally, the question of resilience and reduction of dependencies from other players and nations has become more important than ever which is another political and strategic rationale—and part of long-term system logics.

We are further surprised to learn that in the United States solar panels are booming and continue to be installed en masse on private land and building rooftops, as well as the installation of storage batteries with an increase 2026 compared with 2025. And in January and February 2026, courts made the way free for the completion of the 62nd and final turbine of the Vineyard Wind offshore wind farm south of Martha's Vineyard (Massachusetts).

1.2.2.3 Summary of 1.2.2

The presented use cases—which are a small selection from an overwhelming mass of similar cases—show at least that there is some evidence that the media-driven public perception might be wrong in many cases. We selected these cases to encourage our readers to always question your own perceptions and consider other perspectives as well.

1.2.3 Insights from a Former UN Official[35]

For volume 2 we decided to additionally run an interview with Ovais Sarmad, who served the United Nations Climate Change secretariat (UNFCCC)—COP23–COP 27 (2017–2023). Prior to UNFCCC, he served as Director of Cabinet & Chief of Staff to the Director General at the International Organization for Migration (IOM). At IOM, he worked in several management and policy development capacities over a period of 27 years. Because of the importance of Ovais' statement, we place this interview also in volume unchanged in this book:

Currently, he serves as the Vice Chair of GHG-Protocol Steering Committee and in a number of roles in diverse national and international organizations. His other areas of focus include engaging with and supporting activities and organizations around the world (a) that promise to innovate and accelerate sustainable development and growth without damaging the environment; (b) to contribute to public policy discourse and development; (c) to empower youth through quality education at all levels; and (d) to advise, mentor, and coach individuals and organizations in multilateralism, diplomacy, geopolitical issues on climate change and migration and organizational change and development.

[35] Mainly unaltered excerpt from the previous book

Ovais is a role model for active former UN official after stepping down from the UN, he has developed a worldwide network in all sectors and—as participant in many international conferences and summits—a remarkable overview of "what's really going on around Sustainability and UN SDG Coverage" and "how key organizational principles, procedures, and processes have to be changed to adapt the new developments of the framework preconditions in the last decades." Ovais is very actively and transparently present on diverse platforms like LinkedIn.[36]

We asked Ovais some questions to collect his insights and consideration from the last 2.5 years after leaving UN for the book:

Question: Ovais, after leaving UNFCCC you have been very actively involved outside the official UN formal framework on issues relating to sustainability in a much broader sense than just climate change. What are your key insights so far?

Answer: I left UNFCCC and the UN system in May 2023 after a professional career spanning over 40 years, 33 years of which was with the UN organizations and gained a new outside-in-perspective which is different compared to my previous inside-out-perspective. The current UN system was established after World War II, that means 80 years ago, based on the situation, demands, and insights in the forties of the last century. It has been inevitable that the UN system would become over time "old legacy" in terms of its organizational principles, working styles, bureaucracy, etc. Especially the UN multilateralism concept could not follow the change of the world in the last eight decades. It is thus crucial to achieve a fundamental reform and significant reset for all parts of the UN and its ecosystem and to redefine its multilateralism concept and practice in this context. Here, I would give the perspective in relation to United Nations Climate Change (UNFCCC) organization and its ecosystem where I was for the last six years of my career using a heading of: "Beyond United Nations Multilateralism to a Multilateral Approach that is Practical-Pragmatic-Purposeful Inclusive and Democratic to deal with Global Issues."

I especially see today main objectives for Sustainability and UN SDG coverage as follows:

- **A democratic initiative** in terms of reducing bureaucratic boundaries and rules to get ground-level engagement and demands of society and individuals faster

[36] See https://www.linkedin.com/in/ovais-sarmad/

and more easily translated into action, helping sustainability, preserving resources, etc. In this context it's time for a "Remake-Reimagine-Reform-Reset" of the UN Climate Change process, its COPs, the COP SBs, the negotiations, the climate weeks and all the ancillaries in order to make it leaner and more focused.

- **A better prioritization** of the vast landscape of potential activities to 3–5 focus points (high level of effectiveness and efficiency instead of the hundreds of complex agenda items). This would reduce complexity and allow more focused momentum. Especially picking up the most innovative and transformative initiatives and working out national-regional-global implementation plans for them—in the spirit of the implementation of The Paris Agreement—would help as those initiatives would have the potential that everyone can agree upon in principle and take them away and translate them into their respective transition plans. This would especially allow building more focus on the power of technological innovation—such a plan can and should make climate action a reality at scale and pace that is necessary rather than a political issue of rich versus poor or developed versus developing economies "fight."
- **A better understanding and coverage of public perception** issues of the UN work (for the public, the UN work is often not very visible and understood). An external perspective of the UN system, its working and achievements, and especially the current form of UN multilateralism needs to be created, preferred by people from outside the UN system.

There are different groups of academics and practitioners (incl. Former UN officials) in place which try to schedule realization path ideas and proposals for such a transformation.

Question: Do you experience a critical setback for the numerous activities beyond the narrower formal and political (UN) sphere or has the movement still a strong momentum? What are key examples?

Answer: I would really like to make a difference between those parts of the Sustainability and UN SDG Coverage Activity Landscape which were/are originally initiated and monitored in the formal UNFCCC and COP processes—referred to as Party negotiations—and those which were independently initiated and monitored elsewhere to contribute, referred to as Non-party stakeholder Climate Action. For the first category a constructive critical look at the performance from outside the system is necessary to identify the demands for refinements and redirections—and there are enough events and opportunities in place to present, discuss, and regard the critical perspectives from outside. The second category also contains surprisingly innovative and transformative stand-alone activities and many of them have a very strong momentum which I have been able to verify personally in a large number of cases over the past two years. It is of high importance to have a comprehensive and not a selected view of the overall landscape.

For me, a particularly important key question is how to achieve genuine representation (especially in the first category). Around 70,000 participants, in person and online, were involved in COP30 in Belém, including many government representatives, of course, but also interest groups, NGOs, etc. But how well is civil society and general public represented or involved in the discussion and decision-making process? What platforms for participation are conceivable? Could models such as those used in the context of the change in the Irish constitution (citizen participation via a constitutional convention)[37] also be applied on a global scale to Sustainability and UN SDG Coverage measures? There are many comparable models thinkable. It is in any case necessary to offer more co-creation opportunities for the civil societies.

Question: What is your perspective on the mutual impact of the development of the political sphere and the development of the movement beyond it mentioned above?

Answer: The old dividing lines that used to separate the world—the distinctions between rich and poor, highly developed and less developed countries; first, second and third world, North and South—are, in my opinion, outdated. They are old patterns that are no longer valid (like the old multilateralism concept based on these patterns). During the last Covid pandemic, among other things, we noticed that all parts of the world were affected in their own way, and it is no different with climate change. Covid also highlighted how interdependent, interconnected, and interrelated the global society, economics, safety, security is.

We need another view today and should concentrate on the following (instead of dividing):

- Creation and retainment of **credibility, transparency and integrity**—this is the new "currency" of action to get things done.
- **Co-creation**—the different players must be really and seriously involved, not only UN and its ecosystem, governments, NGOs, etc. but also representatives of the civil societies of every part of the world (see my perspective on the previous question and the Irish example). Co-creation creates co-ownership, stewardship, empowerment—and it does not foster competition but collaboration.
- **Re-imagination of the future**—especially regarding the fast and partly unbelievable technical progress providing new solutions, results, and deliveries in a reasonable amount of time.
- **Focus and prioritization—**concentrate on the most promising new technologies (like reduction of methane, use of biomass management for clean energy production, green hydrogen production).
- **Embrace change—**we learnt in the pandemic what we can change fast, why not transfer the positive experience to the Sustainability and UN SDG Coverage challenge. We need for this an ethics of responsibility (not of conviction) with the focus on purposeful, pragmatic, practical proceeding and credible delivery of results with active involvement of the civil societies. This will guarantee the needed momentum for the Sustainability Movement. Especially vivid civil societies are a success factor—and through co-creation they are strengthened.
- **Accept the multipolar world and learn to manage it**. The key challenges are to understand how the new multipolar world works (especially in terms of inter-

[37] See https://en.wikipedia.org/wiki/Constitutional_Convention_(Ireland)

connectedness and interdependence) in general and in terms of Sustainability, what the impact on modern multilateral concepts is, and how progress can be sustainably achieved under these preconditions. In terms of the results of COP30, this means they are not great and unreservedly satisfactory—but better than commonly expected, and they overall strengthen the movement.[38] We will have to calibrate expectation and interpret results more long term and compared with what is realistically possible.

Question: Which will be the key fields which need most attention in the future?
Answer: I already partly mentioned these fields:

- Focus on "Technological Advancement and Artificial Intelligence (AI) Advancement."
- Focus on "Sustainability of Environment and Resources"—never take these for granted.
- Focus on "Reimaging an Innovative, Transformative Future."

1.2.4 Coping with Perception Biases in General

In general, the human brain prefers clear, simple relationships under stable external conditions that do not change significantly over the long term. This directly results in a preference for easily recognizable patterns, transparent cone-to-one correlation contexts of causes and consequences, linear (in contrast to exponential) developments, and complete clarity at the outset of projects or transformations, rather than clarity that is laboriously achieved over time. Complex situations and (economic, political, or social) systems with multiple causal relationships, dialectical developments, and even the familiar S-curves of the life cycles of products and services are difficult to understand.

Connected with these neurological patterns of thinking we find among others the following thinking and perception traps:

- There is always a pattern we know in what we observe (from the past, which means we have in any case some applicable precedence).
- The development of change is most probably linear (that is the assumption of an illusionary "standard" of predominantly linear change processes).
- One must know and fix everything in detail before one starts (project, transformation, journey; it is the so-called "Waterfall-Model"), no start into uncertainty (too risky).
- Today, we are in the most demanding challenging time ever (the human brain likes to smooth out experiences in retrospect [ex-post] and cuts off extreme peaks

[38] The German magazine Spiegel evaluated the results like a football play with the result 2:1, not an overwhelming success but a winner (see: https://www.spiegel.de/wissenschaft/klimakonferenz-cop30-in-belem-das-taugen-die-ergebnisse-die-analyse-a-05965fb4-dc1a-448a-adc8-e802c52811ed). See also for example: https://www.linkedin.com/feed/update/urn:li:activity:7398272641377386496/

in favor of a more easily imaginable linearity—but it also likes to continue the development of the past linearly into the future and so refuses to learn from past experiences about large transformations).

- The past was—considered ex-post—a lot easier to manage (problem: one understands ex-post developments always easier than in between).
- The future can be derived linearly from the present, using the existing assumptions and paradigms, namely the assumption of an illusionary "standard" of predominantly linear change processes (the alternative option cannot be imagined).
- If a large movement like the UN SDG Coverage has not reached after 50% of the time 50% of the goals it has failed (which neglects the S-Curve-alike development for example of innovation).
- Momentum is nearly always created top down through leaders with power, never bottom up (the hierarchical pattern often even used in our times, the idea of an army-alike character of movements, see controversial statements from Hannah Arendt mentioned above).
- Rate every negative, critical message higher than a positive one—attention for always the worst case happening (the weird delight in failure and ruin).

These perception traps lead to a significant ability lack to realistically assess and evaluate the real status, realization or failure probability, and obstacles of large long-term transformations and change processes with overarching societal, political, economic, environmental, and organizational impacts—especially when their lifecycle exceeds decades or even generations. Thus, a solid knowledge of these perception traps and an efficient coping practice to identify and avoid or reduce them is necessary. As even fundamental decisions are often made based on not-challenged perception, a large degree of discipline and self-reflection is demanded. One should never spontaneously believe what sounds plausible without challenging, also using different perspectives.

1.2.5 Summary of Section 1.2

The summary of the statements and insights from sect. 1.2 leads to the same result as in volume 2: to keep it brief, we should limit again ourselves to two identical statements:

- One should always assume that what one thinks one recognizes may be quite different, especially in the case one does have the full picture and only concentrated on smaller scopes.
- There is—maybe surprisingly—some evidence that the Sustainability Movement (or Sustainability and UN SDG Coverage Activity Landscape) has an unbroken momentum, only slightly irritated by some political developments but nevertheless over the full period irreversible (we will give more evidence in the more than 30 chapters of the book).

We will refer to these statements in the last chapter "Conclusions" again, after appreciating the outcome and the most important takeaways of all chapters.

1.3 Useful Recaps from Volume 1 and 2[39]

As already mentioned, there is a strong general linkage between the first and the second volume, which includes the continuation of using the same terminology, the same underlying thinking styles represented by the used joint "language" for the explorations and evaluations, and the same categorization of the activities and actors of the Movement's activity landscape as in the previous book, which makes a brief recap in this chapter necessary.

1.3.1 The Term "Sustainability"

In the first book "The Sustainable Organization Volume 1"[40] we clarified the term Sustainability in the context we use it also here like this: "'Sustainability'—contextualized in terms of its scope (the world as a whole, regions, single organizations, etc.)—is in place as well in literature as in the daily linguistic usage, and often the terms ESG[41] and CSR[42] are subsumed under the umbrella term 'Sustainability.' In some countries—like Germany for example—the term of 'Sustainability' often has a strong ethical and moralistic dimension and triggers a fundamental dispute between representatives of ethics of conviction and ethics of responsibility whereas other countries have a mainly pragmatic approach."[43] This practice was reconfirmed and continued in volume 2.

As already stressed in Volume 1 and 2, it is unnecessary to go too deeply into the discussion of the precise definitions, because the case studies included as well in Volume 1 and 2 as in this book each present their own definitions tailored to the specific case,[44] so we handle the use of the term in a pragmatic way. UN takes the perspective "Sustainability for the World" and respectively formulates a meta-level target answering the question how we can sustain the world in a human, social, economic, and environmental perspective not only to prevent negative but to generate positive impacts for future generations. Coming from this meta-level target, the UN SDGs (Sustainable Development Goals; see

---

[39] Partly quotation from Volume 1 and Volume 2

[40] https://link.springer.com/book/10.1007/978-3-031-89549-4

[41] ESG = Environmental, Social and Governance

[42] CSR = Corporate Social Responsibility

[43] See chap. 1 of https://link.springer.com/book/10.1007/978-3-031-89549-4

[44] For detailed definitions, see: Helmold, M.; Treu, J.; Fritz, J.; Hummel, F.: "ESG, CSR und SDG als langfristiger Wettbewerbs-vorteil—Nachhaltigkeit durch innovative Konzepte, Methoden und Tools"; Wiesbaden. Springer Fachmedien GmbH, SpringerGabler © 2024

below)[45] are the concrete goals for the world's transformation journey to sustainability recently. The defined SDGs recently were completed by a ground-breaking "Pact for the Future" designed to turbo-charge implementation of the UN SDGs and adopted in the General Assembly in New York on 22 September 2024.[46] On the level of the single organization, Sustainability and its facets have consequently to be concretely defined in detail by each individual player in both the public[47] and the private sectors which includes addressing Sustainability not only within the organizations themselves but also in terms of their impact on the environment and the world as a whole.

1.3.2 The UN SDG

As already touched above the more general term of Sustainability is closely linked to the 17 UN SDGs (see Fig. 1.1), which form so to say the masterplan for the development of the world by 2030 to make it a significantly better place in many perspectives. The history of the ambitious underlying transformation program was described in more detail in chap. 1 and 3 of the first volume[48] in which we also stressed that the landscape of sustainability and UN SDG coverage activities with all its acting players and affected stakeholders is nearly infinite (this was recapitalized in the second volume[49]).

When you consider that each SDG has been broken down into specific targets, which are being pursued in every country on earth by a large number of institutions and organizations with tailored measures involving many stakeholders, you can imagine that the number of acting players and affected stakeholders is nearly infinite. So, the reader might get an idea of the great challenge to generate a halfway

[45] https://sdgs.un.org/goals and https://www.globalcompact.de/en/our-work/sustainable-development-goals

[46] See: https://www.un.org/en/unis-nairobi/press-releaseunited-nations-adopts-ground-breaking-pact-future-transform-global#:~:text=Photo%2FLoey%20Felipe-,PRESS%20RELEASE%3AUnited%20Nations%20adopts%20ground%2Dbreaking%20Pact%20for%20the,a%20Declaration%20on%20Future%20Generations. "The Pact for the Future, the Global Digital Compact, and the Declaration on Future Generations open the door to new opportunities and untapped possibilities," said the Secretary-General during his remarks at the opening of the Summit of the Future. The areas touched are "Peace and Security," "Sustainable Development, Climate and Financing for Development," "Digital Cooperation," Youth and Future Generations," "Human Rights and Gender," and the "Summit Process"

[47] The public sector includes the governmental and also the so-called "Civil Society Organizations (CSOs)": Non-State, not-for-profit, voluntary entities formed by people in the social sphere that are separate from the State and the market

[48] Wollmann, P.; Pemler, D.; Ndrevataj, M. (Eds.): The Sustainable Organization—How Organizations Address the 17 UN SDGs Using the 3-P-Model. Cham Springer Nature, © 2025; (https://link.springer.com/book/10.1007/978-3-031-89549-4)

[49] Wollmann, P.; Pemler, D.; Ndrevataj, M. (Eds.): The Sustainable Organization Volume 2—How Organizations Address the 17 UN SDGs Using the 3-P-Model. Cham Springer Nature, © 2026; (https://link.springer.com/book/9783032238016)

Fig. 1.1 UN SDG (source: https://www.un.org/sustainabledevelopment/), used with friendly permission of UN. The content of this publication has not been approved by the United Nations and does not reflect the views

solid overview of the status of the landscape of activities and projects to reach sustainability for organizations, regions, countries or for the world as a whole and to cover the UN SDGs.

1.3.3 The Common Professional Language or Thinking Style: The Three-Pillar Model

For our endeavor—to continue creating a basic, flexible understanding of the landscape of sustainability and UN SDG coverage activities on a meta-level, to underpin this with some case studies and therefore to generate curiosity and openness for further research and evidence—it is important to introduce a "common professional language"—or perhaps better the applied "analytic thinking style or thinking model." In our case, this "common professional language" is the so-called Three-Pillar (Mindset) Model, based on three key building blocks (pillars). It was developed between 2019 and 2021 for the so-called VUCA[50] world by a global author community of more than 60 professionals from public

[50] VUCA = Volatile, Uncertain, Complex, Ambiguous

and private sector[51] and published at Springer Nature in up to now 6 books (eBook and print).[52] The first three books were already translated into German (by AI) and very successfully published in July 2023 and April and June 2024.[53] The translation of volume 1 is in these days work in progress, the publication is expected for later in 2026. The model has since then been frequently applied—and its applicability and usefulness proven—in various global use cases (so far >100) in the private and public sector. The goal of the Three-Pillar model is to help understanding and navigating the complex world of today's corporate territory in all sectors.

(1) The first Pillar: **Sustainable Purpose**

Sustainable Purpose is the raison d'être of an organization; having a guiding objective means bringing orientation and certainty to the people for their joint endeavor and success. This is especially important in times of crisis and transformation. It is also the key source for the narrative of transformation.[54]

(2) The second Pillar: **Travelling Organization**

Travelling Organization means for an organization accepting to be in a permanent state of flux and transformation while interacting with the external requirements of its journey with rapid learning and adaptivity. Ideally, this is the mindset of loving the unknown and being well able to manage it.[55]

---

[51] The very diverse global author community consists of practitioners, leaders, and academics from all continents and more than 20 countries. They represent knowledge stemming from their roles within more than 20 large global public institutions (incl. UN Agencies) and global enterprises and more than 30 different local public organizations or national companies.

[52] Wollmann, P.; Pemler, D.; Ndrevataj, M. (Eds.): The Sustainable Organization Volume 2—How Organizations Address the 17 UN SDGs Using the 3-P-Model. Cham Springer Nature, © 2026

Wollmann, P.; Pemler, D.; Ndrevataj, M. (Eds.): The Sustainable Organization—How Organizations Address the 17 UN SDGs Using the 3-P-Model. Cham Springer Nature, © 2025

Wollmann, P.; Stricker, L.; Pemler, D. (Eds.): Reading an Unknown Organization—Leaders in New Roles Thriving in Challenging Environments with the 3-P-Model Mindset. Cham Springer Nature, © 2024

Wollmann, P.: Püringer, R. (Eds.): Transforming Public and Private Sector Organizations—Implementing Sustainable Purpose, Travelling Organization and Connectivity for Resilience. Cham: Springer Nature, © 2022Wollmann, P.; Kühn, F.; Kempf, M.: Püringer, R. (Eds.): Organization and Leadership in Disruptive Times—Design and Implementation of the 3-P-Model. Cham: Springer Nature, © 2021

Wollmann, P.; Kühn, F.; Kempf, M. (Eds.): Three Pillars of Organization and Leadership in Disruptive Times—Navigating Your Company Successfully through the twenty-first Century Business World. Cham: Springer Nature, © 2020

[53] https://link.springer.com/book/10.1007/978-3-031-23988-5, https://link.springer.com/book/10.1007/978-3-031-55279-3, https://link.springer.com/book/10.1007/978-3-031-57163-3

[54] Everybody in the organization, but also its key stakeholders in the entire ecosystem and organization environment, must know what the organization stands for and what value and societal contribution it creates.

[55] One has an intuitive idea of a Travelling Organization from the metaphor of an expedition to Africa in former times. The participants of the Travelling Organization must be curious, open to learn, courageous, keen to experiment, and they deal well with uncertainty, stress, unforeseen incidents in a totally new world. This mindset of the Travelling Organization is the core of a Learning Organization. It is impossible in and for the Travelling Organization not to learn. Learning is a question of survival.

(3) The third Pillar: **Connectivity**

Connectivity is all about the ability to interconnect on the journey with all needed parties and resources inside and outside the silos, therefore creating high efficacy and consistency and during this process building also shared learning networks.[56]

Exemplifying the narrative of the Three-Pillar Model to get a deep understanding normally starts with the Travelling Organization. In one's imagination, this term is able to evoke the image of a very agile, flexible team embarking on an unforeseeable journey to explore unknown territories with the underlying expectation that the framing fundamental understanding of the unknown territories might change daily as a consequence of new circumstances, hurdles, and encounters.

The success of the journey of a Travelling Organization especially through unknown territories demands a clear lighthouse, a convincing Sustainable Purpose for the journey as orientation and direction. And without Connectivity survival and success on a journey through unknown territories would not have been possible, the Travelling Organization has to connect with everybody in the team and with all what's important along the chosen path.[57]

1.3.4 Categories Within the Movement's Activity Landscape of Sustainability and UN SDG Coverage

As already stressed above and worked out in volume 1 and 2, we regard the landscape of activities, projects, transformations, actors and affected stakeholders somehow as the world's change and project portfolio on all (environmental, ecological, societal, ethical, organizational, etc.) sustainability and UN SDG coverage topics ("to make the world a better place"), a list with maybe some million entries. This means that even an overview is only possible on a very advanced meta-level or based on sufficiently exemplary and paradigmatic use cases.

In volume 1 we fixed the following categories[58] of the activities, and we will also use them in volume 2 as we have use cases for each category:

- **In the public sector (1): International and national institutional actors**

  These include, for example, UN Agencies like UNFCCC and UNESCO, governmental organizations on the respective topics in the countries or regions or on EU level (partly with significant regulatory tasks), academies, universities, scientific research institutions, think tanks, and consulting governments.

[56] This means that the term organization has to be thought of within a network perspective, including interfaces to adjacent organizations and across organizational borders. In many cases, it might be impossible to learn effectively without collaborating closely with people from outside your silo.

[57] More details in chap. 2 or on: https://www.peterwollmann.com/three-pillars)

[58] See volume 1 and 2

- **In the public sector (2): All sorts of NGOs with respective purposes (a special stake in sustainability topics and the UN SDG coverage)**
  There are numerous global, regional, and local foundations and NGOs working in fields around environmental topics, on inclusion, promotion of women and gender equity, on reduction of poverty, on education, on health topics, etc. The cooperation between NGOs and the other categories of players is getting continuously closer.

- **Larger well-established private sector organizations worldwide**
  In the private sector all enterprises which decided to contribute to sustainability topics named CSR (Corporate Social Responsibility)[59] or ESG (Environmental, Social, Governance)[60] and which kept the engagement in spite of political pressure in some parts of the world are players of interest. Enterprises often advertise their own sustainability effort in their communication[61] reflecting their stakeholders' expectations.

- **Private or public sector start-ups with a special stake in sustainability topics and the UN SDG coverage**
  Often, those start-ups are specialized in one important aspect of sustainability and UN SDGs, for example in the context of a new way to create "green energy" or to implement a circular economy at a special place or to produce clean water or to develop new LifeScience/MedTech products. In many cases such types of start-ups are supported by established enterprises or governmental programs (see, e.g., new legislation in Germany to support R&D[62] projects[63]). How different venture capital is granted for start-ups in USA and Europe regarding the start-up purposes connected with sustainability and UN SDGs was touched in the previous book.[64]

1.4 A Brief Overview of the Book's Structure

Part I of the book ("**Introduction**") contains three chapters. The introducing chapters enlighten the reasons and challenges for this book, its framing and the chosen proceedings and structure. The first chapter also contains a recap of the terminology and other fundamentals, already defined and used in the first and second book, which will be used again in the third book. The second chapter builds on this and additionally describes the well-tailored update of the "User Manual" of the

[59] See for example: https://en.wikipedia.org/wiki/Corporate_social_responsibility#:~:text=Corporate%20social%20responsibility%20(CSR)%20or,volunteering%20through%20pro%20bono%20programs%2C

[60] https://en.wikipedia.org/wiki/Environmental,_social,_and_governance

[61] For example, 10% Profit Pledge for Sacred Forest (for the indigenous people of the Arhuaco in Colombia) by Rituals Cosmetics or Mindful Travelling Campaign by Austrian Airways

[62] Research & Development

[63] See chap. 19 of volume 1: Rosi Hermann on Catalysts for Change—How Research Funding and Financial Incentives Drive Business Transformation and Sustainable Development Goals (SDGs)

[64] See chap. 18 of volume 1

Three-Pillar Model on Sustainability and UN SDG Coverage Following Current Developments which means how the model—or better the thinking style and respective language—has to be best tailored and applied for the topics of the book. In this context the perspective of Transformational Teaching at universities is a bit more intensively touched. The third chapter describes a special historical case how societies might fail from environmental catastrophes and explores what we can learn from this analogy for today.

Part II (“**Academic Institutions to Enable Sustainable Practice and UN SDG Coverage**”) describes in 5 use cases the significant contribution of academic institutions from four countries to enable sustainable practice and UN SDG Coverage (chap. 4–8). The use cases cover sustainable data management by leveraging the data value chain, transformative higher education in line with “The Earth Charter International” and education for sustainable development, purpose-oriented innovation use cases on Sustainability and UN SDG, the impact of perceived stakeholder pressure, benefits, and barriers in Switzerland and Germany, and an essay on reclaiming leadership as a conscious and archetypal way of being.

Part III (“**Cross-Sectoral Approaches to Sustainability from Industry to Fashion**”) is focused on cross-sectoral approaches to sustainability and spans the spectrum from industry to fashion showing six use cases (chaps. 9–14). The use cases cover sustainability efforts at Swiss Post, Thyssen-Krupp activities on resilience as precondition for sustainability, a case study from MedTech, combining sustainability and digital transformation, circular economy and its contribution to sustainability and UN SDG Coverage, collaborative sourcing—a Three-Pillar based approach to meeting the 17 SDGs, and fashioning the future: how luxury brands champion sustainability?

Part IV (“**Conclusions**”) covers the conclusions and contains three chaps. (15–17). It starts with a consideration on the desirable impact of the sustainability books on the consulting industry and for academic education & research, afterwards presents a fundamental consideration on a reconstructive foundation for sustainability—a philosophical deepening of sustainability beyond perception and performance, and is finished by the classic conclusion and takeaways chapter.

Thus, also the Third book offers a significant number of use cases from different fields and thus offers every reader the opportunity to select the chapter that interests them. We want to take our readers on a journey through partly unknown territory extending perspectives and even adding new unexpected key insights and conclusions to those from the first book.

1.5 Recap the Contents of Volume 2

Volume 2 and 3 of The Sustainable Organization were written in parallel and with a joint interconnected author community. Whereas Volume 3 is focusing on especially “Academic Institutions to Enable Sustainable Practice and UN SDG Coverage” and

"Cross-Sectoral Approaches to Sustainability from Industry to Fashion," Volume 2 has "Global Institutions to Foster Sustainable Cooperation and Improve Communication" and "Advancing Sustainable Development in Urban and Rural Areas" in its focus. Together both books give a comprehensive picture. The second volume has the following contents:

Part I: Introduction

Chapter 1: The Framing of the Book: Why a Second Volume is Needed and How It Reflects Current Developments of Sustainability and UN SDG Coverage Activities

Peter Wollmann, Doris Pemler, Mersida Ndrevataj

Chapter 2: An Updated "User Manual" for the Three-Pillar Model on Sustainability and UN SDG Coverage

Peter Wollmann

Part II: Fundamentals

Chapter 3: Considerations on Crucial Institutional, Cooperation and Communication Preconditions for Achieving Sustainability and Coverage of the UN SDG

Peter Wollmann

Chapter 4: A Special Book Focus: Advancing Sustainable Development in Urban and Rural Areas

Mersida Ndrevataj

Part III: Global Institutions to Foster Sustainable Cooperation and Improve Communication

Chapter 5: Impactful Leadership for the Sustainable Development of a Country: Insights from the Unique Model of Qatar

Rabih El-Haddad and Sabrina Illiano

Chapter 6: The Rule of Law and Strong Institutions: The Importance of SDG 16 in post-Conflict Settings

Anna Sabrina Wollmann and Mr. Faisal Aminollah

Chapter 7: Towards a Multilateral State-to-State Dispute Resolution Mechanism within the WIPO Framework to Protect Fundamental Right and Enhance Access to Justice in line with SDG 16

Ahmed Al-Sulaiti

Chapter 8: From Global Goals to Ground-Level Impact: Translating Climate Negotiation into Local and Regional Action

Asif Husain-Naviatti

Chapter 9: From Awareness to Action: The Urgent Role of Education and Communication in Advancing the UN Sustainable Development Goals (SDGs)

Christina D'Asaro

Part IV: Advancing Sustainable Development in Urban and Rural Areas

Chapter 10: Challenges for Sustainable Corporate Management in the Food Supply Chain

Reinhard Grandke

Chapter 11: Sustainability as an Organizational Learning Practice: The Case Study of a Rural Area in Germany

Bianka Renn

Chapter 12: Addressing Sustainability and UN SDG Coverage in Rural Areas of India: A Case Study of Bunker Roy's Endeavors to Develop Rural Regions and its People

Dagmar von Tschurtschenthaler

Chapter 13. Social Innovation in Tourism in Venice: A Three-Pillar Model Analysis of Purpose, Adaptability and Connectivity

Olga Tzatzadaki

Chapter 14. Horizontal Networks for the Implementation of 'Next-door' Welfare Facilities: Case Study from the Third Sector

Giovanna Muzzi

Chapter 15: Sustainable Development in World Heritage Policies and Management: The Italian Context

Remi Wacogne and Chiara Carolina Donelli

Chapter 16: The Presence of Absence: Conflicts in Collaborative Spaces - A Comparative Ethnographic Study in France, Italy, and Czechia

Daphné Reguiessé

Part V: Conclusions and Take-Aways

Chapter 17: Conclusions and Take-Aways

Peter Wollmann, Doris Pemler, Mersida Ndrevataj

1.6 An Outlook

During the creation of the second and the third volume of The Sustainable Organization many valuable ideas for future contributions were placed by existing and new interested authors especially on equality, diversity, and inclusion (EDI) perspectives—with EDI regarded from a Three-Pillar perspective not (mainly) as an ethical task but focusing on performance reasons, or economic and efficiency systemic logic. Such a book could offer a refreshing take on this topic and support an objectification in an atmosphere of a kind of culture war. The realization is planned for 2027.

And at the end of the decade—that means in 2029/2030—we consider a Sustainable Organization retrospective, a potential Fourth Volume of The Sustainable Organization with a focus on a check how the 2025/2026 first and second book volume's discussions, projections, analytical outlooks, etc. for the Sustainability Movement have turned out 5 years later, something like a "checkpoint," a "knowledge update," and "a revaluation" also as a learning point for historical developments, which is necessary for the epistemic evolution of ideas.

Peter Wollmann has been a responsible manager, initiator, mentor, and facilitator in large, predominantly global transformations and strategic developments for almost 40 years in large organizations and since 2017 as independent consultant. He is also a member of an advisory board of an R&D oriented company. Peter holds a degree in mathematics and physics from the University of Bonn. He is the author and publisher of numerous books and articles on strategy, leadership, organization design and project and project portfolio management.

Doris Pemler is an entrepreneur, artist, and researcher based in Zurich. She received her M.A. in Egyptology, Sudan Archaeology and Archaeology of the Ancient Near East from Humboldt University zu Berlin and FU Berlin. Her research focuses on iconography, gender and intercultural communication in Ancient Egypt and Sudan. As a photographer, she contributed to several exhibitions in Germany and Switzerland. As CEO of Ciudad Condal GmbH, she leads an aspiring company focused on import and distribution of high-end sustainable products in Switzerland.

Mersida Ndrevataj is an urban scholar with a background in architecture, urban planning, and environmental psychology. She holds a PhD in Urban Planning and Public Policy from IUAV University of Venice, where her research focused on the implementation of public policies in cultural heritage management. She is currently a lecturer at IUAV University of Venice and collaborates with several COST Action research networks focused on urban studies. Her work brings together academic research and field-based inquiry by adopting audiovisual and sensory approaches to investigate urban spaces, everyday socio-spatial practices, and lived experiences.

A Brief Update on the Three-Pillar Model on Sustainability and UN SDG Coverage

2

Peter Wollmann

Abstract

This chapter is both another recap of the Three-Pillar Model introduced in a series of books before and a tailored reference and adoption, a "user manual" for the readers of the current book. The user manual reflects the current developments "in the world of Sustainability and UN SDG coverage (United Nations Strategic Development Goals, see: https://sdgs.un.org/goals) and the respective ecosystems." The underlying question is how the model can be used and applied to understand, evaluate, judge, and manage the nearly infinite activity landscape striving for Sustainability and UN SDG coverage in general, on the level of a concrete organization and also especially in the context of transformational education in academia (for example, following the Earth Charter). To achieve this crucial understanding of the Three-Pillar Model, we especially focus on the respectively needed capabilities and thinking styles.

Please note: the Three-Pillar Model, which serves as the common professional language and thinking framework for our exploration of activities aimed at achieving sustainability and covering the UN Sustainable Development Goals (SDGs), was already intensively touched in all previous books and it is also briefly described in Chap. 1 (sect. 1.3.3) of this book to provide context for the book's purpose, general direction, and framing. It is repeated here to complete and round out the recap of the Three-Pillar Model. You might also find a similar chapter in Volume 2 of The Sustainable Organization.

P. Wollmann (✉)
Bonn, Germany
e-mail: pw@peterwollmann.com

© The Author(s), under exclusive license to Springer Nature Switzerland AG 2026

P. Wollmann et al. (eds.), *The Sustainable Organization – Volume 3*, Future of Business and Finance, https://doi.org/10.1007/978-3-032-28949-0_2

2.1 Introduction[1]

The Three-Pillar (Mindset and Thinking Style) Model was developed between 2019 and 2021 by a global author community of at that time already more than 60 professionals from the public and private sectors[2] to develop a new option how to cope with the so-called VUCA[3] world. The outcome was published at Springer Nature in up to now 5 books (eBook and print).[4] The first three books were already translated by Springer into German (by AI) and very successfully published in July 2023 and April and June 2024.[5]

The Three-Pillar Model cannot be mechanically applied and is not intended to be used as simple problem-solving recipe set. Instead, it functions more like a meta-level thinking style or a mindset model for the user to tailor application exactly to their unique situation. Therefore, it demands some effort in neutral, unbiased thinking[6] to determine what exactly the Three-Pillar Model can contribute to a specific situation, how it can be effectively and efficiently applied, what concrete outcome might be reached and which detailed measures are requested for long-term success. To help with this process there is a set of questions at the end of the chapter as a backup[7] to animate the discussion and thinking processes needed in each individual situation. In this book, the application of the model is closely connected with the organizations' demands to achieve long-term sustainability and to cover (some of) the 17 UN SDGs[8] and reflects the latest developments in the surrounding ecosystem.

---

[1] quoted almost verbatim from the previous book "The Sustainable Organization" Volume 2, s. https://link.springer.com/book/9783032238016

[2] The very diverse global author community consists of practitioners, leaders, and academics from all continents and more than 20 countries. They represent knowledge stemming from their roles within more than 20 large global public institutions (incl. UN Agencies) and global enterprises and more than 30 different local public organizations or national companies in different industries and sectors.

[3] VUCA = Volatile, Uncertain, Comply, Ambiguous

[4] Wollmann, P.; Pemler, D.; Ndrevataj, M. (Eds.): The Sustainable Organization—How Organizations Address the 17 UN SDGs Using the 3-P-Model. Cham Springer Nature, © 2025Wollmann, P.; Stricker, L.; Pemler, D. (Eds.): Reading an Unknown Organization—Leaders in New Roles Thriving in Challenging Environments with the 3-P-Model Mindset. Cham Springer Nature, © 2024Wollmann, P.: Püringer, R. (Eds.): Transforming Public and Private Sector Organizations—Implementing Sustainable Purpose, Travelling Organization and Connectivity for Resilience. Cham: Springer Nature, © 2022Wollmann, P.; Kühn, F.; Kempf, M.: Püringer, R. (Eds.): Organization and Leadership in Disruptive Times—Design and Implementation of the 3-P-Model. Cham: Springer Nature, © 2021Wollmann, P.; Kühn, F.; Kempf, M. (Eds.): Three Pillars of Organization and Leadership in Disruptive Times—Navigating Your Company Successfully through the twenty-first Century Business World. Cham: Springer Nature, © 2020

[5] See https://link.springer.com/book/10.1007/978-3-031-23988-5, https://link.springer.com/book/10.1007/978-3-031-55279-3, https://link.springer.com/book/10.1007/978-3-031-57163-3

[6] Like Hannah Arendt's "thinking without a banister" (https://medium.com/quote-of-the-week/thinking-without-a-banister-2b68a5120625)

[7] As backup, content taken from the Volume 1, the previous book (see first entry at footnote 3)

[8] https://sdgs.un.org/goals and https://www.globalcompact.de/en/our-work/sustainable-development-goals

The great advantage of the Three-Pillar Model mindset is that by using it and getting used to it, formal, sophisticated, and precisely described operational proceedings are no longer necessary. Instead, engaging in the discussion of the individual questions simply starts every person's and organizations' unique journey and has the potential to lead to meaningful results. Therefore, the demanding challenge is not to cover sophisticated methodical steps but to achieve unbiased thinking and discernment—as the gap between public perception and real situation in topics around Sustainability and UN SDG Coverage and the respective Activity Landscape is huge and the public debate and the formation of personal and organizational opinions are correspondingly difficult. This topic has already been discussed in some detail in the first chapter and will also be tackled among others in the context with transformational education and the Earth Charter in chapters 5 and 15.

2.2 The Three-Pillar Model: An Overview Including Current Application Challenges[9]

The Three-Pillar Model is based on three key building blocks (pillars), defining the Three-Pillar (Mindset or Thinking) Model for organizational design and leadership for the VUCA world. The author community's starting point in 2019 was to establish guiding principles for a world seemingly devoid of them. The model has since been frequently applied—and its applicability and usefulness proven—in various global use cases (so far >80) in the private and public sectors. The goal of the Three-Pillar model is to support understanding and navigating the complexities of today's corporate and institution territory in all sectors.

(1) The first Pillar: **Sustainable Purpose**

Sustainable Purpose is the *raison d'être* of an organization. Having a clear guiding objective provides orientation and certainty to people, fostering a sense of shared effort and success. This is especially important in times of crisis and transformation, as it becomes the key source for the narrative of transformation.[10]

It is obvious and well-known that sustainability, along with the alignment to (some of) the UN SDGs, forms a compelling sustainable purpose for an organization. Sustainability is understood here in two dimensions: sustainability of the organization itself and its contribution to sustainability on a global level. It is clear that achieving sustainability of organizations and their ecosystems as well as the coverage of at least some of the UN SDGs demand significant transformations. It is important that awareness of this issue is already being actively promoted in university education—the chapters in Part II "Academic Institutions to Enable Sustainable Practice and UN SDG Coverage" and also chap. 15 demonstrate that this is already the case in many places.

The ***special challenge with Sustainable Purpose*** linked to a general idea of sustainability and the coverage of selected UN SDGs in detail lies in the perception bias, the gap between the public perception of status, success, reachability of sustainability targets

[9] partly taken from The Sustainable Organization Volume 1 and 2

[10] Everybody in the organization, but also its key stakeholders in the entire ecosystem and organization environment, must know what the organization stands for and what value and societal contribution it creates.

including the judgment of concrete parts of the activity landscape, on the one hand, and the fact-based reality (of the comprehensive big picture), on the other hand. Here, one often finds doubt, discouragement, even despair up to doomsday mood. The perception bias was already tackled in chap. 1—and the whole book's intention is to show at least puzzle pieces of another reality. It is therefore important "to clear the brain" with discipline, try to get a comprehensive, fact-based picture and to start "to think without banister" (that means open-minded and without prejudices—the situation might be different than intuitively expected or from selective perception from media usually consumed. To foster this think style, positive use cases are crucial—they show what is possible and give an idea, what an organization can strive for). In any case, leadership is particularly important in this regard.

(2) The second Pillar: **Travelling Organization**

Travelling Organization refers to an organization accepting to be in a permanent state of flux and transformation while interacting with the external requirements of its journey by means of rapid learning and adaptivity. Ideally, this is the mindset of loving the unknown and being well able to manage it.[11]

It is widely recognized that the transformational journeys of organizations to reach sustainability and align with (some of) the UN SDGs are explorative in nature and often have to pass through unknown territories.

The ***special challenge with a Travelling Organization Mindset and Capabilities*** is today, on the one hand, that the north star of a Sustainable Purpose in the context with Sustainability and UN SDG Coverage is challenged by a pessimistic view and a public perception bias already described above and in chap. 1—which has an impact on motivation and momentum. Going on an explorative journey through the unknown and being confronted with change is always demanding but is even more challenging under these preconditions. On the other hand, there is one more challenging factor: In contrast to the research expeditions of the late nineteenth century through unknown territories in Africa or Asia (mentioned below and in the previous books as a metaphor for Travelling Organizations), which traversed these unknown territory but under relatively stable or linear, albeit unknown external conditions, the journeys today undertaken by organizations are through also uncharted territory, unknown territory but in which external conditions are quickly and unexpectedly constantly changing in a nonlinear manner. This adds another dimension that must be taken into account: fast non-linear changes of the preconditions/contexts. The resulting challenges for leadership are obviously tremendous—but manageable as this book shows in diverse chapters.

(3) The third Pillar: **Connectivity**

Connectivity focuses on the ability to interconnect during the journey with all necessary parties and resources inside and outside the silos, thus creating high effectiveness and consistency and also building shared learning networks[12] during this process.

It is obvious and well-known that an enormous degree of connectivity is needed for transformational journeys of organizations to reach sustainability and align with (some of) the UN SDGs; this is among others a consequence of the high number of stakeholders and affected and/or involved persons.

[11] One has an intuitive idea of a Travelling Organization from the metaphor of an expedition to Africa in former times. The participants of the Travelling Organization must be curious, open to learn, courageous, keen to experiment, and they deal well with uncertainty, stress, unforeseen incidents in a totally new world. This mindset of the Travelling Organization is the core of a Learning Organization. It is impossible in and for the Travelling Organization not to learn. Learning is a question of survival.

[12] This means that organization has to be thought of within a network perspective, including interfaces to adjacent organizations and across organizational borders. In many cases, it might be impossible to learn effectively without collaborating closely with people from outside your silo.

The ***special challenge with Connectivity*** in the context of activities on Sustainability in general and the coverage of selected UN SDGs in detail is the increasingly strong fragmentation among stakeholders in terms of opinions, perceptions, and ethics (ethics of conviction versus ethics of responsibility), all of which are sometimes strongly linked to a certain emotionality. This makes it difficult to agree on a common starting point for cooperation or even neutral exchange. High, often unattainable expectations are met with defeatism, but also with total rejection. The situation often requires the utmost diplomatic skills and special integrative leadership qualities. On the other hand, we experience the strong worldwide connectivity in academia (see Part II "Academic Institutions to Enable Sustainable Practice and UN SDG Coverage") and the connected continuous creation of a global knowledge society (it might make sense to read in this context chap. 3 of the previous book,[13] where this topic is touched in detail.

To exemplify the narrative of the Three-Pillar Model in general, one should perhaps start with the Travelling Organization. In our imagination, this term can evoke the image of a very agile, flexible team embarking on an unforeseeable journey to explore unknown territories. In former times, this meant, as already mentioned above, for example a scientific expedition to Africa or Australia to explore territories previously unknown to us. It is intuitively clear—and supported by various travel reports and even video footage—what kind of team dynamics, personal attributes, and collective capabilities are essential for the successful execution of such an endeavor.

Imagining such an endeavor to succeed, it strongly needs a lighthouse, its own Sustainable Purpose to operate as orientation and direction. In the idea of an explorative expedition, the Sustainable Purpose becomes immediately clear: it is scientific curiosity and the urge to clarify and thus build up lasting knowledge. These sorts of Sustainable Purpose were usually not controversial for the team in any way. Why this might be different in current contexts around Sustainability and UN SDG Coverage was mentioned above. Nevertheless, thinking of a Sustainable Purpose is crucial, even though it might take more efforts today than in former times.

A Travelling Organization does always need its own accepted Sustainable Purpose, leadership has to make this happen and to prepare the stage for it, among others by the mentioned "thinking without banister." And during its journey into or through the unknown, the Travelling Organization also needs the capability to connect within the team but even more so with its surroundings. On a scientific Africa expedition in the nineteenth century,[14] multi-layered Connectivity was an essential skill set—one had to connect between the team members, with indigenous people and their specific cultural traits, with animals, flora and fauna, geography, and many

[13] Chapter 3 of The Sustainable Organization Volume 2, s. https://link.springer.com/book/9783032238016

[14] See also: Wollmann, P.; Püringer, R.: "About Travelling in the Unknown in the 19th Century and Today. A Pattern for Leadership and Management in a 3-P Model Context" in: Wollmann, P.; Kühn, F.; Kempf, M.: Püringer, R. (Eds.): Organization and Leadership in Disruptive Times—Design and Implementation of the 3-P Model. Cham: Springer Nature, 2021. More under: https://www.springer.com/gp/book/9783030630331

other unknowns. Without Connectivity, survival and success would not have been possible.[15]

Applied to the current situation, connectivity must also be established when this is extremely difficult for the reasons mentioned above. The overall success of the nearly infinite Activity Landscape for Sustainability and UN SDG Coverage is dependent on the degree of connectivity realization as the overall movement needs bottom support to be reached through a reasonable degree of participation.[16] This concept is based on the idea that power (as for the activity landscape movement) arises through participation of many from the bottom up. Force—the top-down concept, often confused with power, does not create power according to Hannah Arendt.[17] This is an important statement which has to be linked with the 3-P Model as it stresses the options for action and opportunities to exert influence (a Travelling Organization should always be aware of its scope for action, it is not a passive object but a proactive subject of a journey—and the mindset of the mentioned explorers end of the nineteenth century was exactly like this).

2.3 A Deeper Dive into the Three-Pillar Model Application in Unknown Territories[18]

Living in a VUCA world without many certainties, we increasingly find ourselves navigating through unknown territories. In such situations, we lack precedents to guide us, and our highly developed, sophisticated methods and tools, along with familiar rules and regularities, often become inapplicable—or even entirely obsolete (for example, if stability or slow linear developments turn to become exponential). Furthermore, the carefully collected and analyzed historical data we rely on to predict the future cease to be relevant.

Realizing this has two important consequences. On the one hand, it means that we need another set of thinking models to give orientation more on a meta-level, if it is not possible on a concrete and detailed level. The Three-Pillar Model turns out to be extremely helpful in focusing on meta-capabilities to give a general orientation (see also previous section): how to formulate an exciting and authentic

[15] More details, see: https://www.peterwollmann.com/three-pillars

[16] As a background for Hannah Arendt's concepts see for example: Lederman, S.: Arendt and the Current Participatory Moment, in Lederman, S.: Hannah Arendt and Participatory Democracy—A People's Utopia. Cham. Palgrave Macmillan. 2019. https://link.springer.com/book/10.1007/978-3-030-11692-7

And for a use case see: Ahmad, N.A.; Ismail, K.; Ramli, F.; Ismail, M.: Does Participation in Decision-Making Empower Local People? Understanding Empowerment Through Hannah Arendt's Theory of Power? In: Hamdan, K.R.; Buallay, A.: Artificial Intelligence (AI) and Customer Social Responsibility (CSR), Cham: Springer 2024. https://link.springer.com/chapter/10.1007/978-3-031-50939-1_2

[17] See: https://grattoncourses.wordpress.com/wp-content/uploads/2019/12/hannah-arendt-on-violence-harcourt-brace-jovanovich-1969.pdf, retrieved on June 13, 2023.

[18] Partly reused from The Sustainable Organization, Volume 1 and Volume 2

lighthouse direction (Sustainable Purpose) even in difficult environments, how to embark on agile and flexible journeys as an organization (Travelling Organization), without fearing the unknown territories, and how to effectively connect with all participants needed for the journey (Connectivity). The exact way to nail down and tailor these capabilities in detail has to be developed and decided in the given situation—which is possible with both the meta-capabilities and the right mindset in place. This means that the Three-Pillar Model can be well applied to the Unknown even though or—better because—it demands a lot of effort for the specific tailoring. A crucially needed base is to be able to mind thinking traps like perception biases and to understand that the future can be consciously reshaped by oneself and thus take a different course than predicted by linear thinking[19]—or formulated in another way, that in general future forecasts do only predict what is most likely to happen if people do not intervene and if nothing unexpected occurs—and as already stressed in chap. 1, this is a relevant source for perception biases.

On the other hand, we have a tremendous need to learn about journeys through unknown territories[20] under unclear preconditions even on a detailed level, e.g., with a significant lack of authenticated, reliable facts and knowledge, making necessary a certain level of healthy suspicion to prevent unpleasant surprises. This sort of learning also needs the "thinking without banister," one might extend this to "thinking and learning without banister," without thinking traps and perception biases, always challenging the own perception.

In unknown territories, this specifically means for each organization to learn quickly—it is a matter of survival. Conventional learning approaches are often insufficient in such contexts, making explorative, experimental, and iterative learning with an open mind essential. This must be coupled with intensive, ongoing exchange, including cross-organizational collaboration and mutual peer challenging. Such circumstances demand a new open learning mindset that abandons reliance on rigid specifications and prioritizes adaptability over simply applying lessons from past experiences.

Another valuable reference in uncertain times is Rebecca Solnit and her famous book "A Field Guide to Getting Lost,"[21] in which she works out how important it is to get lost to be able to perceive something important in a completely different and unbiased way. In unknown territories, it is important not to use well-known traditional associations to prevent misperceptions. Or even more directly formulated: The capability to cope with the unknown is based on the ability to get lost. Getting

---

[19] Quote by Hannah Arendt: "Predictions of the future are never anything but projections of present automatic processes and procedures, that is, of occurrences that are likely to come to pass if men do not act and if nothing unexpected happens; every action, for better or worse, and every accident necessarily destroys the whole pattern in whose frame the prediction moves and where it finds its evidence." From Hannah Arendt "On Violence", p.13, Houghton Mifflin Harcourt 1970.

[20] Such unknown territories also might occur in the context of needed innovation to fix issues in the context of Sustainability and UN SDG Coverage

[21] Solnit, R.; A Field Guide to Getting Lost, New York, Viking Penguin, 2005

lost means—according to Rebecca Solnit—"that the world has become larger than your knowledge of it"[22] and you are aware of this.

To paraphrase, a certain openness regarding the Unknown is necessary to be able to gain new knowledge. Rebecca Solnit completes her book with a Greek philosophical perspective: "How will you go about finding that thing the nature of which is totally unknown to you?[23]", which is a fundamental question for each journey through unknown territories. One of her recommendations is to find and use some unusual associations be it from religion, philosophy, literature, arts, or science to introduce radically new perspectives and connected narratives that might help you to cope and understand.

In this context we have to consider three key statements:

- The future[24] is always—despite all efforts to make it predictable and controllable—an "unknown territory." It is a perception bias that one involuntarily tries to continue the present statistically and linearly into the future.[25] This leads to a clash to understand the character of historical developments and the options for shaping the future.[26]
- The best ways to reach sustainability—both for an organization within its ecosystem and for the world as a whole—remain only partly investigated, fixed, and explored concerning their systematic interactions. This in turn means that unknown territories will appear on the journey to reach the sustainability goals.
- Even though there is a system of precise measures and quantification for the UN SDGs for a country and regional level in place, together with some reporting activities, the impact on the concrete organization is the same as can be seen regarding the topic of sustainability: there are so many points to be clarified that the journey will inevitably touch unknown territories.

The exploration of the best ways to secure the organization's future by reaching sustainability and coverage of (some of) the UN SDG will therefore also be a confrontation with a mostly so far unknown world. In a context as such, one must first determine which parts of the existing knowledge and experience, if any, are applicable. In any case, it makes sense to approach the ambitions connected to sustainability and the UN SDGs from a meta-level, using the Three-Pillar Model (defining the sustainable purpose connected with the organization's and the world's sustainability and with the UN SDGs, developing some level of journey mindset for the respective transformation journey and capabilities and the ability to connect with all stakeholders and involved parties) combined with the insights of Rebecca Solnit to

[22] Solnit, R.; A Field Guide to Getting Lost, New York, Viking Penguin, 2005, page 22

[23] Alleged Pre-Socratic philosopher Meno, cited in Solnit, R.; A Field Guide to Getting Lost, New York, Viking Penguin, 2005, page 4

[24] Inclusive future innovation or the long-term impact of current innovation respectively

[25] It is partly a neurological imprint as our brains love special sorts of structure and patterns and have a problem understanding light signals for disruption.

[26] See footnote 17

provide a fresh and unbiased view, always considering to "Leave the Door Open for the Unknown."[27]

In general, it should be noted that unbiased thinking and perception are of utmost importance, especially in unfamiliar territory. The concepts and suggestions mentioned by Hannah Arendt (Thinking Without Railings) and Rebecca Solnit (the ability to get lost and gain a new fresh unbiased view) provide valuable insights in this regard.

2.4 Backup: Helpful Questions for the Application of the Three-Pillar Model[28]

Even though there is an intuitive approach to the three pillars of the Three-Pillar Model, and it quickly generates an image in one mind's eye, there is also the need to get specific and detailed in order to reach a successful application and a fruitful discussion. In this respect, the following questions might be helpful.

Sustainable Purpose
Initial Remarks: It is important to divide the (Sustainable) Purpose of a Travelling Organization into two parts:

- The "noble" Sustainable Purpose, this might be linked with the UN Sustainable Development Goals (UN SDGs), appealing to the value system of society and people (long-term purpose).
- The "basic" (or rather simple and operational) Sustainable Purpose is to cover a special customer need (situational, temporary, and more often changing or rather more often to be newly interpreted purpose).

Additionally, it is crucial to always regard sustainability in two respects: the sustainability of an organization and its long-term survival through continuous development. And on the other hand, the organization's contribution to the sustainability of its environment or the world's sustainability as a whole. Both aspects should be reflected in the formulated sustainable purpose of the organization.

Potential Questions
- Does the Sustainable Purpose aim to "make the world a better place," while also being financially (or reputationally) beneficial for the organization? Is the Sustainable Purpose convincing?
- Does the formulated Sustainable Purpose have the power and convincing force to cope with current doubts and biased public perceptions?
- Does the Sustainable Purpose credibly comprehensively show the organization's social responsibility?

[27] Solnit, R.; A Field Guide to Getting Lost, New York, Viking Penguin, 2005, page 4
[28] Updated version from Volume 1 and Volume 2

- Is it to be expected that the Sustainable Purpose is a key and highly prioritized factor in difficult decision-making processes, or does it always "lose" against profit interests?
- Does the Sustainable Purpose only concentrate on rankings ("we want to be number 1 in ....") and market positions ("we want to grow our market share to ...") or on sense-making future organizational capabilities?
- Is the Sustainable Purpose formulated in a very abstract way ("we want to always satisfy our clients' demands ...")?
- What importance does the Sustainable Purpose of an organization have for you as a person (differentiated, if necessary, between established organizations and start-ups)? How do you evaluate and judge the Sustainable Purpose of a new organization?
- Which sorts of purposes do attract people in your opinion? How do they have to be formulated? How do you evaluate and judge them for a new organization?
- How do you evaluate if the "official" Sustainable Purpose is meant seriously (which also means that it has a high priority) and to what extent does it attract and convince the people in the organization and its stakeholders?
- Do you have best (and worst) practice examples as your benchmarks? And how do you evaluate your Sustainable Purpose in the political current environment?
- What do you do to improve a Sustainable Purpose and its perception in your organization?

Especially for Universities (selection)

- Does the university offer an attractive Sustainable Purpose as an organization, which is attractive as well for students as for employees? Is its Sustainable Purpose convincing? Are there enough events for discussion of the concrete meaning and impact of the defined Sustainable Purpose?
- Does the term Sustainable Purpose play an adequate role in the curricula of the different university departments?
- How does the formulated Sustainable Purpose impact teaching and research in general and in detail? Is Transformational Education part of the university's DNA?

Travelling Organization

Initial Remark: The organizational capabilities of a Travelling Organization are closely linked to the readiness of an organization to transform, its maturity, and its political will to run a transformation journey with high priority. In our days, the readiness to transform is a key (strategic) attribute of an organization, a potential game-changer and competitive advantage.[29] Thus, questioning the existence and maturity degree of a Travelling Organization is crucial. Therefore, a "setting check"

[29] See also: Schmidt, H.: "Psychological Capabilities Required for Continuous Transformations —The Self on Permanent Journeys with a Travelling Organization" in: Wollmann, P.: Püringer, R. (Eds.): Transforming Public and Private Sector Organizations—Implementing Sustainable Purpose, Travelling Organization and Connectivity for Resilience. Cham: Springer Nature, © 2022

of your organization to analyze its ability for journeys also through unknown territory and/or its "readiness (and sustainable ability) to transform" should be mandatory, but kept simple (in application), easily understandable—and certainly built on a process of analytical discussion in a diverse team which includes different expertise, experiences, cultures, geographies, genders, etc. These capabilities should be especially stressed in the context of striving for sustainability and UN SDG coverage.

Potential Questions

- How common is it to start urgent change projects exploratively, that means with a detailed "big picture" and with targets developed during the journey?
- How accepted and easy is it to change the project plan during the journey if needed? How are the decision processes for such changes?
- How high is the trust level in the organization—and the culture of making mistakes (e.g. in basing on assumptions which turn on the journey out to be wrong)?
- How many people in the organization can cope with uncertainty and ambiguity? If you needed several people for a really difficult and demanding project through unknown territory very quickly, would you get a critical mass following?
- Are the colleagues and staff members mostly only able to act if they have a detailed plan (waterfall culture)?
- Is the organization's culture more based on curiosity or more solid-state-oriented?
- What importance does the character of an organization as a Travelling Organization have for you as a person (differentiated, if necessary, between established organizations and start-ups)? What are your first (hard or soft) indicators?
- How do you evaluate in more detail if the organization has the character and mindset/would be able to develop the character and mindset of a Travelling Organization? What are your criteria and checkpoints for evidence?
- Does the organization have a high diversity in thinking styles, leadership styles, experiences, careers? Are there regular exchanges on thinking traps, perception biases, animations through thinker like Hannah Arendt or Rebecca Solnit?
- Is there an adequate learning concept in place (especially in unknown territories)?
- How do you check if there is a critical mass in the organization ready to cope with unknown territory?
- What are your expectations for the organization's HR to foster a Travelling Organization mindset? Do you have best (and worst) practice examples as your benchmarks?
- What do you do to improve your organization's readiness to become a Travelling Organization?
- To what degree is your organization ready and capable of running journeys to reach sustainability and UN SDG coverage?

Especially for Universities (Selection)

- Does the university "walk the talk" as a Travelling Organization, as a role model for the students' idea of a future employer/work environment?
- Does the term Travelling Organization play an adequate role in the curricula of the different university departments? How is the practice evaluated?
- How does the character of university as a Travelling Organization impact teaching and research in general and in detail? Is Transformational Education also from this perspective part of the university's DNA?

Connectivity

Initial Remark: The organizational capability for Connectivity is based among other things on the cooperation and communication culture of the organization and on the personality of key players and their capabilities for Connectivity. In this, a special type of personality is needed, as described in a previous Three-Pillar book.[30]

Potential Questions

- How strongly is cooperation supported across units on all levels (board, senior management level, management level, staff)?
- How intensive are discussions about who exactly is responsible for something vs. how can an issue be fixed in the best and fastest way?
- How is the communication culture: absolutely open inside and outside the units or mostly restricted to the respective unit?
- If you have a quick enquiry: are you allowed to approach the responsible person without asking his/her boss for permission?
- How trustful and vivid are the informal cooperation and exchange channels?
- How are the capabilities to connect and communicate in difficult environments?
- Which importance does Connectivity have within an organization, its ecosystem, and its societal environment for you as a person (differentiated, if necessary, between established organizations and start-ups)? What are your first (hard or soft) indicators for Connectivity?
- How do you evaluate if Connectivity is really in place (e.g., cross-unit cooperation) and not only a not applied official behavior rule?
- How do you check if there is enough Connectivity in your peer network and on the level of your board?
- Which sort of benefit system is necessary to foster Connectivity, especially cross-silo cooperation? How do you check if it is in place? Which priority does this have to have in your opinion?
- Do you have best (and worst) practice examples as your benchmarks?
- What do you do to improve Connectivity and its perception?
- Additional: How do you analyze/realize the difference between official and unofficial org charts, the official and unofficial key influential persons, and their networks? What do the differences mean in your perspective? And what is their

[30] Schmidt, H.: "Personality and Connectivity" in: Wollmann, P. et al.: Three Pillars of Organization and Leadership in Disruptive Times, Springer Nature, 2020

impact on Connectivity but also the Sustainable Purpose and a Travelling organization? And how does your organization cover connectivity in the context of sustainability and UN SDGs?

Especially for Universities (Selection)

- Does the university "walk the talk" as a Connectivity-driven part of the global knowledge society, as a role model for the students' idea of their future style of working?
- Does the term Connectivity play an adequate role in the curricula of the different university departments? How is the practice evaluated? Is it part of the Transformational Education model of the university?
- What does Connectivity mean in detail for the university's teaching and research in general and in detail? Is Connectivity also from this perspective part of the university's DNA?

Peter Wollmann has been a responsible manager, initiator, mentor, and facilitator in large, predominantly global transformations and strategic developments for almost 40 years in large organizations and since 2017 as independent consultant. He is also a member of an advisory board of an R&D oriented company. Peter holds a degree in mathematics and physics from the University of Bonn. He is the author and publisher of numerous books and articles on strategy, leadership, organization design and project and project portfolio management.

3 The End of Meroe: A Surprisingly Modern Lesson in Climate History

Doris Pemler

Abstract

The kingdom of Meroe existed on the Upper Nile in modern Sudan from the 4th century BC until the 4th century CE when it disappeared rather suddenly. In his book "Collapse," renowned Pulitzer Prize-winning scientist Jared Diamond formulated different reasons for the end of civilizations. This text applies his possible factors for failure on the End of Meroe. In this way, it uncovers different aspects of the history of the region from ancient written and archeological sources. We take a look into the enemies of the state, foreign trade, iron production, and climate history. In the end, we discover that the Ancient World was maybe more like ours than we thought: already the Romans used road toll to refinance their infrastructure costs and conducted "globalized" large-scale sea trade to India via the Red Sea. Also, the impact of a changing climate can lead to great stress between different groups of the society about the allocation of diminishing natural resources, a situation very similar in Meroe to that in today's Sudan. One possible solution for these challenges could be the empowerment of the individual along the UN SDGs to make communities big and small more resilient.

D. Pemler (✉)
Ciudad Condal GmbH, Zürich, Switzerland
e-mail: doris@ciudad-condal.ch

© The Author(s), under exclusive license to Springer Nature Switzerland AG 2026
P. Wollmann et al. (eds.), *The Sustainable Organization – Volume 3*, Future of Business and Finance, https://doi.org/10.1007/978-3-032-28949-0_3

3.1 Introduction

And now for something completely different (Monty Python)[1] What will be presented here is a story about the end of the rather unknown ancient kingdom of Meroe in today's Sudan. In the course of the text, we will take a look at some of the original ancient sources surviving the times and their possible interpretation, showing how easy direct access to sources can be and how broad a spectrum for interpretation they leave nevertheless (a lesson also valid in today's media consumption). In the spirit of the Three-Pillar Model, we will not lose our Sustainable Purpose of shedding light and making sense out of our sources on the way, we will be prepared to change course on our journey and keep each other updated at every step of the way.[2] An interpretation for the (in expert circles) much-debated reasons for the end of the Meroitic Kingdom will be sought along with the findings of renowned polymath Jared Diamond and his theory about the collapse of civilizations. Some of the findings will be obscure facts,[3] some will sound surprisingly modern and we will fly the flag in favor of UN Sustainability goals and against authoritarianism (Fig. 3.1).

Let us start with a short introduction to the main subject. The kingdom of Meroe existed as youngest and last part of a succession of civilizations known as the "Kingdom of Kush" from the third century BCE until the fourth century AD. It was situated in the Sudanese Middle Nile region ca. 200 km northeast of Khartoum between Khartoum in the south and the confluence of the river Atbara and the Nile in the north on the east bank of the Nile in a region called the Butana.

The wealth of Meroe originated in the control of abundant gold resources and its mediator role in the trade of goods from Inner Africa highly sought after by the empires of the East and of the Mediterranean such as ebony and ivory, leopard skins and ostrich feathers, skilled workers and slaves. Meroe had its own script which is still not fully known.[4] To reconstruct Meroitic history and society we therefore have to rely heavily on archeological finds and written sources from the outside. One of the outstanding features of Meroitic culture is a very high status of

---

[1] https://en.wikipedia.org/wiki/And_Now_for_Something_Completely_Different

[2] See most recently: Wollmann, P. 2025. A recap of the Three-Pillar-Model and its importance for the Book's Objectives in: Peter Wollmann, Doris Pemler and Mersida Ndrevataj (eds.): The Sustainable Organization. How Organizations Address the 17 UN SDGs Using the 3-P-Model. Springer Nature Cham. 23–32.

[3] https://www.atlasobscura.com/

[4] The letters are known and can be read but it is based on an unknown language which has only recently been identified as part of a language family still used in the region. Claude Rilly as the leading person in deciphering the language gives an overview and also talks about other undeciphered scripts: https://www.youtube.com/watch?v=Myj8ehkVRw8

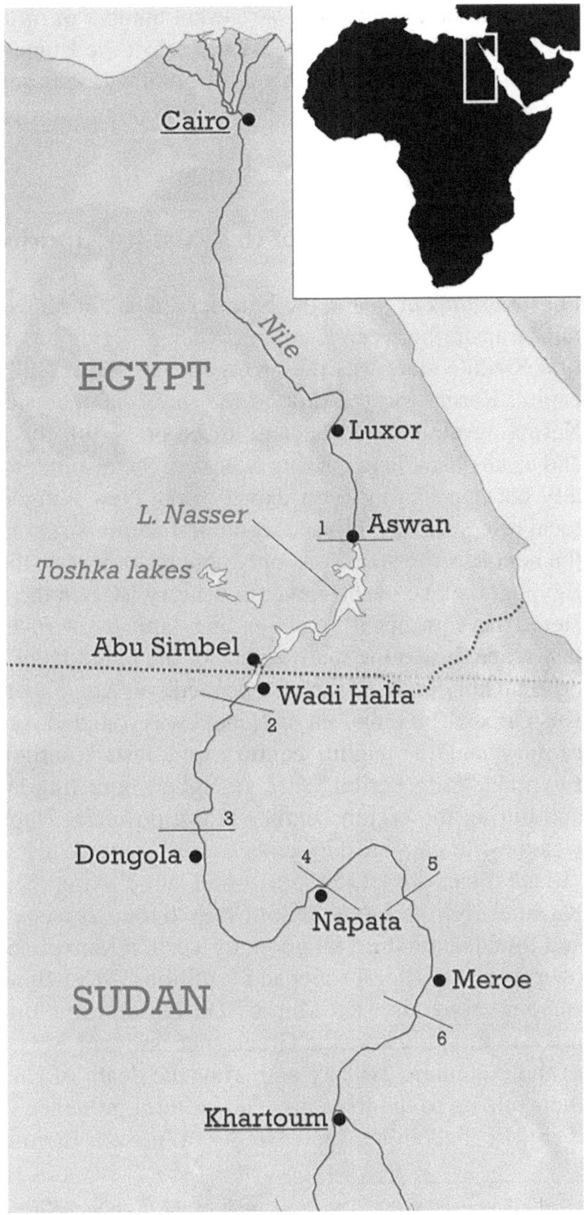

Fig. 3.1 Geographic overview. Source Wikipedia. The original uploader was Mark Dingemanse at English Wikipedia.—Transferred from en.wikipedia to Commons. CC BY 2.5, https://commons.wikimedia.org/w/index.php?curid=2133014

women in society including a relevant number of ruling queens which are reported to have led Meroe into battle, e.g. against the Romans. A fun fact here is that the highest-ranking female title apart from the gender-neutral title of the ruler, "qore," is "candace," which came to be a rather common proper name in the US by a detour via the Bible.

3.2 Short History of the Kushite Kingdoms

For thousands of years, the Nile served as a means of connection and direct cultural transfer between Kushites, Ancient Egyptians, and the Mediterranean.[5] The first Kushite state was the "Kingdom of Kerma" (2500 BCE–1500 BCE) with its capital Kerma located around the third cataract[6] near present-day Dongola. As Kerma developed and became more powerful, the Egyptians saw it as a threat and as an obstacle to gaining access to the resources of Nubia and Kerma eventually became victim to an expansionist New Kingdom Egypt who defeated the local rulers and established a colonial administration. The "Kingdom of Napata," the next Kushite state, was only able to form after the downfall of New Kingdom Egypt at the end of the eleventh century BCE in the region of the fourth cataract. Here, the significant free-standing table rock of Gebel Barkal dominates the landscape. It was the sacred place of the kingdom of Kush and was thought as the original holy place of the ram-headed god Amun, venerated in Ancient Egypt and Kush alike. The kingdom of Napata was founded sometime between the eleventh century and the eighth century and lasted until the third century BCE. The Egyptian "Late Period" sees an Egypt more fragile and ruled by foreign kings and during the eighth century BCE a powerful Napatan Kushite state used this weakness to gain military power over Egypt to rule as 25th dynasty, the so-called "Black Pharaohs" (754 BCE—656 BCE). After nearly 100 years of rule, the Kushites lost the Egyptian kingship to the Assyrians, retreated to their homeland and founded the third kingdom of Kush in Meroe, near the sixth cataract, with its own specific cultural traits and traditions. Most significant today are perhaps the steep royal pyramids at Meroe.[7] During Meroitic time, a great cultural shift took place in Egypt, where the Greek successors of Alexander the Great began to rule as the Ptolemaic dynasty and, after the death of Cleopatra VII in 30 BCE passed their rule on to the Romans. The cultural influence from the North into Meroe is therefore becoming increasingly a mixed Egyptian/Greek/Roman one. The

[5] For a short overview and some pictures of the history of Kush from an African perspective, see:
Kingdom of Kush—History Of Africa with Zeinab Badawi (Episode 4) https://youtu.be/CwaP1kyAqqo?si=HmQIrrF-cs3C8dIl where Zeinab Badawi from BBC returns to her home country Sudan looking for its ancient history.

[6] The Nile between Assuan in the north and Khartoum in the south passes through six areas of cataracts. These are numbered from north to south, beginning with the first cataract near Assuan up to the sixth cataract south of Meroe: https://en.wikipedia.org/wiki/Cataracts_of_the_Nile

[7] The archaeological sites of the region of Meroe are UNESCO world heritage: https://whc.unesco.org/en/list/1336/

Fig. 3.2 Illustration by Cailliaud, Frédéric in 1826—Voyage à Méroé, au fleuve Blanc, Public Domain, https://commons.wikimedia.org/w/index.php?curid=91555519

contact zone in Lower Nubia between the first and second cataract remained contested but was too far away from their heartland to be a threat for Meroe proper. Meroe at its zenith must have been a sight to behold (Fig. 3.2):

3.3 Jared Diamond and the Collapse of Civilizations

The Meroitic kingdom came to an end in the fourth century CE. There seems to have been a multivariate set of reasons for this. To illuminate these further, let us have a look at the work of Jared Diamond and his ideas about the Collapse of Civilizations.

Jared Diamond (*1937) has a long and very successful career as one of the leading generalists of the historical "Zeitläufte." As an American with eastern European roots, he skillfully both analyzes and explains the big picture of human civilization. He can tap into vast resources of several successful academic careers from physiology to environmental studies, including fieldwork in New Guinea. In his works, he combines his own everyday observations with archeological and scientific results from a wide range of other areas to come to far-reaching theories about the whys and ways of history including winning a Pulitzer Prize in 1998. In his 2005 book "Collapse" he researches and describes the fates of different societies in history and today and defines a set of factors contributing to their collapse. He starts with defining collapse:

> By collapse, I mean a drastic decrease in human population size and/or political/economic/social complexity, over a considerable area, for an extended time.[8]

The main question leading his research is:

> …the real question is why only some societies proved fragile, and what distinguished those that collapsed from those that didn't.[9]

As for the reasons for collapse, he came to understand it in most of the cases due to environmental causes and here often caused by an overuse of resources. The main factors contributing to societal collapse are according to him:

> Eventually, I arrived at a five-point framework of possible contributing factors that I now consider in trying to understand any putative environmental collapse. Four of those sets of factors—environmental damage, climate change, hostile neighbors, and [loss of] friendly trade partners—may or may not prove significant for a particular society. The fifth set of factors—the society's responses to its environmental problems—always proves significant.[10]

This leaves us with the tentatively positive outlook that collapse does and did not have to be inevitable but can be moderated or averted by rational decisions of a society.

Regarding Meroe, it appears that collapse did indeed happen. At some point around the middle of the fourth century AD, the archeological remains of Meroitic culture stopped, royal burials as well as the burials of the elite came to an end, temples were no longer built and maintained, the production of an exquisite kind of ceramic associated with the temples, the "Meroitic Fine Ware" ceased, as did the administrative inscriptions. The last known name of a Meroitic ruler known to us is Yesebokheamani, who is generally dated to the end of the fourth century.[11]

We will now take a closer look at how the factors defined by Jared Diamond relate to the end of the Meroitic kingdom. Fortunately, research interest in the study of Ancient Sudan shifted in recent decades from an archeological and epigraphic focus to align with Diamond's questions and research interests. As Diamond chose a different set of examples in his book, I will in the following only use his reference system to check against our current knowledge about the End of Meroe.

---

[8] Diamond, Jared 2005. Collapse: How Societies Choose to Fail or Survive. Penguin Books London. 3.

[9] Diamond 2005.10.

[10] Diamond 2005. 11.

[11] T. Eide, T. Hägg, R. H. Pierce, L. Török (eds.) 1998 Fontes Historiae Nubiorum, Vol. III. Textual Sources for the History of the Middle Nile Region between the Eighth Century BC and the Sixth Century AD: From the First to the Sixth Century AD. University of Bergen, Bergen. 1049–51.

3.4 Hostile Neighbors

When the archeological remains of Meroe vanish around the middle of the fourth century, two different protagonists were made responsible for that, the kingdom of Aksum and the Noba.

The kingdom of Aksum (second century CE—eighth century CE) was situated in the Ethiopian highlands. The Aksumites, who had already for a long time entertained close ties to the kingdoms of South Arabia[12] on the other side of the Red Sea and were therefore well versed in seafaring, took part in the international Indian-Mediterranean trade via land routes and through precursors of the Suez Canal, which were known to be trade routes already in the second millennium BCE.[13] Already in the second century, Aksum was described by classical authors as a metropolis with a royal palace.[14] The Aksumites used their own script Ge'ez (next to Greek and South Arabian script) and their own coinage.[15]

There are only a few sources for the End of Meroe. We are in fact lucky to have some written sources from the Aksumites. The most famous of these is probably the victory inscription of king Ezana from Aksum. It is written in three different languages[16] and tells the story of king Ezana's military campaign against Noba and Kushites.

In the Ge'ez version of his victory stela,[17] the Aksumite king Ezana claims for himself the titles of *"...Ezana,... king of Aksum, Himyar, Raydan, Saba, Salhin, Tsiyamo, Beja and of Kasu...never defeated by the enemy..."* He continues by describing his campaign: *"By the might of the Lord of all I campaigned against the Noba when the Noba peoples revolted and boasted "They will not dare to cross the Takaze (river)" said the Noba people. When they had oppressed the Mangurto, Hasa and Barya peoples, and when the black fought the red people and they broke their word for the second and third times and put their neighbors to death without*

---

12 The most famous of the South Arabian kingdoms was probably the kingdom of Saba, from where the legendary Queen of Sheba set out to visit Jerusalem bringing gifts for king Solomon. South Arabia was also one of the main centers for frankincense trade.

13 The range of the trade with India is quite spectacularly illustrated by the recent archaeological works in Egypt at the harbor site in Berenike, where, e.g., Buddha statues and a large amount of black pepper were found, see https://honorfrostfoundation.org/2021/07/02/berenike-excavations-at-a-red-sea-maritime-hub-linking-the-mediterranean-world-to-the-indian-ocean-ongoing/

14 Munro-Hay, S. 1991. Aksum. An African Civilization of Late Antiquity. Edinburgh University Press Edinburgh. 41.

15 for an overview expert discussion including trade routes and early Christianity, see the BBC podcast the Forum episode: The kingdom of Aksum: Africa's trading empire: https://www.bbc.co.uk/programmes/w3cszjwh

Aksum is also UNESCO world heritage site: https://whc.unesco.org/en/list/15/

16 For a detailed discussion of the sources, see Munro-Hay 1991 and Hatke, George. 2013. Aksum and Nubia. Warfare, Commerce, and Political Fictions in Ancient Northeast Africa. New York University Press New York. 71–137.

17 The stela is trilingual, written in Ethiopian Ge'ez, the South Arabian Sabaic and in Greek. https://en.wikipedia.org/wiki/Ezana_Stone

mercy, and pillaged our envoys and the messengers whom I sent to them to admonish them, and they plundered them of what they had including their lances; when finally having sent new messengers to whom they did not wish to listen but replied by refusal, scorn and evil acts; then I took the field. I set forth by the might of the Lord of the Land and I fought at the Takaze and the ford Kemalke. Here I put them to flight, and, not resting, I followed those who fled for twenty-three days during which I killed some everywhere they halted. I made prisoners from others and took booty from them. At the same time those of my people who were in the field brought back captives and booty. At the same time I burnt their villages, both those with walls of stone and those of straw. My people took their cereals, bronze, iron and copper and overthrew the idols in their dwellings, as well as their corn and cotton and threw them themselves in the river Seda (Blue Nile). Many lost their lives in the river, no-one knows their number. At the same time my people pierced and sank their boats which carried a crowd of men and women... (Description and names of spies and notables killed)...Thus five notables and a priest fell. I arrived at the Kasu, fought them and took them prisoner at the confluence of the rivers Seda and Takaze. And the day after my arrival I sent into the field the... (different troops)... along the Seda going up to their cities with walls of stone and of straw; Their cities with walls of stone are Alwa and Daro. And my troops killed and took prisoners and threw them into the water and they returned home safe and sound after terrifying their enemies and vanquishing them thanks to the power of the Lord of the Land. Next I sent ... (other troops) along the Seda, going down towards the four towns of straw of the Noba and the town of Negus. The towns of the Kasu with walls of stone which the Noba had taken were Tabito(?), Fertoti; and the troops penetrated to the territory of the Red Noba and my peoples returned safe after taking prisoners and booty, and killing by the might of the Lord of Heaven. And I erected a throne at the confluence of the rivers Seda and Takaze opposite the town with walls of stone which rises on this peninsula....[18]

Keeping in mind that this text is composed as victory inscription of the king and therefore a royal propaganda text including an (alleged) justification for the campaign and not a historical report, we can nevertheless learn several things out of it: In his titulary, Ezana assumes overlordship over the Kushites and calls himself the king of Kasu (Kush).[19] He describes a time of heightened tensions with the Noba who did not think he would cross the Takaze river and who oppressed other people who called for Ezana for help.[20] The behavior of the Noba described as a revolt does suggest a former agreement of overlordship by the Aksumites, probably with the river Takaze forming a border of some sort. As diplomatic endeavors did not bear fruit, Ezana had no other choice but to campaign against the Noba. (This is his justification for the campaign). He fought them at the Takaze river and followed them into their territories, describing what seems separate parts of the campaign which

[18] Translation after Munro-Hay 1991. 227–228

[19] This is a tradition which is known already from his predecessors in similar monuments, for the documents, see Munro-Hay 1991. 223–227. The standard formula was Beja and Kasu.

[20] A pattern of warfare not unknown in other historical periods including today.

led himself and other divisions of his army into different regions. An interesting fact is the passage where Ezana talks about the *"towns of the Kasu with walls of stone which the Noba had taken"* in contrast to the *"towns with walls of straw."* It is therefore tempting to assume a setup in which the Noba had towns with walls of straw (round huts) whereas the Kushites lived in towns with walls of stone, an assumption which can be supported by the archeological finds. Some of these towns had been taken over by the Noba already before the Aksumite campaign. From the text it is also probable that Meroitic and Noba towns were at this time situated side by side in the same region at the same time.[21] Ezana is however talking about fighting the Kushites which would mean that they did still have some kind of organized military at that time.

Another text fragment from the city of Meroe further helps interpret the possible historical events. It is a fragmentary Aksumite victory text written in Ancient Greek, which was part of one of the thrones which were erected by the Aksumites in the place of their victory (also described in the Ezana text).[22] Unfortunately, the names of the kings are not part of the fragment; we know neither the name of the Aksumite king nor the name of the (possibly last) Meroitic king. However, in this fragment, the god of war is mentioned in its Greek form as Ares, in contrast to the Ezana stele, which was erected after Ezana converted to Christianity and where he speaks of the Lord or Lord of Heaven as his protector. This gives us a rough timeline, and we can tentatively reconstruct a chain of events where a predecessor of Ezana campaigned against Meroe and left several victory inscriptions which were not destroyed or removed by the Meroites. After these events, which would also have included the Noba if they were already living inside Meroitic territory, an agreement or peace treaty would have been established with the river Takaze as border. Sometime between this inscription and the Ezana text, the Noba conquered at least some of the Meroitic "cities of stone" and began hostilities against their neighbors to challenge the Aksumite sphere of influence, which finally led to Ezana's campaign.

The Noba themselves did not have their own script at that time. We do not know if their communications took place via texts in a common language like Greek, or functioned strictly via oral messages, that is through messengers. Their archeological material shows a distinct material culture which we know mainly from graves. These graves have a different architecture from Meroitic ones, and they had a different set of ceramics buried with them. We know from classical author Strabo citing Eratosthenes that in the third century BCE the Noba settled on the opposite side of the Nile from Meroe, that they were a large group divided into different tribes and that they were not subject to the Meroitic kings.[23] Accordingly, archeology shows that the large graves of the Nubian rulers containing many iron objects were situated

[21] From the topography described it is unlikely that the Aksumites actually came to the capital Meroe during this campaign.

[22] Fontes Historiae Nubiorium III 1998. 1066–1070. Another even more fragmentary inscription was found on one of the temples of Meroe. Ibid 1070–1072.

[23] T. Hägg, I. Hjortsjö, M. Pernler, N. Török (eds.) 1996. Fontes Historiae Nubiorum: Textual Sources for the History of the Middle Nile Region between the Eighth Century BC and the Sixth

on the other side of the river from Meroe, suggesting that the Noba of the fourth century still regarded the Western Bank of the Nile their homeland as well as presumably having access to iron smelting technology and using it for their weaponry as well as for other prestige goods.[24] We only get a closer look into the Noba from the sixth century onward when they became Christians and adapted writing systems in Old Nubian and Greek. During that time three distinct Christian kingdoms of Nubia emerge, which last until the end of our Middle Ages.

To sum up, textual sources and archeological finds from the fourth century CE both show a weak Meroitic state. Beside the end of practically all Meroitic elite culture, royal victory texts from Aksum describe different campaigns against Meroe, one of them with a victory over Meroe, the other with a victory against the Noba and the Kushites. To celebrate and remember their victories, the Aksumites erected inscribed thrones at the site of the victory and at home. We know of at least one other in Meroe (from which we have a fragment) and another in an unnamed town (known from another text). Due to the fragmentary state of the source for the campaign in Meroe, we are neither given a reason of the first campaign nor can tell of the situation of the state of Meroe at that time or name the rulers on both sides. The Ezana stele then tells about another campaign during which the Noba took advantage of the Kushites by conquering some of their cities. After the first campaign, the Meroites did not remove the Aksumite victory inscriptions from their capital, which could very reasonably be interpreted as a sign of weakness.

3.5 Loss of Friendly Trade Partners

In order to understand the state and organization of Meroitic trade it might be of use to take a look at the general setup of the Meroitic state. The Meroitic State as seen from the sources exuded hegemony over distinct parts of territory and distinct groups of people and worked as a redistribution economy without monetary system. Geographically, there were four distinct regions along the Nile, Lower Nubia, the border zone to Egypt between the first and second cataract, the region around the third cataract around former Kerma, the fourth cataract region including Napata and Gebel Barkal, and the "Meroitic Heartland" between the Nile and its confluence with the Atbara in the North and the Blue Nile in the South.[25] The eastern regions further away from the Nile were inhabited by non-sedentary herders summarized as

Century AD, Vol. II: From the Mid-Fifth to the First Century BC. University of Bergen Bergen. 557–561.

[24] *«Es ist anzunehmen, dass in den grossen Tumulusgräbern von El Hobagi lokale Fürsten begraben waren, die aufgrund der reichen Waffenausstattung als «Warlord» anzusprechen sind."* Lohwasser, A. 2013. Das Ende von Meroe. Gedanken zur Regionalität von Ereignissen. In: Feder, F; Lohwasser A. (eds.). Ägypten und sein Umfeld in der Spätantike. Vom Regierungsantritt Diokletians 284/285 bis zur arabischen Eroberung des Vorderen Orients 635–646. Akten der Tagung vom 7.-9.7.2011 in Münster (Philippika 61). Harassowitz Verlag Wiesbaden. 275–290.

[25] There are indeed ideas that three of the different regions of Meroe did find a different end and finally resulted in the three longstanding Christian kingdoms of Nubia. Lohwasser 2013. 276

Blemmyes in the classical sources and Beja of the Ancient Egyptian and Aksumite sources, whereas the inner reach of the Nile Bend, the Bayuda, was Noba territory.[26]

The different regions were held together by the sovereignty of the ruler and their power to redistribute centrally acquired luxury goods. The connection between international trade and the Meroitic economy can therefore be summarized as follows:

> …the rather patchy geographical evidence of royal investment implies that territorial sovereignty was limited to the state's core areas in the Meroitic heartland and central places in Upper Nubia. Here, the kingdom's territorial sovereignty was realized by controlling population and subsistence resources as well as by direct accumulation, redistribution, and organization of the exchange of agro-pastoral surplus, whereas in between these core areas and in the periphery, direct coercive power was rather exercised by local elites. To maintain royal suzerainty over their local sovereignty, the Meroitic crown was forced to legitimate its authority by means of inter-marriage, ritual performance, and hegemony as well as by a prestige-goods economy, providing valuable commodities like wine, oil and honey from Ptolemaic and Roman Egypt and fine tableware, furniture, jewelry and luxury items of Mediterranean style or origin… Apart from local production of Hellenizing-style items in the kingdom's core areas, the royal house was able to acquire such commodities through diplomatic gifts and exchange against slaves, gold, iron tools, gum Arabic, and natron, as well as inner-African exotica. The control over craft production in the core areas, a certain military force for raids and the monopoly of long-distance exchange were therefore vital to sustain economic and political power. The loss of particularly the latter monopoly finally resulted in dramatic political consequences at the end of the Meroitic period.[27]

According to this point of view, the Meroitic rulers were dependent on long-distance trade of prestige goods to keep the goodwill of the regional elites supporting them. Part of the social cohesion of Meroe seems therefore to have relied on a functioning import–export system and on the constant availability of luxury goods. Another part was realized by means of prestigious building projects like temples and water reservoirs.[28] The nonmaterial part of Meroitic rule would have been realized by means of state religion and superregional ceremonies where people from a

[26] Already in Napatan times the newly crowned king had to pay a visit to the main temples of the different regions to see his coronation confirmed.

[27] Wolf, P; Nowotnick, U. 2021. The Meroitic Heartland. In: Emberling, G; Williams, B.B.(eds.). The Oxford Handbook of Ancient Nubia. Oxford University Press New York. 511–544. 522

[28] The Meroitc state built several known water resources (hafirs), often near Meroitic temples. https://de.wikipedia.org/wiki/Hafir

see also Weschenfelder, P. 2018. An Ethno-archaeology of Mobile Networks on the Island of Meroe. In: Tipper, S; Tully, G. (eds.). Current Research in Nubian Archaeology. Gorgias Press Piscataway, New Jersey. 89–108. Her ethno-archaeological survey in the Butana illuminates possible forms of cooperation and mutual trust building between different groups. The central element here is the access to water, another possibility is lending out animals for a certain time for rearing or transhumance. To keep up the long-distance trade networks of antiquity but also of today, there had to be infrastructure like watering stations, local guides and local markets… and mutual trust in keeping to the respective agreements.

larger area came together and celebrated on a regular basis. This kind of setup is inherently more stable during prosperous times than in crises.[29]

During the later part of Meroitic history, several problems seemed to have arisen for the trade in different parts of the country as well as internationally.

Internal territorial problems could have made long-distance trade less secure and limited in its routes. From the lack of Meroitic archeological material from that time, there is reason to assume a takeover of a local Noba group in the region between the third and fourth cataract sometime during the third century CE.[30] How much this scenario influenced the Nile trade remains unclear—was a significant part of the trade negotiated via the river or along it? If so, did the Noba simply establish their own taxation system or make the river route impassable for Meroitic trade? However, the situation, it did not make trade and internal communication to Lower Nubia impossible, only limited the possibilities to an inland route shortcutting the Nile bend from halfway between the fourth and the fifth cataract to Lower Nubia that could be used to keep trade and communication going and had already been in use for a long time.[31] Besides possibly limiting the choice of communication and trade routes, the reduction to only one passage to the North also made this route more exposed to taxation or raids by the Blemmyes/Beja people living in the area. A formerly presumably easy route along the Nile through Meroitic territory had now ceased to be Meroitic territory and the only remaining desert route also went through territory which cannot reliably be described as friendly.

Another problem was the rise of Aksum as a competitor in international trade. Understanding trade from archaeological records holds its own challenges:

> Archaeological evidence for these (trade) contacts is derived mainly from artefacts and materials recognized as having originated far from their areas of distribution may be inadequately reflected in the archaeological record as a result of uneven coverage of research. Evidence for political relations—whether military, religious or diplomatic—comes primarily from oral and written sources; these present remarkably little overlap with the archaeology and are usually subject to varied emphasis and distortion according to the sympathies of the societies where they were created and/or transmitted.[32]

[29] A typical scenario here is a gradual shift of power from the center to the periphery and from the royal family to the elites or other powerful segments of society. Such a shift of power is also assumed by Wolf/Nowotnik 2021. 531–532

[30] "*Ich meine aus der Zusammenstellung der archäologischen Belege der früheren Zentren... zu lesen, dass diese vom Königshaus bereits spätestens im 3.Jh. aufgegeben wurden.* Lohwasser 2013. 282.

[31] The Korosko road, which had the additional advantage of passing through gold mining territory and also branched off westwards to the third cataract. «*Trotzdem ist eine Verbindung zwischen dem Kernland im Süden und Unternubien ganz deutlich,...Da die Verbindung entlang des Nils aber durch den Abfall von Prä-Makuria unterbrochen war, musste auf eine Inlandsroute ausgewichen werden. Diese sehe ich in der Korosko Road.*" Lohwasser 2013. 282.

[32] Phillipson, D.W. 2012. Foundations of an African Civilisation. Aksum & the northern Horn 1000 BC—AD 1300. James Curry Woodbridge. 195.

This is especially true when archeological evidence is scarce and the few remaining text sources from Aksum and from classical authors hold inherent bias.

Our knowledge about Aksumite sea trade already in the first century CE comes from a unique source, the "Periplus of the Erythrean Sea." In this, the unknown author describes the different ports of the Indian trade from the Red Sea as far as India including distances from each other, special topographic features, their people and their possible trade goods in amount and quality as well as the best season to sail there.[33] The passage about the Aksumite trade goods reads:

> There are imported into these places, undressed cloth made in Egypt for the Berbers; robes from Arsinoe; cloaks of poor quality dyed in colors; double-fringed linen mantles; many articles of flint glass, and others of murrhine, made in Diospolis; and brass, which is used for ornament and in cut pieces instead of coin; sheets of soft copper, used for cooking-utensils and cut up for bracelets and anklets for the women; iron, which is made into spears used against the elephants and other wild beasts, and in their wars. Besides these, small axes are imported, and adzes and swords; copper drinking-cups, round and large; a little coin for those coming to the market; wine of Laodicea and Italy, not much; olive oil, not much; for the king, gold and silver plate made after the fashion of the country, and for clothing, military cloaks, and thin coats of skin, of no great value. Likewise, from the district of Ariaca across this sea, there are imported Indian iron, and steel, and Indian cotton cloth; the broad cloth called monachê and that called sagimtogênê, and girdles, and coats of skin and mallow-colored cloth, and a few muslins, and colored lac. There are exported from these places ivory, and tortoiseshell and rhinoceros-horn. The most from Egypt is brought to this market from the month of January to September, that is, from Tybi to Thoth; but seasonably they put to sea about the month of September.[34]

This excerpt nicely shows the variety of the goods traded on a daily basis, which would range from luxury to everyday items. Aksumite interest is described more in metal than in wine and olive oil and its main export is ivory, tortoise shell and rhinoceros's horn. At that time, much of the import consisted of manufactured items while raw material was exported.[35] Also, the export of ivory brought Aksum into direct competition with Meroe already in the first CEntury CE, surely on the market but probably also with the suppliers.[36] Iron was a possible export product of Meroe and as such (although it would not be visible in the archeological material) either was not traded to Aksum, was more expensive than the iron coming from India or did not come in enough quantities. Aksum seemingly did not have direct access to the gold resources of the Meroitic kingdom[37]

---

[33] For a short overview: https://en.wikipedia.org/wiki/Periplus_of_the_Erythraean_Sea

[34] https://en.wikisource.org/wiki/Periplus_of_the_Erythraean_Sea paragraph 6.

[35] Which also constituted one of the problematic aspects of colonial trade.

[36] From the third century onwards the supply of ivory to Rome has risen and *"a corresponding fall in value was recorded in the price-control edict issued by the Emperor Diocletian c.301"*. Phillipson 2012. 196.

[37] *"Rapid debasement of the coinage... could imply that gold itself was in short supply, at least from the mid-fourth century onwards, despite continuation of its use for embellishments of base-metal issues"* Phillipson 2012. 165. During the fourth century we can assume a certain amount of gold coming from Meroe to Aksum as spoils of war as well as in tributes.

There is in fact not a lot of evidence regarding a direct relationship between Meroe and Aksum. As direct neighbors they must surely have been aware of each other during all this time. However, we know of no written sources apart from the Aksumite victory inscriptions and only small and isolated finds hint to an exchange of goods. A lack of mutual interest in trade would be surprising but not impossible. More likely, the traded goods simply are not visible in the archeological material. Animals such as cows, goats and sheep, textiles and leather as well as raw materials like gold and iron would not have left any traces in the archeological material.[38] On the other hand, a substantial and direct trade between Meroe and Aksum should have left more traces. Meroe and Aksum trade could de facto even have worked parallel to each other, not trying to access the same regions for resources and serving different routes, the Meroites via the Nile, the Aksumites via the Red Sea,[39] although there certainly existed routes between Aksum and Meroe.[40] A possible parallel trade ended during the third century at latest when we know of the construction of a road by a once again unknown king between Aksum and Egypt.[41] From this time on, Aksum is also in direct competition with Meroe in trading goods overland with Egypt.

Looking at the recipient's side of the trade, the internal crises of the Roman Empire and also of Roman Egypt could also be felt in the trade:

> Roman society was haunted by the specter of civil war since the first century BCE. The imperial monarchy was essentially based on the promise that only the almighty emperors could establish and maintain inner peace. But already the second princeps, Tiberius, is reported to have said that being emperor meant to hold a wolf by the ears. For centuries, the Roman monarchy resembled the dance on a volcano. Countless emperors were being killed, and from the third century on it became increasingly difficult to control this problem, culminating in the disintegration of imperial rule in the West.[42]

[38] This kind of information can only be obtained by provenience analysis. Jacke Phillips has compiled a list of the known pieces linked to Meroitic—Aksumite relations: Phillips, J. 2014. The Foreign Contacts of Ancient Aksum: New Finds and some Random Thoughts in: Angelika Lohwasser and Pawel Wolf (eds.) Ein Forscherleben zwischen den Welten. Zum 80. Geburtstag von Steffen Wenig (Der Antike Sudan. Mitteilungen der Sudanarchäologischen Gesellschaft zu Berlin e.V. Sonderheft) 253–268.

[39] Phillipson 2012. 200. Aksum profited from its direct connection to South Arabia and seems to have actively taken part in the India trade. For knowledgeable and easy access to information regarding the Red Sea trade, see also: https://www.ancientportsantiques.com/ancient-sailing/red-sea-versus-nile-ancient-sailing/

[40] For a recent survey along the ancient caravan route between the Nile and the Red Sea, see Bashir, M.S. 2017 Archaeological Survey along the Berber-Suakin Caravan Route: Preliminary Report in: Sudan & Nubia. Bulletin of the Sudan Archaeological Research Society London 21. 204–210.

[41] As learned from another victory inscription in Greek from the throne of an unknown Aksumite king, only known of by its copy by classical author Cosmas Indicopleustes: https://en.wikipedia.org/wiki/Monumentum_Adulitanum

[42] Börm, H. 2022. The End of the Roman Empire: Civil Wars, the Imperial Monarchy, and the End of Antiquity. In: Gehler, M; Rollinger, R; Strobl, P. (eds.) The end of empires. Springer Fachmedien Wiesbaden. 191–212. 199–200. For a short introduction, see, e.g., Patrick Wyman The Fall of Rome Podcast. Episode 5 "Just how screwed up was the later Roman Empire" (from every major podcast app). The central element of Roman rule was trust in the Roman model. When this trust

The political situation in Rome was not reliably stable and military costs rose. In contrast to the political struggles of Roman elites and military commanders, international trade was carried out by private individuals, the state profited mainly in the form of taxes. The trade goods coming from India were often loaded onto caravans in the Red Sea harbors like Berenike because northerly winds made sailing difficult north of there. Infrastructure like roads and watering stat*wards*ions had been built and maintained by the Roman state and secured by the military. Using it would have required payment of a road toll.[43] To understand the scale and cultural impact as well as the volume of the Indian trade in the second century we can rely on the testimony of Papyrus Muziris:

> Customs dues on cross frontier trade, where we have direct evidence for their levels on the eastern frontiers, were 25% at least until sometime in the third century CE. The so-called 'Muziris papyrus' values the cargo of the ship Hermapollon, which sailed from Muziris in India to the Red Sea in the mid-second century CE, at 1,151 talents, 5,852 drachmae of silver, or nearly 7 million sestertii. (To put this in perspective, to qualify as a senator one had to own property of at least 1 million sestertii.) This valuation was made after the deduction of the 25% customs tax, so that the pre-tax value of the cargo was 9,215,803 sestertii, and the customs dues were 2,303,951 sestertii on this cargo alone… We do not know whether the Hermapollon was typical or exceptional, though statistical probability is in favour of the former; but over a hundred years earlier Strabo (Geography, 2.5.12) says that 120 ships per year left Myos Hormos for India, and if we imagine just 100 such cargoes each year, then the customs revenues on imports from India alone would total c.230 million sestertii, one-third of Rome's estimated annual military budget of 643–704 million sestertii.[44]

This made the Indian trade a highly profitable business for the income of the Roman State. To further encourage the business of overseas transport we know of significant tax exemptions for shipowners who put a ship of over 350 t capacity or several ships of 70 t at the disposal of the state.[45]

Roman roadbuilding would surely also have been an example for the roadbuilding of our unknown Aksumite king in the third century between Aksum and the Nile valley (see above with footnote 39).

got lost, a lot more was falling into pieces. Another popular interpretation blames the adoption of Christianity for much of the Empire's crises, Melvyn Bragg discusses with history experts in the BBC programme "In our time" The Roman Empire's Collapse in the fifth century: https://www.bbc.co.uk/programmes/p00547ds

43 Wilson, A. 2024. Trade in the Roman Empire. In Jeremy Tanner and Andrew Gardner (eds.) Materializing the Roman Empire UCL Press London. Chap. 4. 99–121. 113.114.

44 Wilson 2024 Trade.116.

45 Wilson 2024 Trade. 108.

3.6 Environmental Damage

Meroitic iron production has long been held in suspicion for environmental damage from overuse of wood resources. Already the first archeologists working in the Sudan noticed great slag heaps surrounding Meroe and it got a reputation as "Birmingham of Africa." From today's perspective, Meroe still holds *"some of the most extensive evidence for iron production in Africa."*[46]

Invention and widespread use of iron technology came about during the first half of the first century BCE and changed the political landscape as iron weapons became a significant military advantage. The process of iron production needed much more skills and material than bronze production. To reach the high temperatures of over 1000 degrees Celsius needed for smelting iron, a special kind of kiln fired by charcoal was necessary.

> All stages of the production of an iron object required various inputs and resulted in specific outputs and required specialist technological knowledge and skill. At the end of a complete iron production episode, the iron slag (the waste product containing various amounts of parent materials including the iron ore, clays, and fuel…), was discarded along with other waste such as broken or failed fragments of the furnace structure and of the blowpipes (tuyères) used to supply air to the furnace, as well as unreduced ore and wasted fuel. Over time, this discarded waste was piled up, creating for example the famous slag heaps of Meroe.[47]

> With the adoption of iron production by society, a specialized and particularly resource and labor-thirsty technology joined the local craft scene, producing a valuable material with properties "far in advance" of anything that had gone before. Despite the technical skill, effort and materials required to produce iron, this production was viable and, in some cases, became of paramount importance to society, because it provided a material that could be worked to produce tough, durable objects as well as ornamental and prestige items suitable for local consumption and for external trade. From rings, pins, bracelets and bells to hoes and axes, to spears and arrowheads, the malleability of heated iron meant that a smelted bloom could be easily forged and shaped into desired objects with good tensile strength. Thus, the practical significance of iron lay in the vast array of objects that could be produced, as well as the superior mechanical properties of these objects.[48]

Reaching genuine conclusions from the archeological material of the slag heaps is very difficult as we lack practically all relevant information. On the iron production side, much of the extensive material remains unexcavated and even excavated exploratory trenches are often not meaningful because of the variety of the material

---

[46] Humphris, J. 2021. Iron Production in Meroe. In: Geoff Emberling and Bruce Beyer Williams (eds.) The Oxford Handbook of Nubia. Oxford University Press New York. Chap. 48,975–994. 975. We are talking about 5000–10,000 tons of slag material here. Humphris, J. and Rehren, T. 2014. Iron Production and the Kingdom of Kush—an Introduction to UCL Qatar's research in Sudan in: Angelika Lohwasser and Pawel Wolf (eds.) Ein Forscherleben zwischen den Welten. Zum 80. Geburtstag von Steffen Wenig (Der Antike Sudan. Mitteilungen der Sudanarchäologischen Gesellschaft zu Berlin e.V. Sonderheft) 177–190. 182.

[47] Humphris, J. 2021. 976.

[48] Humphris, J. and Rehren, T. 2014. Iron Production. 177.

and an uneven layering of the finds. The excavated material afterward requires extensive and cost-intensive scientific analysis. On the consumer side we lack any kind of demographic data of the time and the knowledge of how widespread the use of iron objects would have been. We also do not know much about the amount of iron produced, which, together with the previous information, could give us an idea of possibilities and amount of trade of iron.[49]

What we do know is that we are talking about a timespan of over 1000 years of iron production at the same place with a period of heightened productivity between the second half of the first millennium BCE and the mid of the second century AD.[50] During that time, societal requirements as well as fashions probably changed more than once. The archeological iron material is not extensive, most of it coming from graves and a majority being weapons.[51] The main reasons for the scarcity of actual iron finds are probably corrosion and recycling of the precious material. For most of this timespan, almost all of the known iron production has been taking place in the environs of Meroe City. Smaller slag deposits found in several other places such as Hamadab at the outskirts of Meroe and at Musawwarat es-Sufra can be dated to Post-Meroitic time.[52] At Hamadab, there is evidence of fewer smelts per year in kilns not suitable for repeated uses.[53] It is therefore possible to assume a centralized iron industry in Meroitic times and a decentralized smaller scale iron industry using a slightly different kind of kilns in Post-meroitic times. The general impression about iron production in Meroe considering the longtime of production is therefore that it was still small to medium scale in contrast to the first impression of the amount of slag remains.

During all this time, the type of ore used, and the wood used for charcoal production remained the same. The ore was mined from easily accessible deposits on hilltops ca 9 km outside the city[54] and the wood used was consistently Acacia Nilotica.[55] Iron production in general needs a significant amount of wood for producing charcoal and is therefore often connected to an overuse of resources. The evidence from Meroe shows the use of a single type of wood for charcoal production over a timespan of over 1000 years without signs of a shortage. Acacia Nilotica was available abundantly in the area. With its dense wood and high calorific value, it is a perfect

[49] Humphris 2021. 977–978.

[50] Humphris 2021. 982.

[51] For an overview of the iron objects from Meroe see Abdelrahman, M.F. 2011. A New Study Concerning Kushite and Post-Meroitic Iron Objects in: Vincent Rondot, Frédéric Alpi & François Villeneuve (dir.) La Pioche et la Plume autour du Soudan, du Liban et de la Jordanie: Hommages Archéologiques à Patrice Lenoble. PUPS Paris. 391–402.

[52] For an overview see Abdelrahman 2011. 393–395.

[53] Humphris 2021. 981;983; 986.

[54] Humphris 2021. 986.

[55] Consistently about 86% of the charcoal samples proved to be of Acacia Nilotica. Jane Humphris & Barbara Eichhorn (2019) Fuel selection during long-term ancient iron production in Sudan in: Azania: Archaeological Research in Africa, 54:1, 33–54. 43–46.

choice for charcoal production.[56] It is also a fast-growing resistant pioneer plant able to survive floods as well as extreme temperatures. Even when cut down almost to the ground, it is still able to regrow quickly and even become a weed. Furthermore, it had been in use to produce gum Arabic for a long time, a substance both used as a binder, in medicine and for tanning leather. This means the plant was well known in its properties and used already for a long time. From these findings, the first impression of an overwhelmingly big iron production has to be reevaluated. Taking into account the long period of iron production at the same place of over 1000 years, we should talk about a small or medium-sized iron production. The study of charcoal remains shows the continuous use of Acacia Nilotica for coal production, a species very well suited for the purpose. Acacia Nilotica is a fast-growing and resistant plant which does not show signs of overuse during the centuries, at least for iron production, either because it was always growing in abundance or because a conscious wood management.[57]

3.7 Climate Change

The next in our set of questions to the material deals with the impact of climate change in Ancient Sudan.

Findings in that field have been facilitated by advanced technology and scientific analysis. A recent study used isotopic analysis of dental enamel samples of individuals from Upper Nubia from different periods of Meroitic and Post Meroitic times to determine their main sources for nutrition. Additionally, human samples outside the given space and time as well as samples from different animals have been analyzed to calibrate the findings. The results show a shift in the change of diet from the so-called C3 crops like wheat and barley to C4 crops like sorghum and millet, that is a shift toward plants which are more suited to a dryer climate. This hypothesis is also confirmed by the isotopic data regarding the water consumed.[58]

This study fits in other studies regarding diminishing flood levels in Roman Egypt, the "breadbasket of the Roman Empire." The general situation of flood-based agriculture in Egypt was the following:

> Before the twentieth century damming projects, the monsoon rains in Ethiopia caused flooding in the summer months. Every year in July the waters of the River Nile started to rise on Egypt's southern borders in Syene. Week by week the waters rose higher and slowly started to burst the riverbanks and cover the adjacent flood plains along the river from south to north. The highest levels were reached in August and September. While the plains were inundated, villages along the river turned to islands, and voyaging was only possible by boat. In September/October the waters then slowly receded and left rich alluvial deposit on

[56] Humphris and Eichhorn 2019. 48.

[57] Humphris and Eichhorn 2019. 48–50.

[58] Kozieradzka-Ogunmakin, I.; Sołtysiak, A. 2023. Isotopic evidence of an environmental shift at the fall of the Kushite kingdom of Meroë, Sudan in: Antiquity 97 (396): 1501–1515.

> the fields along the river. In mid-November the waters in northern Egypt had receded so far that sowing could begin.[59]

Such a high reliance on the Nile flood as the single constituent of the agricultural supply of the regions along the Nile was often problematic. Too little water for several years in a row would pose grave problems for the food supply of the people but on the other hand, also floods could prove fatal in this aspect.[60]

Our data is once again patchy and has to be collected from many different sources and different fields. Around 260 CE, ice core records show the biggest volcano eruption of the first 500 years CE. Correlation with international tree ring data from Siberia, Scandinavia, and the Alps show normal years before 265 so that the unknown volcano likely erupted in 265 CE.[61] Records of that time in Egypt show a correlating series of problematic floods. Other relevant data from the fourth century comes from sediment cores in the Nile Delta that suggest a drought in the fourth century,[62] coinciding with the vanishing of Meroitic civilization.

> I argue that we witness a transition to a century-length episode of greater aridity and lower river flow starting at the beginning of the third century CE. Paleoclimate proxies indicate the onset of a general shift in the East African monsoon pattern and a switch to a drier climate during the third century CE. Reduced monsoon rainfall in the northern Ethiopian highlands would have affected summer Nile floods, leading to gradually lower Nile levels.[63]

In our sources, we only see the consequences of environmental changes such as an orientation of the non-sedentary population toward the Nile as their traditional regions did not provide sufficient subsistence. This can be seen in the Noba overtaking of the region of the third and fourth cataract as well as a conflict between Blemmyes/Beja and Noba to take over Lower Nubia which was deserted by the

---

[59] Huebner, S.R. Climate Change in the Breadbasket of the Roman Empire—Explaining the Decline of the Fayum Villages in the Third Century CE in: Studies in Late Antiquity (2020) 4 (4) 486–518. 488.

[60] As Huebner puts it: "*...any change in the flood timing or quality would have had a significant impact*

on the agricultural year. A lower-than-normal flood meant that not all fields were covered by the floods and needed to be artificially irrigated if this was possible. In addition, less fertile silt was deposited in the fields. Moreover, a series of insufficient Nile floods would mean drying out of fields and dune invasion resulting in a poor harvest and associated famines. A strong flood, on the other hand, would destroy any infrastructure such as dikes that had required several years to construct, wash away any adjacent buildings, and even endanger the villages themselves. Breaches in dams and the lake bursting its banks would also result in contamination of the soil with excessive salt. If the waters left too soon, the fields were not sufficiently watered. If the waters remained too long, sowing had to be postponed and the season for harvesting was cut short as a consequence. Furthermore, stagnating water favored the development of bacterial disease and plant parasites in the soil. Any changes in the annual Nile flood thus had a direct impact on Egyptian agricultural economy, harvest size, and tax revenues." Huebner 2020. 505–506.

[61] Huebner 2020. 512–513.

[62] Tahir, Y.F. 2017. Environmental Archaeology of the Greater Meroe Region in: Beiträge zur Sudanforschung 12. 229–252. 237 with Fig. 4.

[63] Huebner 2020. 517.

Romans struggling with their own crises. Further evidence for a perceptible shift in the subsistence base of Meroe can be seen in the building of the Hafir water reservoirs. The size of some of them and their location next to Meroitic temples suggest a state enterprise in critical times undertaken either for the benefit of its own people doing seasonal transhumance, for the benefit and goodwill of the non-sedentary people of the Blemmyer/Beja of the Eastern desert or for a mixture of both.

3.8 Society's Response to the Problems

The last and often decisive criterion in Jared Diamond's theory about the collapse of societies is the reaction of societies to the given problems. Do they recognize the problem, do they search for solutions, and does the given structure of society have the means needed for the necessary changes? Examples of different outcomes to similar sets of problems are the fate of Norse settlement in Greenland versus in Iceland[64] or the markedly different development of the states of Haiti and the Dominican Republic on the same island.[65] It is important both in reality and in psychology not to negate the agency of a state or an organization to change. On the other hand, substantial and lasting change in often already challenging situations—to turn the tide—is often hard or even impossible as each society and other organizational leaderships are dealing with their own complex rather than complicated situation and require the active participation of relevant parts of society to succeed in fundamentally changing their situation.[66]

At the end of these considerations, a short recapitulation of the different markers of societal collapse outlined by Jared Diamond and its application to the End of Meroe shows a clear picture: The impact of a changing climate set the non-sedentary people in motion toward the Nile where there ensued conflicts with the Meroitic people. This also had an impact on the trade routes along the Nile and the territory under Meroitic control. At the same time, the Indian trade of the Roman empire expanded massively. Rome hugely profited from taxes and goods imported. They built ports and roads to facilitate the trade and established local partners like Aksum, which was well suited to serve as a waypoint for the Indian trade, whereas Meroe got left behind and did not have any possibilities to change the trade. Noba hostilities from within the territory of Meroe and Aksumite military action finally put an end to a kingdom already weakened from the other circumstances. Against first impression, there seems not to have been an environmental impact of the local iron industry overusing the wood resources. As far as we can tell, the Meroitic rulers were aware of the problems which arose from changing trade routes and a dryer

[64] Diamond 2005. 211–276 (Norse Greenland;179–206 (Iceland).

[65] Diamond 2005. 229–257.

[66] As shown nicely in the example sustainable transformation in the insurance sector from another book in this series: Stricker, L. 2022. The Sustainability Transformation in: Peter Wollman und Reto Püringer (eds.): Transforming Public and Private Sector Organizations. Springer Cham. Chap. 18,325–344.

climate with less water available also from the Nile floods by building water reservoirs to counter the drought. We do not know of other measures taken but the obvious one would have been to better the relationship to Roman Egypt and Rome in order to stabilize the trade and maybe get reinforcements along the Nile route. On the other hand, we still see the surplus of Meroe going into elite infrastructure and the upkeep of religious institutions. The Meroites could also have introduced another language as lingua franca or try to change their economy into a monetary-based one.

3.9 Conclusion: Modern Sudan and UN Sustainability Goals

This article started in the spirit of the Three-Pillar Model as exploration of the end of the Meroitic Kingdom along the ideas proposed by Jared Diamond about the collapse of civilizations. I assumed that this would be a straightforward endeavor of putting together some scientific articles and coming to the conclusion that the Meroites most probably overused their resources just as we are doing today and which would have been one of the main reasons for their own demise. As I graduated my MA in Egyptology with Sudan Archaeology as one of my minor subjects at Humboldt University zu Berlin quite some time ago, this would give me the possibility to read up on the recent research and once again follow my passion for history and archeology. However, honoring the scientific method, I had to challenge and change my hypotheses on the way. In this, my findings hugely benefited from the use of an external theoretical model. Using the ideas of Jared Diamond as a blueprint for my questions to the sources gave me a fresh perspective and forced me to expand my reading to scientific research outside the "usual suspects" of the discipline. My first hypothesis of man-made catastrophe would have aligned nicely with the UN sustainability goals as an example from far-away places and times not to be repeated today and we are actually living it. For the reasons explained in the text, this did not work out. I therefore had to think about the connection of my article to UN Sustainability Goals in a new way and came up with two things, first thinking about the general use of applying UNSDGs on historical societies and then, taking Goal 13 "Climate Action" and weaving some thoughts around this.

The UN Sustainability goals[67] are outlines for organizations as well as individuals about different fields and measures to engage in in order to build and preserve a world worth living for future generations. They are the stabilizing measures of a rule-based world order meant to empower and encourage free and equal individuals to work together toward this goal. These goals are, however, modeled along our modern understanding of an ideal society based on equality and separation of powers. They are very much a late product of enlightenment and most of them cannot be transferred to a pre-modern world in which catastrophes were products of divine doing and in which status and well-being of individuals relied on destiny more than on a benign ruler.

[67] Graphic illustration: https://sdgs.un.org/goals

Therefore, the most likely connecting factor from UNSDGs to the Sudan in Antiquity and today is SDG 13 "Climate Action," also in relation to the political situation today in the Sudan and worldwide. In Ancient Meroe, one of the two main reasons for the frailty of the society was the impact of climate change. The climate must have changed perceptibly as there was less rain and the monsoon patterns at the sources of the Nile changed and led to unreliable yearly Nile floods. Consequently, there arose struggles over the distribution of basic resources like water and food as traditionally nomadic parts of the population came to the Nile and came into conflict with sedentary farmers. This sounds remarkably modern today as we are also in the middle of a very fast and far-reaching climate change[68] where one of the most affected countries is once again Sudan. The Darfur Conflict in 2003 was labeled as the "world's first climate change conflict"[69] (which seems slightly misleading as many of the conflicts in world history came about in a similar way when climate changes triggered struggles over more scarce basic resources). The political situation in Sudan since its independence from Egypt in 1956 has been the story of much internal conflict. However, the country mastered the independence of South Sudan 2011 and the end of authoritarian rule under Omar al Bashir, which ended 2016/17 from a big democracy movement. The military involved did not keep their promises of a peaceful transition to democracy but instead started another huge crisis. Since 2023, a gruesome civil war left over 13 m people displaced and about double the number are seriously threatened by hunger. Two rival generals lead the regular army Sudanese Armed Forces (SAF) and the Rapid Support Forces (RSF). This conflict once again plays out as conflict between the sedentary (SAF) and the nomadic (RSF) part of the Sudanese population and is fired by the interests of outside players as RSF is supported by the UAI whereas the regular army SAF is supported by Egypt, Turkey, Saudi Arabia and Qatar. The still rich gold resources of Sudan serve as additional trigger for the conflict.[70] As the victory of the people in the peaceful revolution is seen as a threat to further authoritarian rule, both of the conflict parties do not care about the well-being of the civilians, and the eyes of the international community are on other problems. Another recent victim of this civil war are the cultural goods of Ancient Sudan. The most extensive collection of Ancient Sudanese Archaeological finds in the Khartoum National Museum was looted by paramilitary RSF and many of its priceless artifacts were destroyed or sold illegally.[71] This situation is the same with many regional museums painstakingly built and preserved by local people and international archeologists alike during many decades. Sudan has once again become the playing field for a multitude of international interests

[68] https://www.unhcr.org/publications/no-escape-frontlines-climate-change-conflict-and-forced-displacement - interactive data story.

[69] See also from the BBC podcast program "The Compass" the episode: Climate Wars: Darfur: https://www.bbc.co.uk/programmes/w3ct165c

[70] DW News: How gold drives the Sudan war in "Mapped Out": https://www.youtube.com/watch?v=CNzC2U-P5kM

[71] From prized artworks to bullet shells: how war devastated Sudan's museums: https://www.bbc.com/news/articles/c5y4lxxj055o

including geopolitical strategic interest in the countries of the Horn of Africa to keep international maritime trade through the Red Sea running.

The modern Sudanese conflicts therefore mirror closely the ancient ones: The situation becomes violent when a changing climate triggers longstanding conflicts between sedentary and non-sedentary people or ethnicities over scarce resources, international interest for the region is bound elsewhere or centers around the safety of the maritime trade routes to the East and exploitation of gold resources and authoritarian rule shows no interest in empowering of the people. In fact, this main goal of UNSDGs puts them in direct opposition to any authoritarian rule, which can only see them as a threat, a fact which can very much be seen today not only in the Sudan but everywhere around the world.

Regarding history, let me end this text with one last citation for the road:

> As for most periods of Middle Nile history, the historian's view of the Late Meroitic and early Post-Meroitic periods is based, for the dearth of textual evidence, principally on archaeological material. He is not only, directly or indirectly, dependent on the current ideologies which strongly influence the interpretation of archaeological finds, but is also expected to change his mind with every significant new archaeological discovery in a more radical way than he would be willing to do in the case of a culture which is documented in a more balanced manner.[72]

Doris Pemler is an entrepreneur, artist, and researcher based in Zurich. She received her M.A. in Egyptology, Sudan Archaeology and Archaeology of the Ancient Near East from Humboldt University zu Berlin and FU Berlin. Her research focuses on iconography, gender and intercultural communication in Ancient Egypt and Sudan. As a photographer, she contributed to several exhibitions in Germany and Switzerland. As CEO of Ciudad Condal GmbH, she leads an aspiring company focused on import and distribution of high-end sustainable products in Switzerland.

[72] Török, L. 1999. The end of Meroe in Welsby, D.A. Recent Research in Kushite History and Archaeology. Proceedings of the eighth International Conference for Meroitic Studies (British Museum Occasional Paper Number 131) British Museum Press London. 133–156. 133.

Part II

Academic Institutions to Enable Sustainable Practice and UN SDG Coverage

4 Sustainable Data Management: Leveraging the Data Value Chain for UN SDG Achievement Through the Three-Pillar Model

Sebastian Fischer, Claudia Lemke, and Benjamin Rausch

Abstract

In the global pursuit of the UN Sustainable Development Goals (SDGs), data is often hailed as the solution. Yet, its exponential growth poses a paradox: the *intangible* resource of the digital age carries a massive, rapidly expanding material footprint. This chapter argues that sustainable data management is no longer optional but a strategic imperative. By applying the Three-Pillar Model—Sustainable Purpose, Travelling Organization, and Connectivity (see chap. 2)—we provide a framework for organizations to decouple business growth from their digital carbon emissions. We dissect the Data Value Chain to expose hidden CO_2 hotspots, from energy-intensive model training to the indefinite storage of *Dark Data*. Moving beyond theory, the chapter offers actionable strategies, highlighting Data Virtualization as a critical lever to reduce redundancy and infrastructure load. We equip leaders with a new set of sustainability KPIs, transforming data management from a silent environmental liability into a transparent catalyst for the Green Digital Economy.

S. Fischer (✉) · C. Lemke
Berlin School of Economics and Law (Hochschule für Wirtschaft und Recht Berlin), Berlin, Germany
e-mail: sebastian.fischer@hwr-berlin.de; claudia.lemke@hwr-berlin.de

B. Rausch
Virtual Data Platform GmbH, Gröbenzell, Germany
e-mail: br@virtualdataplatform.com

© The Author(s), under exclusive license to Springer Nature Switzerland AG 2026

P. Wollmann et al. (eds.), *The Sustainable Organization – Volume 3*, Future of Business and Finance, https://doi.org/10.1007/978-3-032-28949-0_4

4.1 The Tangible Effects of an Intangible Resource

The global pursuit of the United Nations' 17 Sustainable Development Goals (SDGs) by 2030 (https://sdgs.un.org/goals) presents an urgent mandate. In our digital age, data, as an intangible resource, is both the raw material and the service outcome of any value-creation process; therefore, data management becomes a strategic necessity with a direct impact on these goals. The enormous global increase in data volume, from creation, storage, and processing to redistribution, is now generating significant interactions with measurable real-world effects, carrying a substantial material footprint. While data centers alone consumed about 1–1.3% of global electricity in 2022, the combined impact of data centers and data transmission networks accounts for nearly 1% of all energy-related greenhouse gas emissions.[1] That directly challenges SDGs like SDG 7 (Affordable and Clean Energy), SDG 9 (Industry, Innovation and Infrastructure), SDG 12 (Responsible Consumption and Production), and SDG 13 (Climate Action). This paradox gains urgency when we consider the foundational 1987 Brundtland Commission report, "Our Common Future," which defined sustainable development as "meeting the needs of the present without compromising the ability of future generations to meet their own needs."[2]

The interplay between Information Technology (IT) (the infrastructure powering data management) and sustainability is twofold.[3] First, there is Sustainability *in* IT (often referred to as the "Greening of IT"). It encompasses the entire IT lifecycle, from production to implementation, use, and operation, ensuring that natural, material, and human resources are utilized sustainably. Second, there is Sustainability *through* IT (known as "Green by IT"), which means that structures, processes, products, and services in other sectors are enabled or optimized by IT to reduce their resource consumption. While the ecological footprint of the digital world involves both perspectives, this article focuses specifically on the former: Sustainability *in* IT, particularly within the context of data management. To navigate the complexity of reducing this intangible resource's tangible footprint, we apply the Three-Pillar Model.[4]

---

[1] International Energy Agency: Data Centres and Data Transmission Networks. IEA, Paris (2022). https://www.iea.org/energy-system/buildings/data-centres-and-data-transmission-networks

[2] World Commission on Environment and Development: Our Common Future. Oxford University Press, Oxford (1987).

[3] Lemke, C. (2025). Information Technology for Sustainability: The Hidden Cost of the Digital Era. In *The Sustainable Organization: How Organizations Address the 17 UN SDGs Using the 3-P-Model* (pp. 195–215). Cham: Springer Nature Switzerland.

[4] Wollmann, P., Kühn, F., & Kempf, M. (2020). Three pillars of organization and leadership in disruptive times: Navigating Your Company Successfully Through the twenty-first Century Business World. Springer.

4.2 Applying the Three-Pillar Model to Data Management

The first pillar, *Sustainable Purpose*, demands that organizations align data practices with sustainability objectives and UN SDGs. It means stewarding data responsibly for long-term societal, environmental, and economic well-being, not just profit. For instance, a company defining a *Sustainable Purpose* to actively reduce its data-related carbon emissions by 30% by 2028 directly supports SDG 13 and provides a clear mission. The second pillar, *Travelling Organization*, emphasizes continuous learning, agility, and adaptation. As sustainable technologies evolve, organizations must explore, learn, and adapt, understanding that the path is rarely linear.[5] A *Travelling Organization* might pilot emerging, greener data storage solutions or adopt agile project management for its sustainability initiatives to iterate and learn quickly. The third pillar, *Connectivity*, focuses on breaking internal silos (e.g., between IT and business) and fostering external collaboration. It could involve IT working closely with procurement to source energy-efficient hardware or partnering with cloud providers to ensure workloads run on renewable energy, thereby enhancing ecosystem-wide sustainable practices.

4.3 Understanding and Mapping the Data Value Chain

Effectively managing data sustainability requires understanding its lifecycle—the Data Value Chain (DVC)—to identify CO_2 hotspots and optimization opportunities. Data, unlike physical goods or tangible assets with depletion and significant marginal production costs, can be replicated infinitely at near-zero marginal cost. However, each copy, process, and transmission consumes energy, creating a tangible footprint from an intangible asset. The DVC has six critical stages, each offering chances for impactful optimization.

4.3.1 Data Generation

Data Generation is where vast data streams originate from IoT sensors, smartphones, and enterprise systems. An autonomous vehicle, for example, can generate terabytes daily. The CO_2 hotspot here is the aggregate energy for sensor operation and immediate local storage. A key optimization is *Data-by-Design*—generating data only if a clear, sustainable purpose exists, embodying the *Sustainable Purpose* pillar. This requires a *Travelling Organization* mindset and *Connectivity* with product designers and engineers to ensure data generation aligns with actual needs and avoids unnecessary data creation from the outset. For instance, reducing unnecessary high-frequency sensor readings in a large industrial IoT deployment by just 10% can lead to corresponding energy savings in data generation and all subsequent lifecycle stages.

[5] See Wollmann, P., et al., op. cit. (Note 4).

4.3.2 Data Collection & Transmission

Data Collection and Transmission involve moving data across networks. Global data transmission consumes significant energy, a figure expected to grow with 5G and high-bandwidth content.[6] Energy use in network equipment is the primary CO_2 hotspot. Optimization can be achieved by using edge computing to process data closer to its source, thereby reducing transmission volumes. Employing efficient data compression techniques before transmission can also yield substantial benefits; compressing data by 50% effectively halves the energy used for that data's network transit.

4.3.3 Data Storage

Data Storage presents a significant challenge. *Dark data*—collected, processed, and stored but never used for meaningful insight—continuously consumes energy for powered disks and cooling systems.[7] It includes embodied carbon from manufacturing storage hardware and operational carbon from power use. Implementing strict data retention and archival policies is crucial. Regularly identifying and deleting Redundant, Obsolete, and Trivial (ROT) data can directly reduce storage energy costs by a similar percentage to the data eliminated and defer capital expenditure on new hardware. A *Travelling Organization* would proactively experiment with automated data classification tools to identify and manage ROT data. A *Sustainable Purpose* drives this proactive experimentation to minimize waste and resource consumption. Effective ROT data management also requires strong *Connectivity* between IT, legal (for defining retention policies), and business units (to determine ongoing data value).

4.3.4 Data Processing and Analysis

Data Processing and Analysis is where energy consumption can vary dramatically. Training large AI models is exceptionally energy-intensive. Training the GPT-3 model, for instance, was estimated to consume approximately 1287 MWh,[8] equivalent to the annual electricity use of over 120 average US homes. The CO_2 hotspots are CPU/GPU cycles for complex analytics, AI model training, and inefficient queries. Optimizing database queries and analytical algorithms for efficiency is vital. For AI, exploring techniques like model pruning, quantization, federated- and

[6] See International Energy Agency, op. cit. (Note 1).

[7] Mersy, G., Krishnan, S.: Toward a Life Cycle Assessment for the Carbon Footprint of Data. In Proceedings of the 2nd Workshop on Sustainable Computer Systems (HotCarbon '23). ACM, New York, Article 14, 1–9 (2023).

[8] Patterson, D., Gonzalez, J., Le, Q., Liang, C., Munguia, L., Rothchild, D., So, D., Texier, M., & Dean, J. (2021). Carbon emissions and large neural network training. arXiv (Cornell University).

transfer-learning can reduce computational load. Scheduling energy-intensive, non-critical processing for times of high renewable energy availability is another impactful strategy. Optimizing a frequently run, inefficient query that consumes 100 kWh/month to use only 50 kWh/month saves 600 kWh annually.

4.3.5 Data Usage and Distribution

Data Usage and Distribution involve disseminating insights and data products. Inefficient data access patterns and unnecessary data replication across platforms are CO_2 hotspots. Designing efficient APIs and data delivery mechanisms, alongside using caching strategies to reduce redundant data retrieval and processing, are key optimizations. Implementing a caching layer for frequently accessed reports can reduce database load and associated energy significantly.

4.3.6 Data Archival and Deletion

Data Archival and Deletion manages data's end-of-life. A lack of systematic approaches often leads to indefinite storage of obsolete data, which continues to consume energy. Establishing and automating clear data destruction protocols for data that has reached the end of its mandated retention period or business value is essential. It requires *Connectivity* between legal, business, and IT departments to define and enforce these policies effectively. This *Connectivity* ensures that the *Sustainable Purpose* of responsible data end-of-life management is met. Successfully implementing and automating these processes often requires a *Travelling Organization* approach, willing to adopt new tools and refine workflows to ensure efficient data disposition. Automating the deletion of data past a 7-year retention policy can free up significant storage capacity and reduce associated energy costs annually (Table 4.1).

4.4 From Theory to Practice: Implementing Sustainable Data Management

Transitioning to solutions requires strategic governance, technical excellence, and a link between data security and sustainability.

4.4.1 Strategic & Governance Foundations

Embed sustainability at the core of IT strategy. It begins with top-management commitment to a *Sustainable Purpose* that redefines the organization's relationship with data. Rather than setting isolated technical targets immediately, the strategy must first establish a guiding narrative: that data is a resource to be stewarded, not

Table 4.1 The sustainable data value chain—hotspots & optimization. Created by the authors. All rights reserved, used with permission

Value Chain Stage	CO_2 Hotspot	Optimization Lever	Strategic Link (Three-Pillar / SDG)
1. Generation	Sensor operation & indiscriminate local storage.	*Data-by-design:* Minimizing data creation at the source.	*Sustainable purpose:* Defining why data is needed prevents waste.
2. Collection & Transmission	Network energy (5G, bandwidth) & continuous streaming.	*Edge computing:* Processing data locally to reduce transfer.	*SDG 9:* Building resilient, resource-efficient infrastructure.
3. Storage	"Dark data" & embodied carbon of hardware (disks/cooling).	*Deduplication:* Removing redundant, obsolete, trivial (ROT) data.	*Connectivity:* Legal & IT aligning on strict retention policies.
4. Processing	AI model training (GPU cycles) & inefficient queries.	*Green AI:* Model pruning & carbon-aware scheduling.	*SDG 13:* Directly reducing the energy intensity of computation.
5. Usage	Redundant copies & inefficient API calls.	*Data virtualization:* Accessing data without replicating it.	*Connectivity:* Breaking silos to share data without copying.
4. Archival & Deletion	Indefinite retention ("data hoarding").	*Automated deletion:* Enforcing end-of-life policies.	*Travelling organization:* Adopting agile disposal workflows.

hoarded. The strategic goal is to decouple business growth from data-driven carbon emissions. Once this purpose is established as the organization's "raison d'être" regarding IT, it provides the orientation needed to define specific operational KPIs in the subsequent execution phase. This approach ensures that metrics, such as data center compute efficiency, storage utilization rates, or hardware durability, are not just compliance tick-boxes, but indicators of a genuine transformation.

This stewardship mindset reveals that data protection and sustainability are effectively two sides of the same coin. Robust security prevents breaches, thereby avoiding energy-intensive recovery operations. Moreover, data minimization—a core GDPR principle—directly supports the environmental strategy: by minimizing personal data, leaders reduce the organization's "attack surface" and cut the volume of data that requires energy for storage, backup, and security analytics. Thus, security and sustainability become mutually reinforcing goals.

4.4.2 Technical & Operational Excellence

Specific technical interventions are key. Optimizing Data Lifecycle Management (DLM) starts with data minimization and using tiered storage, which can move archival data to energy-efficient tiers, potentially reducing energy for that data by up to 70–80% compared to high-performance tiers. Automating deletion policies for

ROT data is also critical, as eliminating 30–50% of ROT data is an achievable target leading to direct energy savings.

Successfully implementing a comprehensive DLM is a clear manifestation of the organization's *Sustainable Purpose* to reduce its environmental footprint. It necessitates a *Travelling Organization* willing to adopt new tools and processes for data classification, tiering, and automated deletion. It relies on strong *Connectivity* between IT and business units to accurately define data value, access needs, and retention requirements.

Consider modern technologies like *Data Virtualization* not only to drive better decision-making based on insightful analytical content and cost efficiencies, but also to reduce the CO_2 footprint along the DVC. *Data Virtualization* offers a powerful way to minimize data redundancy and its environmental impact. Instead of physically replicating data into central data repositories, *Data Virtualization* can access, combine, and manipulate data regardless of where it is stored or how it is structured. Standard approaches usually rely on placing data from different sources, stored in heterogeneous file formats, into a (one) central data warehouse, thereby requiring energy-intensive Extract, Transform, Load (ETL) processes.

Green Software Engineering involves designing and developing inherently efficient software applications. Key principles include writing energy-efficient code by optimizing algorithms and data structures, maximizing the energy efficiency of underlying hardware, and creating "carbon-aware" applications. Carbon-aware applications can adapt their behavior based on grid carbon intensity, for example, by shifting non-critical, energy-intensive tasks to times when renewable energy is abundant, potentially using APIs that provide real-time grid information. Actively measuring the energy and carbon footprint of software is crucial for identifying optimization opportunities.

Understanding the Carbon Divide between classical IT and modern workloads is crucial. While classical applications often have moderate energy profiles, modern workloads, especially training large AI/ML models[9] or HPC simulations, consume vastly more energy. The International Energy Agency notes that AI's share of data center energy is growing rapidly, with Google estimating that Machine Learning accounted for 10–15% of its total energy use in 2019–2021.[10] Organizations investing heavily in AI must prioritize energy efficiency in models (both training and inference), algorithms, and hardware. It requires a *Travelling Organization* approach to allocate resources effectively, focusing optimization on these high-impact areas.

Sustainable Infrastructure Choices are vital. When using cloud services, evaluate providers on their commitment to 100% renewable energy, transparency in sustainability reporting (e.g., tools to track the carbon footprint of workloads), and efficient data center designs (low Power Usage Effectiveness, PUE).[11] Actively use provider tools to select greener regions. For on-premises data centers, continuously improve PUE, invest in modern cooling, and consolidate servers. Adopting these requires a

---

[9] See Patterson, D., et al., op. cit. (Note 8).

[10] See International Energy Agency, op. cit. (Note 1).

[11] See International Energy Agency, op. cit. (Note 1).

Travelling Organization mindset, willing to invest in and adapt to new, greener infrastructure and software development practices.

4.5 Deep Dive: How to Optimize Data Management along the Data Value Chain Through Data Virtualization

At the Data Generation stage, it is crucial to understand that one core principle of Data Virtualization is that it does not generate data on its own. Data Virtualization relies on the availability and integrity of data stored in existing data sources. It provides complete flexibility to use data from any data source for multiple purposes without owning the data in the sense of having a physical copy. Newly created information, generated by virtually using or analyzing existing data (e.g., analytical insights from combining data from different sources to prepare new content), can be stored for further use, time-series analysis, or auditing, based on an "*if-needed-principle*."

Data Virtualization does not require any Data Collection before data can be analyzed or used. It facilitates easy, lean Data Transmission. Even within large organizations, data can be spread across individually secured networks (e.g., for geographical or regulatory reasons, or following a recent merger or acquisition) in data silos. A complete integration and bidirectional access to secured networks would not only be time-consuming but also critical from a security standpoint. Modern Data Virtualization solutions securely transmit, access, and use all or only pre-defined data based on the "*need-to-know-principle*" from secured networks. As with Data Virtualization, the data volume can be precisely limited to what is actually needed, significantly reducing the CO_2 footprint of network equipment.

Large organizations regularly grapple with data silos and redundant physical databases of harmonized data, mainly built for single-purpose needs (e.g., combined reporting). Identifying and deleting ROT data can be challenging in such cases. Data Virtualization avoids redundant and trivial data, sometimes freely available on the Internet, so that it can be accessed and used efficiently. Data Virtualization also delivers insightful information from executed data queries, making ROT data transparent. Data that has not been accessed or used for a well-defined time period might have become obsolete. This enables organizations to actively avoid and reduce ROT data and its associated energy, hardware, as well as ETL maintenance and governance costs.

Data Virtualization results from Data Processing and Analysis can be used by any authorized person within the organization, subject to the organization's Identity and Access Management (IAM). Newly created analytical results can, but do not have to, be stored as a repeated execution of a virtual data query, and the application of the same virtual logic (virtual model) will provide the same results (reproducible results). It enables a digital and data-driven steering of the organization as data virtualization offers a logical, unified view of data from disparate sources in real time.[12] Minimizing physical copies leads to significant reductions in storage requirements, thereby directly cutting energy costs for storage and cooling and reducing the need for energy-intensive ETL jobs.

In a simplified view, Data Virtualization can be seen as a universal connector or (configurable) API that can apply any logic, data operations, or even complex algorithms to data. While using Data Virtualization, any Data Usage and Distribution does not require data replication or dissemination of analytical insights. Real-time data and analytical results are available, and intelligent caching reduces database queries and associated energy costs.

Data Virtualization itself does not generate any data and provides reproducible results. Generally, there is no need for Data Archival and Deletion. In addition, Data Virtualization supports ROT data. Taking both into account, Data Virtualization minimizes data storage in a hybrid data strategy and can significantly reduce energy costs and CO_2 footprint while improving analytical insights.

Beyond optimizing Data Management, Data Virtualization also serves the Security-Sustainability Nexus by reducing or limiting data redundancies and thereby reducing the attack surface for network or data breaches. With every redundant data persistence comes the requirement to maintain data governance policies for personal access of joiners, movers, or leavers, and the need to timely monitor and update data persistence with the latest security patches.

4.6 Measuring What Matters: Sustainability Metrics for Data Management

What gets measured gets managed. Integrating sustainability into data management requires expanding the traditional Total Cost of Ownership (TCO)[13] model to include environmental externalities throughout the IT lifecycle, such as energy consumption, carbon emissions, water usage, and e-waste liabilities. This extended perspective—often referred to as the Total Cost of Sustainability (TCS) approach[14][15]—aligns financial decision-making with responsible resource stewardship and UN SDGs. It acknowledges that the **cost of data** is not limited to hardware, licenses, and labor, but also includes its ecological footprint throughout three key stages:

[12] For sales management use cases, see: Rapp, M.S. and Rausch, M. (2024). "Digitale Steuerung mithilfe von Datenvirtualisierung," Controlling & Management Review, 68(4), pp. 40–47.

[13] David, J. S., Schuff, D., & St. Louis, R. (2002). Managing your total IT cost of ownership. Communications of the ACM, 45(1), 101–106.

[14] G. Callou et al., "Estimating sustainability impact, total cost of ownership and dependability metrics on data center infrastructures," Proceedings of the 2011 IEEE International Symposium on Sustainable Systems and Technology, Chicago, IL, USA, 2011, pp. 1–6.

[15] Parmar, K. (2024). Total Cost of Ownership to Total Cost of Sustainability, https://www.intelligentcio.com/me/2024/11/05/total-cost-of-ownership-to-total-cost-of-sustainability/

- **Procurement phase:** Embodied energy and carbon of hardware (Scope 3 emissions), percentage of renewable materials, and supplier sustainability ratings.
- **Operation phase:** Energy efficiency, carbon and water performance (PUE, CUE, WUE), data efficiency (dark data reduction), workload optimization, and share of renewable power.
- **End-of-life phase:** Hardware reuse, recyclability, and e-waste recovery rates.

By embedding these indicators into TCO analyses, organizations can make data-driven decisions that align cost efficiency with environmental stewardship. Table 4.2 shows selected KPIs that enable a total-cost-of-sustainability view.

Table 4.2 Key Performance Indicators (KPIs) for sustainable data management. Created by the authors. All rights reserved, used with permission

Metric Category	KPI Example	Why It Matters	Standardization Status	SDG Link
Energy efficiency	**PUE** (power usage effectiveness)	Measures how efficiently a data center uses energy (computing equipment) vs. cooling overhead as Total Facility energy / IT equipment energy. Ideal is 1.0; good practice is 1.1 to 1.4 for high-efficiency data centers.	**Yes** (ISO/IEC 30134–2)	7, 9, 12
Carbon impact	**CUE** (carbon usage effectiveness)	Measures the carbon emissions produced for every unit of IT energy consumed.	**Yes**(ISO/IEC 30134–2)	7, 13
Water usage	**WUE** (water usage effectiveness)	Quantifies water intensity of data center cooling systems; highlights resource trade-offs.	**Yes** (ISO/IEC 30134–9)	6, 12
Software efficiency	**SCI** (software carbon intensity)	A standardized way to measure the carbon efficiency of software, algorithms, and AI workloads.	**Yes** (De-facto standard)	9, 12, 13
Data efficiency	**kWh per terabyte**	Links data volume growth to energy demand, enabling data-efficient architectures.	**No** (indicative non-standardized)	7, 12
Operational	**Carbon per query**	Makes digital services comparable in carbon terms; supports green software engineering.	**No** (use-case specific)	13
Infrastructure	**Server utilization rate**	Higher utilization reduces idle energy consumption and stranded capacity.	Partially standardized (ITIL / ISO/IEC 20000)	9, 12
Energy source	**% renewable energy**	Measures progress in decarbonizing IT workloads; key for ESG alignment.	**Yes** (GHG protocol)	7, 13

(continued)

Table 4.2 (continued)

Metric Category	KPI Example	Why It Matters	Standardization Status	SDG Link
Circularity	**E-waste recovery rate**	Encourages circularity and responsible end-of-life asset management (recycling/reuse).	**Yes** (EU WEEE / ISO 14001)	12

Each of these KPIs reflects a different dimension of the data management lifecycle.[16] **Infrastructure Efficiency** (PUE, WUE) indicates how effectively a facility converts resources into computing power. **Climate Impact** (CUE, SCI) moves beyond efficiency to account for real-world emissions. **Data Efficiency** (Dark Data, kWh/TB) recognizes that data itself drives sustainability outcomes—more data is not inherently better. Finally, **Circularity** extends sustainability to the IT hardware lifecycle, ensuring that the physical substrate of data does not become a pollutant.

Based on these KPIs, organizations can establish a framework to measure the impact of **Data Virtualization specifically**. While estimating the precise contribution requires detailed knowledge of the IT landscape, data virtualization inherently simplifies this measurement by providing granular visibility into data usage patterns. Although individual user behavior is traditionally complex to quantify, data virtualization offers the transparency needed to identify inefficiencies (such as redundant queries) and drive behavioral change. Ultimately, while exact attribution remains complex, the evidence suggests that data virtualization acts as a significant lever for improving the sustainability metrics outlined above.

[16] For detailed discussions on these specific metrics, see the following sources. The relevant standards can be found in the respective publication of the standardization institution (ISO/IEC, EU WEEE) or standardization body (ITIL).

Banijamali, P., et al.: Unveiling key performance indicators for the energy efficiency of cloud data storage. In 2024 IEEE 21st International Conference on Software Architecture Companion (ICSA-C) (pp. 222–229). IEEE;

Dodge, J., et al.: Measuring the carbon intensity of AI in cloud instances. In Proceedings of the 2022 ACM conference on fairness, accountability, and transparency (pp. 1877–1894);

Gandhi, A., et al.: Metrics for sustainability in data centers. ACM SIGENERGY Energy Informatics Review 3(3), 40–46 (2023);

Greenhouse Gas Protocol (GHP): We set the standards to measure and manage emissions, https://ghgprotocol.org/ (2025).

Green Software Foundation: Software Carbon Intensity (SCI) Specification, https://sci.greensoftware.foundation/, (2025).

Rongali, L. P.: Green DevOps Metrics for Utility Operations. Authorea Preprints (2025).

SIXFOLD Group: Improving Power Usage Effectiveness (PUE) in Data Centres: Poven Tips. https://sixfoldgroup.com/2024/09/23/power-usage-effectiveness-pue/ (2025).

4.7 Actionable Recommendations: Your Roadmap

Immediate Actions (0–3 months): Conduct a data audit to baseline energy/data volumes and identify dark/ROT data. Refine lifecycle policies for automated archiving and secure deletion. Optimize power management features on existing infrastructure. Evaluate cloud providers for renewable energy commitments.

Medium-term Initiatives (3–12 months): Deploy data virtualization to reduce physical copies and ETL workloads. Redesign key applications using Green Software Engineering principles. Implement carbon accounting for data operations. Invest in team training on sustainable data practices to foster the *Travelling Organization* mindset.

Long-term Transformation (12+ months & ongoing): Fully align data strategy with corporate sustainability goals and UN SDGs, making it a C-Level priority. Participate in industry collaborations like the Green Software Foundation, an example of *Connectivity*. Innovate business models based on sustainable data use. Transparently report progress on KPIs and SDG targets.

4.8 Conclusion: Data as a Force for Sustainable Transformation

Transforming data management from an environmental liability into a force for sustainable development is crucial. As global data volumes double roughly every 2 years, today's choices will profoundly impact our collective contribution to the UN SDGs.

The Three-Pillar Model—Sustainable Purpose, Travelling Organization, and Connectivity—provides a robust framework. A clear *Sustainable Purpose* aligns data practices with SDG ambitions. A *Travelling Organization* cultivates agility to navigate evolving sustainable technologies. Enhanced *Connectivity* breaks silos and fosters collaboration for systemic change.

This path requires courage to challenge data hoarding, wisdom to balance performance and sustainability, and persistence. Organizations embedding sustainability into their data practices will not only contribute to global goals but also lead the sustainable digital economy. Every data choice has environmental consequences. Through strategic purpose, organizational agility, and collaborative connectivity, we can transform data management into a catalyst for achieving the UN SDGs and a sustainable future.

Sebastian Fischer is Professor of Business Informatics at the Berlin School of Economics and Law (HWR Berlin), where he focuses on the intersection of Data Science, Business Intelligence, and organizational transformation. As a member of the Institute for Data-Driven Digital Transformation (d-cube), he researches applied Machine Learning and AI. Sebastian bridges the gap between academic rigor and industry practice, drawing on his experience leading AI initiatives at Deutsche Telekom AG. He holds a PhD in Innovation Economics from TU Berlin and has pursued advanced specialization in Data Science at Harvard and Stanford.

Claudia Lemke holds a Professorship in Business Information Systems at the Berlin School of Economics and Law (HWR Berlin). She is also the Academic Director of the Master's Program in Digital Transformation at the Berlin Professional School, the Professional Institute at the HWR Berlin, and a member of the HWR Berlin Harriet Taylor-Mill Institute (HTMI) of Economics and Gender Studies. She has over 25 years of research, teaching, and consulting experience in different countries. Her most recent work focuses on the impact of digitalization on the economy and society, driven by emerging technologies.

Benjamin Rausch is Managing Director of Virtual Data Platform GmbH. He helps organizations leverage data virtualization to drive digitalization and automation, and to grow their business. His aim is to access and integrate all relevant data to provide meaningful insights for better decision-making. Thereby, he focuses on minimizing data redundancy and footprint while maintaining compliance and security. He brings many years of experience in finance, valuation, and capital markets in the investment banking and insurance sectors.

5 Transformative Education and the Earth Charter: Enabling Sustainable Development Through Transformative Learning in Higher Education

Maike Land, Claudia Lemke, Jacobus Bracker, and Sam Crowell

Abstract

Amidst the planetary polycrisis, it is perplexing how slowly significant transformation appears to occur. Despite growing awareness of planetary boundaries and social inequalities, the move toward just and sustainable futures remains disjointed. This indicates a deeper shortfall—not in knowledge, but in shared values based on common understanding and holistic perspectives, serving as an ethical foundation for sustainable development. Therefore, we question: why and how can we embed such an ethical foundation in education to promote a fundamental shift toward a sustainable future?This article presents the Earth Charter as a fundamental ethical framework for our thinking, actions, and coexistence. We argue from the perspective of the ideas of transformative learning and show approaches to how transformative education can succeed based on the principles of the Earth Charter. In doing so, we discuss some possible starting points for implementation, including in relation to the application of the three-pillar model concept underlying this edited volume.

M. Land · J. Bracker
German Rectors' Conference/Hochschulrektorenkonferenz (HRK), Berlin, Germany
e-mail: land@hrk.de; bracker@hrk.de

C. Lemke (✉)
Berlin School of Economics and Law (Hochschule für Wirtschaft und Recht Berlin), Berlin, Germany
e-mail: claudia.lemke@hwr-berlin.de

S. Crowell
Earth Charter International c/o University for Peace, San José, Costa Rica
e-mail: idyllsam@gmail.com

© The Author(s), under exclusive license to Springer Nature Switzerland AG 2026

P. Wollmann et al. (eds.), *The Sustainable Organization – Volume 3*, Future of Business and Finance, https://doi.org/10.1007/978-3-032-28949-0_5

5.1 Introduction

Amidst the planetary polycrisis, it is perplexing how slowly significant transformation appears to occur. Despite growing awareness of planetary boundaries and social inequalities, the move toward just and sustainable futures remains disjointed. This indicates a deeper shortfall—not in knowledge, but in shared values based on common understanding and holistic perspectives, serving as an ethical foundation for sustainable development. Therefore, we ask: why and how can we embed such an ethical foundation in education to promote a fundamental shift toward a sustainable future?

Following John Dewey, we do not view education as a formal act of preparation for life; instead, we see education as life itself, shaped by our cultural and social belonging, and as the ability to share in humanity's intellectual and moral resources.[1]

The Earth Charter, adopted in 2000, presents an ethical vision for human and ecological well-being rooted in culture and values. Its principles encourage us all to envision and build just and sustainable futures based on respect and care for the community of life, ecological integrity, social and economic justice, and democracy, nonviolence, and peace.[2]

Our approach is based on the theoretical principles of transformative education or transformational learning, aiming to create pathways for individuals and organizations to actively initiate and accelerate value-driven change. Transformative education aligned with the Earth Charter encourages educators and learners to turn ethical visions into practical action.

Simply put, transformative education helps individuals see the polycrisis in a new, empowering way: their actions and decisions—such as being ethically grounded and responsible business leaders—affect communities and ecosystems, and they can actively change systems, attitudes, and power relations. Therefore, it supports sustainable development for both current and future generations.

Building on a research perspective on transformative education and a practice perspective grounded in the Earth Charter, our approach advocates a shift from thinking to feeling when identifying and implementing pathways for transformative change. It offers recommendations for action on how we can rethink education for future decision-makers in business and society. This can help achieve the UN Sustainable Development Goals.

---

[1] Dewey, J. (1958). *My pedagogic creed (1897).*

[2] Earth Charter International (2020). The Earth Charter Booklet.

5.2 Transformative Learning, Transformative Education

Transformative learning is a concept that can change the way people see themselves and their world, thereby creating a changed worldview, often triggered by their own powerful experiences.[3]

> Transformative Education ... empowers learners of all ages with the knowledge, skills, values and attitudes to address the interconnected global challenges we are facing, including climate change, environmental degradation, loss of biodiversity, poverty and inequality. It builds a sense of belonging to a common humanity and help them become responsible and active global citizens in building inclusive, peaceful and sustainable societies.[4]

Transformative education helps learners clarify their own and others' worldviews through dialogue. By reflecting on how they and others relate to the world, learners identify common actionable pathways toward just and sustainable futures. Learners "come to see the world not as a static reality, but as a reality in the process, in transformation." In a swiftly changing world, where many of us feel like bouncing from one crisis to the next, education must prepare learners to navigate uncertainty, complexity, and ambiguity. However, education is still too often viewed primarily as a process of knowledge transfer.

Transformative education challenges this perspective. It empowers learners to critically analyze and address the economic and cultural roots of inequality, structural injustice, unsustainability, and entrenched power structures. It offers a learning environment where individuals can feel safe and be courageous, creating a space designed to turn conscience into value-based action. It deliberately fosters critical awareness about the polycrisis while helping to alleviate feelings of being overwhelmed by it.

Transformative education pertains to pedagogical frameworks such as Global Citizenship Education (GCE) and Education for Sustainable Development (ESD). All three share a commitment to empowering learners to critically engage in societal challenges and to take meaningful action. They emphasize reflective, participatory learning that questions assumptions and promote an understanding of interconnected social, economic, and environmental systems.

Transformative education is not just about imagining a perfect world; it's emotionally demanding work that critically examines societal structures and reveals patterns of discrimination. This learning process helps students become aware of oppressive systems and see themselves as part of an interconnected and interdependent world.[5] Drawing on Freire's pedagogy of the oppressed, we view

[3] Clark, M. C. (1993). Transformational learning. *New directions for adult and continuing education, 1993*(57), 47–56.; Mezirow, J. (1997). Transformation theory out of context. *Adult education quarterly*, 48(1), 60–62.

[4] UNESCO (2025). Transformative Education. https://www.unesco.org/en/fieldoffice/beirut/transformative-education,December 2025

[5] Freire P. (2005). Pedagogy of the oppressed (30th Anniversary Edition). The Continuum International Publishing Group, New York (Original work published 1970).

transformative education as emancipatory—encouraging students to question how societal structures influence them. It raises awareness of systemic injustice and empowers learners to see the world—and their role in it—from a new perspective.

Freire's (2005) pedagogy of the oppressed supports learning through dialogue rather than hierarchy. He sees dialogical pedagogy as a way to shift learning from vertical transmission to horizontal co-creation. The concept of conscientization—derived from the Portuguese term *conscientização*—is central to his philosophy. *Conscientização* can also be understood as "consciousness raising" or "critical consciousness." In simple terms, conscientization is the process by which individuals develop a profound awareness of social, political, and economic inequalities, enabling them to critically reflect on their reality and take informed action to transform it.

While some fear that becoming critically conscious of intersecting inequalities may lead to uprisings and destruction, the opposite is true: "by making it possible for people to enter the historical process as responsible Subjects, *conscientização* enrolls them in the search for self-affirmation and thus avoids fanaticism."[6] In that sense, transformative education helps victims of injustice to recognize oppression and develop the critical awareness and collective capacity for action necessary to constructively challenge it.

Freire's pedagogy mirrors what de Andreotti outlines as the general objective of Critical Global Citizenship Education (CGCE). Namely, "that injustices are addressed, more equal grounds for dialogue are created, and people can have more autonomy to define their own development."[7]

While, ideally, transformative education and CGCE alike promote critical thinking and ethical action, it could also be that learners feel guilty when they discover their own complicity in injustices or feel helpless when they become aware of the extent of injustices and global crises.

However, according to Mezirow, it is precisely these kinds of experiences that turn our understanding of the world upside down that are essential for transformative learning. Inspired by Freire's (2005) *pedagogy of the oppressed*, Mezirow pinpoints the starting point of transformative learning experiences as *disorienting dilemmas*—moments of cognitive dissonance—that stimulate self-reflection. It is not uncommon that this process generates pressure and anxiety,[8] i.e., self-reflection is linked to feeling afraid, angry, guilty, or ashamed.[9] To progress from this

[6] Freire P. (2005). Pedagogy of the oppressed (30th Anniversary Edition). The Continuum International Publishing Group, New York (Original work published 1970), p. 36.

[7] de Andreotti, V.O. (2014). Soft versus Critical Global Citizenship Education. In McCloskey, S. (Ed.) *Development Education in Policy and Practice* (pp. 21–31). Palgrave Macmillan, London.

[8] Mezirow, J. (1978). Perspective transformation. *Adult education*, *28*(2), 100–110.

[9] Fleming, T. (2018). Mezirow and the theory of transformative learning. In Wang, V. (Ed.) *Critical theory and transformative learning* (pp. 120–136). IGI Global Scientific Publishing.

often-paralyzing emotional state to action, learners must adopt new perspectives and relationships and develop new skills and self-confidence in new roles.[10]

Educational researchers distinguish between different ESD "levels" to categorize ESD approaches. However, the boundaries between these levels are not clearly defined and vary across the literature. Some scholars conceptualize learning "about" sustainable development and change as ESD I; learning "for" sustainable development and change as ESD II; and learning "as" sustainable development and change as ESD III.[11] Others, by contrast, conceptualize learning "for" sustainable development and change as ESD I; learning "as" sustainable development and change as ESD II[12]; and learning in relation with a changing world as ESD III.[13]

While the labels assigned to the ESD approaches vary, the underlying conceptual progression is largely consistent: At the most basic level ESD is mainly about acquiring knowledge related to sustainable development. More advanced approaches move from merely knowing things to the critical reflection and deconstruction of expert knowledge as well as the development of action competencies. At its most comprehensive, ESD is understood as a learning process that engages with uncertainty, complexity, and ambiguity. A dialogical space that is "sensitive to power asymmetries, responsibilization practices and structures, processes and mechanisms of inequality"[14] and "where a change of mindset occurs toward one which is more holistic, connected, agile, and open in outlook."[15]

This paper does not seek to delineate different ESD approaches, as this has been addressed by other authors comprehensively. Instead, we proceed from a shared understanding that ESD—as we understand it—is a fundamentally transformative educational approach. One that is consistent with the idea of CGCE and gives learners a space

> to reflect on their context and their own and others' epistemological and ontological assumptions: how we came to think/be/feel/act the way we do and the implications of our

---

[10] Mezirow, J. (2009). An overview on transformative learning. In Illeris, K. (Ed.) *Contemporary Theories of Learning: Learning Theorists ... In Their Own Words* (pp. 90–105). Routledge, London.

[11] Sterling, S. (2014). Separate tracks or real synergy? Achieving a closer relationship between education and SD, post-2015. *Journal of Education for Sustainable Development, 8*(2), 89–112.

[12] Vare, P., & Scott, W. (2007). Learning for a change: Exploring the relationship between education and sustainable development. *Journal of education for sustainable development, 1*(2), 191–198.

[13] Pettig, F., & Singer-Brodowski, M. (2024). Learning in relation with a changing world: thinking beyond ESD 1 and ESD 2 towards ESD 3. *Journal of Education for Sustainable Development, 18*(2), 176–201.

[14] Pettig, F., & Singer-Brodowski, M. (2024). Learning in relation with a changing world: thinking beyond ESD 1 and ESD 2 towards ESD 3. *Journal of Education for Sustainable Development, 18*(2), 176–201.

[15] Sterling, S. (2014). Separate tracks or real synergy? Achieving a closer relationship between education and SD, post-2015. *Journal of Education for Sustainable Development, 8*(2), 89–112.

> systems of belief in local/global terms in relation to power, social relationships and the distribution of labor and resources.[16]

In that sense, transformative education disrupts learned belief systems and shakes up our worldviews: the attitudes, values, stories, and expectations that shape how we interpret the world around us and how we act.[17] Our observable actions are only the tip of the iceberg; beneath them lie our worldviews. If we want to change how we interact with one another, education must address these underlying worldviews across multiple dimensions: cognitively, by fostering critical reflection of the interconnectedness and shared responsibility for global crises; behaviorally, by encouraging actions toward challenging and changing future trajectories; and emotionally, by supporting learners to navigate discomfort, empathy, and moral judgment. To do so, transformative education must empower learners to develop ethical orientations.

5.3 The Earth Charter

The Earth Charter, adopted in 2000, is an ethical framework that can be integrated into education curricula to engage educators and learners cognitively, behaviorally, and emotionally in co-creating future-fit economic models and organizational strategies rooted in its four pillars: Respect and Care for the Community of Life, Ecological Integrity, Social and Economic Justice, and Democracy, Nonviolence, and Peace.[18] The principles can serve as ethical guideposts during *conscientization* and *disorienting dilemmas*. They open up new ways of seeing the world and one's place in it. Importantly, the Earth Charter is a living document. Its meaning emerges through the way people who work with it interact with each other and their environment. The following section provides an overview of these four pillars, their corresponding principles, and the history behind the Earth Charter Initiative (ECI).

The idea for the Earth Charter dates back to 1987, when the World Commission on Environment and Development (known as the Brundtland Commission) published its report "Our Common Future." This report called

> for a common endeavor and for new norms of behavior at all levels and in the interests of all. The changes in attitudes, in social values, and in aspirations that the report urges will depend on vast campaigns of education, debate and public participation.[19]

---

[16] de Andreotti, V.O. (2014). Soft versus Critical Global Citizenship Education. In McCloskey, S. (Ed.) *Development Education in Policy and Practice* (pp. 21–31). Palgrave Macmillan, London.

[17] Sire, J. W. (2014). *Naming the elephant: Worldview as a concept.* InterVarsity Press.

[18] Earth Charter International (2020). The Earth Charter Booklet.

[19] United Nations General Assembly. (1987). *Report of the World Commission on Environment and Development: Our common future* (UN Doc. A/42/427). United Nations. https://digitallibrary.un.org/record/139811?v=pdf (accessed December 18, 2025)

Based on this call for new shared norms that could guide us toward a just and sustainable future, an ethical framework emerged: the Earth Charter. Initial discussions took place in the run-up to the 1992 Earth Summit in Rio de Janeiro. In 1994, the idea of the Earth Charter was launched as a civil society initiative by Maurice Strong (then Secretary-General of the Rio Earth Summit) and Mikhail Gorbachev (President of Green Cross International) with the support of the Dutch government.

In 1997, an independent Earth Charter Commission was established, marking the beginning of a decade-long, worldwide, cross-cultural dialogue involving thousands of individuals and organizations from the fields of science, law, religion, human rights, environmental protection, and civil society. In 2000, the Earth Charter was presented at the Peace Palace in The Hague in the Netherlands.

Since then, the Earth Charter has evolved beyond a document. It has become a living document, i.e., a global movement. It is an ethical foundation, an educational framework, and a reference for policy, governance, community development, peacebuilding, and sustainable development around the world.[20]

The Earth Charter consists of a Preamble; followed by its four pillars: Respect and Care for the Community of Life, Ecological Integrity, Social and Economic Justice, and Democracy, Nonviolence, and Peace; and concludes with The Way Forward. The following provides an overview of Earth Charter pillars and selected principles.

Pillar I, Respect and Care for the Community of Life, highlights that ethical responsibility begins with an attitude of respect. The corresponding principles focus on interdependence of all life (1), love and responsibility (2), democracy and freedom (3), and justice across generations (4).

The Preamble of the Earth Charter introduces the principle of universal responsibility, stressing that we all are accountable for our actions and have a shared responsibility to care for the Earth community. However, subprinciple 2b states "that with increased freedom, knowledge, and power comes increased responsibility to promote the common good." In that sense our responsibility is differentiated depending on our place in the world. Those who have historically benefited most from damaging the planet and have accumulated powerful positions in doing so have a greater moral obligation to restore ecological integrity for the benefit of future generations.[21] The idea of transgenerational justice is also emphasized in principle 4.

Pillar II, Ecological Integrity, outlines principles that restore and protect interconnected and interdependent ecosystems. The corresponding principles focus on protecting Earth's diversity (5), preventing ecological harm (6), sustainable lifestyles (7), and sharing knowledge (8). This pillar focuses on nature and biosphere reserves; responsible resource management; and pollution prevention. Central to

[20] Earth Charter International. *History*. https://earthcharter.org/about-the-earth-charter/history/ (accessed December 18, 2025)

[21] Brenes, A. (2005). Universal and differentiated responsibility. In: Corcoran, P. B., Vilela, M. & Roerink, A. (Eds.), *The Earth Charter in action: Toward a sustainable world* (pp. 35–37). KIT Publishers.

this principle is a precautionary approach: environmental damage should be prevented before it occurs, and decisions should be made with consideration for long-term ecological consequences.

Pillar III, Social and Economic Justice, contains guiding principles for human actions to promote social and economic justice, systemically. The corresponding principles focus on eradicating poverty (9), equitable human development (10), gender equality and equity (11), and dignity, inclusion, and well-being (12). While most of us will intuitively agree with principle 10, namely, to "ensure that economic activities and institutions at all levels promote human development equitably and sustainably" in our everyday lives, this ethical consideration sets us up for facing disorienting dilemmas regularly. These dilemmas are rooted in the illusion that unlimited growth is possible on a finite planet. While the Earth Charter does not call for zero growth, since growth and change are part of nature, it stresses that growth is not only about quantity, but also about quality. This ties in with the preamble to the Earth Charter, which states: "We must realize that when basic needs have been met … human development is primarily about being more, not having more."

Pillar IV underscores that caring for people and building just and peaceful societies are central concerns of the Earth Charter. The corresponding principles focus on transparency and participation (13), integrating values into education (14), respecting all living beings (15), and nonviolence and peace (16). Principle 13 emphasizes the importance of creating democratic societies that are "participatory." The understanding here is that all human beings have the right to participate in making the environmental, political, and social decisions that affect them. Participation requires education to ensure that everybody involved is empowered to actively contribute.[22] The mutual interdependence of participation and qualification is also reflected in Earth Charter Principle 14, which emphasizes the importance of education and life-long learning. This closely aligns with SDG 4.7, which considers education as the driver for transformation and for the implementation of all the SDGs.

The Earth Charter and the SDGs share many core ideas and can be seen as complementary documents. While the SDGs formulate clear targets, the Earth Charter provides the ethical framework for achieving them. It helps learners to critically examine the lines on accountability and system-immanent shortcomings that make achieving the SDGs so difficult now. While the SDGs leave the current power relations economic system largely unquestioned, the Earth Charter provides concrete entry points for challenging and reshaping the mindsets that sustain unsustainable practices.

Another complementary approach to achieving just and sustainable futures is the Inner Development Goals (IDGs). The IDGs are a framework that defines 25 dimensions of human inner development and links inner growth with outer change. These

[22] Bundesministerium für Familie, Senioren, Frauen und Jugend, Jugendstrategie der Bundesregierung, & deutscher Bundesjugendring (2023). Mitwirkung mit Wirkung—Qualitätsstandards für Kinder- und Jugendbeteiligung. Impulse zur Weiterentwicklung in Theorie und Praxis. https://www.bmbfsfj.bund.de/resource/blob/204010/affa15d77c0b4a8da7e6c476 (accessed December 18, 2025)

25 skills are grouped into five pillars (being: relationship to self; thinking: cognitive skills; relating: caring for others and for the world; collaborating: social skills; and acting: driving change). They are designed to address the complexity of the SDG agenda by empowering individuals to link personal transformation with systemic transformation, thereby bridging self-awareness and societal well-being. The framework positions inner development as a critical enabler for collective and systemic transformation. Like Earth Charter Pillar I "Respect and Care for the Community of Life" the IDGs emphasize "Caring for Others and the World." While the Earth Charter outlines principles that paint a picture of how a world in which we all care for other people and the planet would look like, i.e., provide an ethical vision, the IDGs paint a picture of how we personally would be in such a world. Key skills are, e.g., humility and compassion.

In that sense, the Earth Charter serves as a foundational guide for the SDGs and IDGs by providing a shared ethical vision for individuals, organizations, and educational systems working toward a more sustainable world. They form a complementary trio: The Earth Charter provides the ethical vision for just and sustainable futures; SDGs serve as measurable targets for achieving these futures, and the IDGs outline the skill set we all need to effectively realize these targets together.

Emerging research on transformative learning suggests that our emotions shape our trajectory of deep personal transformations.[23] While *disorienting dilemmas* centered around guilt, fear and shame can promote transformative experiences; another path toward deep personal transformations is *reorienting connections*. Such learning experiences evoke emotions like curiosity, enthusiasm, and joy.[24] They are moments of discovery, moments in which we reconnect with nature and each other. Connections we cannot unsee once we become aware of them and that inspire us to undertake actions. Working with the Earth Charter can foster such reorienting connections by bringing people together and uniting them around a shared ethical framework.

[23] Grund, J., Singer-Brodowski, M., & Büssing, A. G. (2024). Emotions and transformative learning for sustainability: A systematic review. *Sustainability Science, 19*(1), 307–324.

[24] Hathaway, M. (2022). Fostering reorienting connections via ecological practices. In: Nicolaides, A., Eschenbacher, S., Buergelt, P. T., Gilpin-Jackson, Y., Welch, M., & Misawa, M. (Eds.). *The Palgrave handbook of learning for transformation* (pp. 279–300). Cham: Springer International Publishing.

5.4 Transformative Learning in Higher Education Institutions

Higher Education Institutions (HEIs) are key players in implementing the UN SDGs.[25] According to the latest OECD report, "Education at a Glance 2025,"[26] in Western industrialized nations, 70% of 25- to 34-year-olds with academic parents hold a college or university degree, compared with 26% of those without an academic background. This underscores once again that educational attainment plays a decisive role in democratic societies and in the labor market. HEIs are thus becoming the place where future employability is shaped. For years, business and politics have called for fundamental systemic changes, especially regarding study formats and curricula, organizational structures, and the reform of higher education as a whole. In Germany, as in other industrialized nations, this pressure repeatedly runs up against the limits of Humboldt's ideal of a university education, encompassing the unity of research and teaching, the autonomy of institutions, and other systemic factors. In the European context, reforms aligned with the Bologna Process have already expanded these ideals, but often with little focus on transformative education as we define it (see sect. 5.2).

However, if we see HEIs as the places where the creators and architects of our future are educated, these institutions should be more strongly oriented toward the principles of transformative education—not only in line with ecological and social sustainability, but also in the sense of the shift toward labor marketability and economic sustainability, as already explained. Nevertheless, this transformation requires a fundamental paradigm shift in how teaching and research are viewed at HEIs. If these institutions are to educate future decision-makers, they must first develop a "sustainability mindset," which can help shape social, ecological, and economic developments worldwide.[27] Studies, such as that by Žalėnienė & Pereira (2021), which examine 120 countries, show that higher education participation correlates with positive indicators, including lower infant mortality, lower job insecurity, lower air pollution, higher access to clean energy, and higher women's participation. HEIs act as transformative agents through teaching, research, governance, and leadership; they particularly support SDGs 1, 3, 5, 8, 12, 13, and 16, and are related to all 17 SDGs.

HEIs should see themselves as morally responsible institutions that help secure a just, peaceful, and ecologically sustainable future through teaching, research, and institutional practice.[28] The Earth Charter provides a normative baseline of values as

---

[25] Žalėnienė, I., & Pereira, P. (2021). Higher education for sustainability: A global perspective. Geography and Sustainability, 2(2), 99–106.

[26] OECD (2025). Education at a Glance 2025, https://www.oecd.org/en/publications/education-at-a-glance-2025_1c0d9c79-en.html, Dec 2025

[27] Žalėnienė, I., & Pereira, P. (2021). Higher education for sustainability: A global perspective. Geography and Sustainability, 2(2), 99–106.

[28] Weakland, J. P., & Corcoran, P. B. (2009). The Earth Charter in higher education for sustainability. Journal of Education for Sustainable Development, 3(2), 151–158.

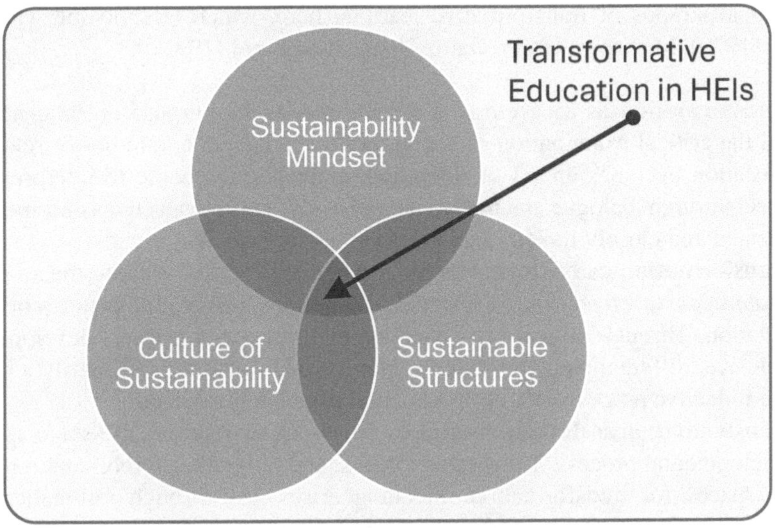

Fig. 5.1 Transformative Education in HEIs, according to Weakland & Corcoran (2009) and Žalėnienė & Pereira (2021). Developed by the authors, used with permission, all rights reserved

an ethical framework for HEIs to reference when discussing and aligning more closely with curricular and extracurricular offerings.[29] Transformative education forms the central pillar of value orientation, critical reflection, participatory learning, inter- and transdisciplinarity, and institutional reform, and aims to bring about profound changes in mindsets, structures, and culture (see Fig. 5.1).

If we follow the fundamental definition of transformative learning,[30] we can see it as a profound process of meaning and significance formation that goes beyond instrumental learning and aims at personal and social transformation.[31] Learning is not understood as mere knowledge acquisition but as an existential process in which individuals reflect on, question, and reconstruct their understanding of themselves and the world.[32]

In the context of higher education, this formation of meaning and significance also needs to be anchored in the areas shown in Fig. 5.1. Dirkx (1998)[33] sees four

[29] Žalėnienė, I., & Pereira, P. (2021). Higher education for sustainability: A global perspective. Geography and Sustainability, 2(2), 99–106; Weakland, J. P., & Corcoran, P. B. (2009). The Earth Charter in higher education for sustainability. Journal of Education for Sustainable Development, 3(2), 151–158.

[30] See also Baumgartner, L. M. (2001). An update on transformational learning. New directions for adult and continuing education, 2001(89), 15–24.

[31] Dirkx, J. M. (1998). Transformative learning theory in the practice of adult education: An overview. PAACE journal of lifelong learning, 7, 1–14.

[32] Ibid.

[33] Ibid.

theoretical strands of transformative learning here, which refer to the works of Freire (1970), Mezirow (1991), Daloz (1986), and Boyd (1991)[34]:

- **Transformation as awareness:** Learning serves the purpose of emancipation and the critical examination of social, political, and economic power relations. Education becomes an act of liberation, empowering people to interpret their world through dialogue and to change it actively. This perspective is normatively oriented and closely linked to questions of social justice.
- **Transformation as critical reflection:** The focus is on changing the so-called *meaning perspectives*—deeply rooted assumptions, values, and patterns of interpretation. Through critical self-reflection and discourse, learners develop more inclusive, differentiated worldviews. Here, transformation is primarily a cognitive-reflective process with developmental psychological depth.
- **Transformation as development:** Learning is understood as a life-stage-specific developmental process. Transitions, breaks, and biographical reorientations create spaces for transformation. Meaning arises less through rationality than through holistic meaning-making, supported by mentoring and supportive learning relationships.
- **Transformation as individuation:** This perspective emphasizes the emotional, symbolic, and spiritual dimensions of learning. Transformation occurs through the integration of unconscious elements (e.g., through narratives, images, and imagination). Here, learning is a deeply internal process of self-actualization.

If HEIs meet these requirements to cultivate a sustainability mindset, a culture of sustainability, and sustainable university structures, transformative learning can be lived integrally[35] and empower the institution, teachers, and learners to bring about a desired, emotionally driven change in research and knowledge creation, transfer, and representation. In doing so, they make a demonstrable contribution to achieving the UN SDGs in line with the values of the Earth Charter.

[34] Freire, P. (1970). *Pedagogy of the oppressed.* New York: Seabury Press; Mezirow, J. (1991). *Transformative dimensions of adult learning.* San Francisco: Jossey-Bass; Daloz, L. *(1986). Effective teaching and mentoring: Realizing the transformational power of adult learning experiences.* San Francisco: Jossey-Bass; Boyd, R. D. *(1991). Personal transformation in small groups: a Jungian*

perspective. London: Routledge.

[35] O'Sullivan, E. (2002). The project and vision of transformative education: Integral transformative learning. In Expanding the boundaries of transformative learning: Essays on theory and praxis (pp. 1–12). New York: Palgrave Macmillan US.

5.5 Recommendations on Transformative Learning for Sustainable Development in Higher Education

In the following, selected aspects of implementing transformative learning in HEIs are presented to show how universities can contribute to sustainable development. Our recommendations rest on a cultural-semiotic premise: Cultures can be conceptualized as webs of meaning[36]: historically grown, socially shared systems of meaning through which individuals and groups make sense of the world and of their own actions. These webs are not static. They are continuously reproduced, modified, and reoriented through communication.[37] Culture is not merely a background variable of education; it is the *ongoing outcome* of communicative processes—of what is said, shown, practiced, rewarded, normalized, and silenced.

Education constitutes a **specific, institutionalized form of communication** that deliberately designs communicative situations (curricula, learning settings, media, assessments, institutional rituals) to shape meaning-making over time. These specific characteristics of education allow for a distinctive transformative function: to deliberately *dynamize* cultural webs of meaning. Transformative educational communication does not merely transmit existing knowledge or values; it introduces new perspectives, challenges habitual interpretations, and creates structured opportunities for reflection, dialogue, and reorientation. In this sense, education acts as a cultural medium that enables societies to respond to changing conditions without losing coherence.

In times of global socio-ecological crisis, this dynamizing function becomes particularly significant. Digitalization, globalization, and accelerating flows of information increase the density and speed of communication, multiplying competing interpretations of sustainability, responsibility, and the future. Under such conditions, meaning is not lost because culture collapses, but because it becomes *unstable* and contested. Education therefore plays a crucial stabilizing–dynamizing role: it helps learners navigate complexity, relate new information to ethical frameworks, and integrate knowledge into coherent worldviews.

From this perspective, the central argument of this chapter can be sharpened: **the sustainability mindsets required for global transformation cannot emerge automatically from information availability or technological innovation. They depend on transformative Education for Sustainable Development (ESD)**. Transformative ESD enables learners to actively reconfigure their meaning structures in relation to ecological integrity, social justice, and intergenerational responsibility. It supports not only cognitive learning, but also ethical orientation, emotional engagement, and agency.

---

[36] C. Geertz, The Interpretation of Cultures. Selected Essays (New York 1973) 5.—Following more-than-human approaches, these webs are not necessarily spun by humans only, cf. P. Cobley, The Cultural Implications of Biosemiotics, Biosemiotics 3, 2010, 225; D. Abram, The Spell of the Sensuous. Perception and Language in a More- Than-Human World (New York 1996) IX.

[37] Cf. M. Sturken/L. Cartwright, Practices of Looking. An Introduction to Visual Culture (New York 2001) 4.

Sustainability Mindset

A sustainability mindset can be understood as a relatively stable configuration of meaning: how individuals and groups *perceive* and *interpret* socio-ecological realities, what they treat as *valuable*, what they consider *possible*, and how they locate *responsibility and agency*. Such mindsets are cultural achievements, shaped through repeated communicative experiences, happening in environments, rather than isolated learning events.[38]

In recent years, commercialization and technologization have narrowed educational communication: teachers are pushed toward "content delivery," learning becomes "Netflix-like," and questions of meaning and shared sense-making are displaced—often alongside declining occupational well-being for educators.

This is not merely a didactic problem; it is a cultural one. If the dominant communicative codes of the university reduce education to consumption, optimization, and individualized performance metrics, then sustainability mindsets remain structurally unlikely.

Higher education therefore should not simply "produce competencies"; it has to provide for **communicative environments** in which sustainability-related meanings can be explored, negotiated, and internalized. Teaching formats, curricula, assessment practices, and institutional norms all communicate—explicitly or implicitly—what counts as relevant knowledge, legitimate concern, and responsible action.

Transformative ESD must intentionally use this communicative power to dynamize existing meaning structures without dissolving them. This includes:

- Shifting **from** a **focus on** *knowledge about sustainability* to ***meaning-making for sustainability***. Design learning explicitly around the move from *what we know* to *how we know*—and how knowledge relates to values, emotions, identities, and action.
- Use **dialogic formats as core educational media**, not as add-ons. Transformative learning depends on dialogue that surfaces assumptions and enables learners to test and revise interpretive frames in relation to others.
- **Acknowledging ambiguity and uncertainty** as integral to sustainability challenges rather than as deficits. Because communication is inherently ambiguous and context-dependent, transformative ESD should not promise "the" correct sustainability worldview. Instead, it should cultivate competencies for working with contested meanings while keeping an ethical orientation visible.
- **Create "disorienting dilemmas" and "reorienting connections."** Transformative learning often begins with disruption—moments where old meaning structures fail—but it must also provide relational and experiential anchors that reconnect learners with purpose, care, and agency (including connection with community and more-than-human life).

[38] Cf. also the concept of ecocultural identities: T. Milstein/J. Castro-Sotomayor, Ecocultural identity. An introduction, in: ibid., Routledge Handbook of Ecocultural Identity (New York 2020) xviii.

- **Strengthen epistemic resilience in the age of generative AI.** When AI-generated content increases and misinformation spreads, the educational task becomes doubly essential: fostering sustainability mindsets while protecting scientific integrity and evidence-based reasoning—without collapsing into technophobia or naïve "AI-as-solutionism."
- **Treat educators as ethical communicators, not delivery systems.** If education is a culturally formative practice, teachers inevitably function as role models and "carriers" of meaning. HEIs must therefore empower educators with time, support, and institutional legitimacy to teach sustainability as ethical and emotional work—not merely as content.
- **Assess what you actually value.** If assessment rewards only speed, individual competition, and narrow reproducibility, it communicates an unsustainability code. Assessments should include reflective and collaborative dimensions (e.g., deliberation quality, ethical reasoning, systems thinking, relational accountability, iterative learning from practice). This reduces the risk that measurement regimes undermine the very mindsets transformative ESD seeks to build.

Through such approaches, education enables learners to reorient their understanding of progress, success, and responsibility in ways that are compatible with long-term sustainability.

Cultures of Sustainability
If culture is a web of meaning, a culture of sustainability within HEIs emerges from the **patterned ways in which sustainability is communicated and practiced** across the whole institution through teaching, research, governance, and everyday institutional life. Mission statements alone are insufficient; what matters is the consistency of communicative signals—what is rewarded, what is marginalized, and how conflicts are addressed.

A culture of sustainability is therefore not created by adding sustainability as a topic, but by **embedding sustainability-oriented meaning structures into institutional communication**. This includes

- **Making the ethical narrative explicit and shared.** Use the Earth Charter or similar frameworks not only as content but as a *common language* for interpreting tensions and dilemmas (e.g., ecological integrity vs. economic pressure; inclusion vs. performance metrics). The point is not moralizing, but enabling a shared ethical reference system that stabilizes meaning amid pluralism.
- **Institutionalizing dialogue across differences (including power asymmetries).** Sustainability controversies are not just "opinions"; they are structured by unequal access to resources and voice. A sustainability culture requires formats that are sensitive to inequality and capable of producing more equal grounds for dialogue—rather than reproducing existing hierarchies.
- **Protect plural ways of knowing—without abandoning epistemic standards.** Transformative learning must integrate cognitive, emotional, relational, and experiential forms of knowing while maintaining robust evidence practices.

Otherwise, either technocratic reductionism or relativistic fragmentation will "blow up" the shared meaning resources needed for collective action.

- **Beware of "learning analytics" as hidden cultural messaging.** If student success is framed primarily through quantification and prediction, the institution communicates meritocratic and epistemic narrowing. Sustainability cultures need evaluation systems that recognize relational, collective, and long-term capacities—especially where transformative learning is intentionally non-linear.
- **Create psychologically safe spaces for ethical-emotional work.** A culture of sustainability must make it legitimate to feel grief, anger, shame, hope, and joy in relation to the polycrisis—and to translate these emotions into ethical reflection and action rather than cynicism or paralysis.

Sustainable Structures

Finally, meaning-making is stabilized—or destabilized—by structures. HEIs cannot expect transformative ESD to flourish if governance, funding logics, workloads, promotion criteria, and campus operations communicate contradictory values. **The Whole-Institution Approach** (WIA) can be understood as the organizational manifestation of an integrative view on sustainability: sustainability is not a "topic," but a structural condition of how the institution learns, decides, and relates to society. It is recommendable to:

- **Align governance with ethical orientation**. Establish decision procedures that operationalize, for example, Earth Charter–consistent commitments (participation, justice, ecological integrity) in budgeting, procurement, partnerships, and risk management—so the institution communicates sustainability not only in lectures, but through its institutional behavior.
- **Integrate sustainability across teaching, research, operations, and transfer**. Treat the campus as a living lab: connect curricula to real institutional sustainability challenges (energy, mobility, food systems, inclusion), enabling iterative learning and visible agency.
- **Build capacity for educators and staff**. Provide professional development, time allocations, and recognition structures for transformative pedagogy—because sustainable transformation cannot be an unpaid "extra" carried by a few committed individuals.
- **Link the Whole-Institution Approach to a "Whole Society Approach."** Sustainability mindsets are cultural achievements that must resonate beyond campus boundaries. HEIs should cultivate long-term partnerships with civil society, public institutions, and businesses to co-create communicative spaces for shared learning—mirroring the Earth Charter's orientation toward "right relationships" across self, others, other cultures, and the more-than-human world.

In short: Understanding culture as a web of meaning highlights that sustainability transformation is fundamentally a **cultural and communicative process**. The struggle for sustainability is also a struggle over meaning—over which stories, values, and interpretive frames become socially binding. Communication can fragment these webs, especially under commercial and algorithmic pressures; but education, as a deliberately designed form of communication, can also reweave them. Because the mindsets required for transformation are not produced by knowledge transfer alone, HEIs must treat transformative ESD as a central institutional responsibility—one that integrates ethical orientation, dialogic meaning-making, and whole-institution structures into a coherent communicative ecology.

Education *dynamizes* ecocultural identities and empowers societies to reinterpret themselves in the face of new challenges. Transformative ESD is thus not an optional add-on, but a core societal mechanism for cultivating the sustainability mindsets required for global transformation.

Application of the Three-Pillar Model for the Implementation of Transformative Learning in HEI

Regardless of some further explanations in Chap. 15,[39] we would like to provide an overview of how the concepts fit into the Three-Pillar Model.

Even though this model addresses organizations of all kinds, we tend to think of it as something that is driven by economic considerations. It is therefore worth examining how the model can be adapted to educational institutions, which, according to an idealized view shaped by history, should not be subject to economic constraints. As mentioned, Chap. 15 covers this in more detail, so we will only address one basic idea here.

Basically, we are seeing a global trend toward the commodification of education, nevertheless driven by the technologization of education. As a result, education is increasingly driven by economic considerations, which, based on experience, leads to increased pressure for efficiency and effectiveness. This way of thinking is therefore inherently opposed to the idea of transformative learning. However, in our view, there are opportunities here to use the concept of transformative learning to counterbalance the economization of education. The application of the three-pillar model can be seen as an opportunity to establish ideas of sustainable education in educational institutions as well.

Transformative education can be the lever, enabler, and catalyst, as well as the goal, of creating a sustainable institution that is responsible, has clearly defined guidelines (sustainable purpose), and is thus enabled not only to respond resiliently and agilely to changes, but also to actively shape them (travelling organization), thereby integrating all actors in the sense of a holistic view of education (connectivity).

[39] See Chap. 15

5.6 Conclusion

The implementation of the idea of transformative learning begins with the shift from what we know to how we know[40] (Baumgartner, 2001). Transformative learning in higher education with the Earth Charter is not about adding content to the curriculum; it is not about including a sustainability checklist to module plans. Rather, it helps educators and students to place the entire learning process within an ethical and transformative framework. This foundation helps to implement the concept of emergent teaching which means to address the "whole person" through a sustainable teacher or lecturer mindset, a culture of sustainability and sustainable structures in the institutions.

By embedding the principles of the Earth Charter, learning becomes a reflective and action-oriented experience. It promotes dialogic learning, in which students and teachers engage in open, critical discussions, and fosters a mindset characterized by responsibility, empathy, and global citizenship. This approach transforms education into a space where knowledge, values, and action are interconnected, preparing students to address complex social and environmental challenges with insight and ethical commitment.

The Earth Charter helps learners to hold the tension of trade-offs between different Sustainable Development Goals. For instance, advancing industrial growth (SDG 8) may conflict with climate action (SDG 13), or promoting economic development can challenge responsible consumption (SDG 12). These conflicts often appear as visible "problems at the tip of the iceberg," but addressing them effectively requires looking deeper—into the underlying structures, values, and mental models that shape decision-making. By providing an ethical and systemic framework, the Earth Charter supports students and educators in developing a both/and mindset, enabling them to navigate complexity, uncertainty, and ambiguity without reducing challenges to simple solutions.

In the AI era, we are increasingly confronted with the question of how to fundamentally shape the feasible and desired boundaries of human–machine interaction. According to widespread opinion, only the concept of "human-in-the-loop" (HITL) should be applied, according to which humans are ultimately responsible or actively involved in data preparation, model training, and in a supervisory role in automated decision-making.

Nevertheless, we are increasingly seeing a shift in the concept of epistemic injustice,[41] in which AI systems are achieving positions of power that are leading to a change in their role in the process of knowledge sovereignty. AI systems are increasingly taking control of knowledge generation processes and their representation in the education sector. We run the risk here of creating, as humanity as a whole, a **generative epistemic injustice**[42] that will not only fundamentally change future economic, social,

[40] Baumgartner, L. M. (2001). An update on transformational learning. New directions for adult and continuing education, 2001(89), 15–24.

[41] Fricker, M. (2017). *Epistemic Injustice. Power and Ethics of Knowing*, Oxford University Press.

[42] Anneborg, Raymond. (2020). Computing Machinery and Knowledge. 10.48550/arXiv.2012.06686.; Origgi, G., Ciranna, S. (2017). Epistemic Injustice, The Case of Digital

and ecological structures, but may also systematically reinforce existing inequalities. For these reasons, the concept of transformative learning should be applied and institutionally anchored, especially in the context of academic education.

Additionally, we can also emphasize again that it is important in transformative learning journeys to include different forms of knowing, because otherwise: "When those shared epistemic resources are biased, this transformation prevents students from retaining aspects of their personal epistemic resources which it is strongly in their interests (as well as in the interests of the broader epistemic community) to retain."[43]

We see transformative education and learning as a fundamental prerequisite for counteracting these developments and contributing to the achievement of the SDGs: There is an emphasis on capacity building, empowerment, and action competence. Learners face unpleasant emotions, so-called "edge emotions,"[44] when confronted with information that challenges their assumptions. Educators can open a space to discuss emotions and engage in perspective change to support transformative learning processes during which mindsets change and learners reconstruct their assumptions adapting a more interconnected worldview.[45] This expanded and interconnected perspective reflects the Earth Charter's emphasis on interdependence, sustainability, and global responsibility. Co-created in response to global ecological and social challenges, the Earth Charter aims to promote a shared ethical consciousness. As such, it can serve as a guiding framework for educators and learners committed to transformative change, supporting learning processes that engage not only cognitively and behaviorally, but also emotionally.

Maike Land is an ambassador for climate justice and ESD. She facilitates educational projects and creates transformative learning spaces that enable people to actively engage in socio-ecological transformation. She is a community space holder with a strong sensitivity to collective, discrimi-nation-aware, and self-organized structures. With a background in environmental economics (M.Sc. in Resource and Environmental Economics) and cultural research focusing on Southeast Asia (B.A. in International Culture and Business Studies), she brings an interdisciplinary outlook to her work. She is committed to educational approaches that bridge diverse perspectives and foster intercultural understanding. Within the academic system, she connects stakeholders and supports dialogue as well as collaboration to strengthen democracy and sustainable development.

Environments. Kidd, I.J., Medina, J., Pohlhaus, G. (Eds). *The Routledge Handbook of Epistemic Injustice*, Routledge, 303–312.

[43] Fairbairn, F. (2023). Epistemic injustice through transformative learning, *Journal of Philosophy of Education*, Volume 57, Issue 4–5, August–October 2023, Pages 964–982, https://doi.org/10.1093/jopedu/qhad082

[44] Mälkki, K. (2019). Coming to grips with edge-emotions: The gateway to critical reflection and transformative learning. In: Fleming, T., Kokkos, A., Finnegan, F. (Eds.) *European perspectives on transformation theory* (pp. 59–73). Cham: Springer International Publishing.

[45] Sterling, S. (2014). Separate tracks or real synergy? Achieving a closer relationship between education and SD, post-2015. *Journal of Education for Sustainable Development, 8*(2), 89–112.

Claudia Lemke holds a Professorship in Business Information Systems at the Berlin School of Economics and Law (HWR Berlin). She is also the Academic Director of the Master's Program in Digital Transformation at the Berlin Professional School, the Professional Institute at the HWR Berlin, and a member of the HWR Berlin Harriet Taylor-Mill Institute (HTMI) of Economics and Gender Studies. She has over 25 years of research, teaching, and consulting experience in different countries. Her most recent work focuses on the impact of digitalization on the economy and society, driven by emerging technologies.

Jacobus Bracker is an archeologist and interdisciplinary educator working at the intersections of visual culture, narrative, and cultural heritage in the context of ecological transformation. He holds an M.A. in Classical Archaeology from the University of Hamburg. Jacobus brings experience from environmental humanities teaching as a lecturer at the Hamburg University of Technology (TU Hamburg), and from science management at the German Rectors' Conference, where he supports the mainstreaming of Education for Sustainable Development (ESD) in higher education contexts. As a certified ESD Educator and founder of "Planetary Ways," he designs and facilitates learning spaces for the Anthropocene.

Dr. Sam Crowell is professor emeritus at California State University, San Bernardino, where he founded the Masters Program in Holistic and Integrative Education. He served as a doctoral faculty member at Cal State and is currently an affiliate faculty of the UNESCO Chair for Education for Sustainable Development with the Earth Charter and serves as a senior researcher and Chair of the Earth Charter International Council. Sam was a Fulbright Specialist in ESD and is the recipient of both the Outstanding Teaching Award and the Outstanding Professional Accomplishments Award. Dr. Crowell has experience in elementary and secondary settings as a teacher, principal, alternative charter school director, and federal program administrator. He also served as director of university programs in both elementary education and educational administration. He continues to accept invitations as a visiting professor, keynote speaker, and international consultant in the areas of Education for Sustainable Development, Holistic Pedagogy, and Neuroscience Applications to Learning.

Professor Crowell writes in the field of education, the implications of socio-neurobiology, social values and worldviews, and sustainability with the Earth Charter where he explores the implication of a transmodernist worldview toward an ecological civilization. He is interested in holistic perspectives, Earth-based traditions, and heart-centered ways of living and being. Among his recent books are *The International Handbook of Holistic Education*, *Emergent Teaching: a Path of Significance, Creativity and Meaning*, *The Re-Enchantment of Leaning*, *Earth Charter Pedagogy: New Understandings of Emergence Applied to ESD* as well as many chapters on the Earth Charter. His latest book is *The Tao of Spiritual Development*.

Educating Purpose-Driven Innovators for Circularity: Insights from Global Industry-Academia Collaboration

6

Albena Björck, George Toye, and Cedric Gysel

Abstract

Sustainability challenges confront organizations in the life science industry with a structural paradox: a sector dedicated to protecting human health relies on linear production and disposal regimes that generate substantial environmental harm. Although research on corporate purpose, ecosystem governance, and the circular economy specifies what systemic transformation requires—alignment across actors, orchestrated collaboration, and circular business models—it offers limited insight into how the necessary capabilities are developed and enacted.

Drawing on three interdisciplinary collaborations between Stanford University, ZHAW, and industry partners, this chapter demonstrates that purpose-driven system innovation is not primarily a technological and material science challenge, but a learning and coordination challenge. It shows how design-oriented educational architectures—rooted in the Stanford ME310 tradition and extended through international business perspectives—can function as micro-foundations for circular transformation by cultivating ambidexterity, boundary-spanning, and ecosystem governance.

A. Björck (✉)
ZHAW School of Management and Law, International Management Institute, Winterthur, Switzerland
e-mail: bjoe@zhaw.ch

G. Toye
Stanford University, ME Design Group, Stanford, CA, USA
e-mail: toye@stanford.edu

C. Gysel
CEO Be Circular, Winterthur, Switzerland
e-mail: cedric@be-circular.com

© The Author(s), under exclusive license to Springer Nature Switzerland AG 2026

P. Wollmann et al. (eds.), *The Sustainable Organization – Volume 3*, Future of Business and Finance, https://doi.org/10.1007/978-3-032-28949-0_6

6.1 Why Purpose-Driven Innovation in Life Sciences Requires New Learning Capabilities

The life sciences face a structural paradox: while their mission is to protect health, their products and practices contribute substantially to environmental harm. Healthcare systems account for a sizeable share of global emissions, and medical and pharmaceutical waste streams remain largely linear—incinerated or landfilled despite substantial embedded value.[1,2] If healthcare were a country, it would rank among the top five global emitters. Across the life sciences value chain, medical technologies and pharmaceuticals are major contributors, accounting for roughly one-third of healthcare-related emissions and the majority of healthcare waste.[3] Linear "use-and-dispose" logics dominate, driven by safety requirements, and regulatory complexity in clinical environments. As a result, high-value materials embedded in medical devices, pharmaceutical products, and delivery systems are routinely incinerated or landfilled, while downstream emissions and waste flows remain poorly measured, fragmented, and weakly governed.[4]

Despite growing awareness and policy ambition, progress toward circularity in the life sciences remains slow. Regulatory approval processes prioritize patient safety and risk minimization, often limiting experimentation with alternative materials, reuse, or recovery systems. At the same time, fragmented global supply chains, limited traceability, and misaligned incentives across manufacturers, hospitals, recyclers, and policymakers inhibit coordinated action. Empirical indicators suggest that less than 10% of healthcare materials are currently reused or recycled, with most medical and pharmaceutical waste still thrown away due to safety and regulatory constraints.[5]

Addressing such system-level challenges requires more than technological innovation; it demands capability development and collective learning. Empirical evidence from collaborative projects conducted between Stanford University ME310, ZHAW's Master in International Business and industry partners shows that circularity challenges are experienced by participants as ambiguous, evolving, and highly constrained by regulation and organizational boundaries.[6] Participants reported limited data availability, fragmented responsibility, and persistent tensions between

---

[1] Health Care Without Harm. (2020). *Health care's climate footprint: How the health sector contributes to the global climate crisis and opportunities for action.* Health Care Without Harm & Arup.

[2] Metzke, S. (2022). *Sustainability in healthcare: Environmental impact and transformation pathways.*

[3] Boston Consulting Group (BCG). (2022). *The green transformation of healthcare: Reducing emissions while improving outcomes.* Boston Consulting Group.

[4] Health Care Without Harm Europe. (2020). *Circular economy and the health sector.* HCWH Europe, Brussels.

[5] Health Care Without Harm Europe. (2020). *Circular economy and the health sector.* HCWH Europe, Brussels.

[6] Chauhan, S. (2024). *Purpose-driven innovation in interdisciplinary university–industry collaboration: A qualitative study in life sciences.* Master's thesis, ZHAW School of Management and Law.

usability, patient safety, regulatory compliance, and economic viability. Traditional, discipline-based educational formats proved insufficient to prepare participants for this complexity. Progress depended on the ability to collaborate across domains, negotiate competing logics, and sustain engagement under uncertainty—capabilities rarely cultivated explicitly in conventional curricula.

Against this backdrop, this chapter positions education as a central micro-foundation of purpose-driven system innovation in the context of circularity. In the projects studied, a shared project purpose served as a coordination and legitimization mechanism that enabled alignment, engagement, and experimentation under uncertainty. Collaboration supported learning processes that integrate systems thinking, reflexivity, and collective problem framing—capabilities increasingly recognized as essential for addressing sustainability challenges.[7, 8]

The chapter examines how such collaborative learning environments can be deliberately designed and facilitated. Drawing on three interdisciplinary projects linking education, research, and industry practice, it analyzes how design-oriented didactic approaches, such as open-ended problem framing, iterative prototyping, stakeholder engagement, and structured reflection, enable actors to enact purpose-driven innovation in practice.

The chapter is structured as follows. Section 6.2 situates the study within the relevant literature. Section 6.3 examines the educational architecture and didactic design co-developed between ZHAW and Stanford University. Section 6.4 presents the three empirical use cases—circular pharmaceuticals, circular medical devices in hospital operating rooms, and circular delivery systems—followed by an analysis of identified challenges and success factors in 6.5. Section 6.6 synthesizes the findings into a practice-oriented framework. The chapter concludes by outlining implications for future research, education, business and policy.

6.2 Theoretical Foundations and Definitions: Purpose-Driven Innovation as a Logic for System-Level Circular Transformation

Recent research on purpose-driven innovation, open innovation, and the circular economy documents a fundamental shift away from firm-centric models toward more systemic and collaborative forms of value creation.[9,10] In highly regulated and

[7] Schön, D. A. (1983). *The reflective practitioner: How professionals think in action.* Basic Books.

[8] Bushe, G. R. (2021). *Clear leadership: Sustaining real collaboration and partnership at work.* Berrett-Koehler.

[9] Björck, A., Pregmark, J., Brandin, K., & Schoch, D. (2025). From purpose to circularity: Unpacking the strategic and systemic role of corporate purpose. *Swiss Journal of Business*, 25/3. doi.org: 10.5771/2944-3741-2025-3-262.

[10] Tushman M., Volberda H., Di Minin A., Bushe G., Björck A., Pregmark J., Gionfriddo G. / Fredberg T. (2024). Purpose-driven Startegic Renewal, Open Innovation & Generative Change: Models, Governance, Practice. *Academy of Management Proceedings*. https://doi.org/10.5465/AMPROC.2024.12103symposium

complex sectors such as the life sciences, innovation increasingly depends on the ability to coordinate heterogeneous actors—manufacturers, healthcare providers, regulators, recyclers, and users—around shared objectives that extend beyond short-term economic returns. Across these streams of literature, there is growing consensus that sustainability transitions require not only technological advances, but new forms of coordination, governance, and collective learning.[11]

To integrate these insights, we adopt a conceptual hierarchy: We treat sustainability challenges in the life sciences as *system innovation problems*; they are addressed through *design-oriented innovation* processes; they are coordinated and legitimized through *purpose-driven innovation*; and they materialize in the form of *circular innovation outcomes*. This hierarchy establishes focus, scope, mechanism, and outcome and provides a structured lens for the case analysis.

System Innovation Problems

Sustainability challenges in the life sciences—such as medical waste—are deeply embedded in healthcare systems, global value chains, and regulatory regimes that no single actor controls. They exhibit the defining characteristics of wicked problems: they are ill-defined, evolve over time, involve multiple stakeholders with divergent interests, and lack definitive solutions[12, 13]). They also display tragedy-of-the-commons dynamics, where individually rational behavior leads to collectively suboptimal outcomes due to delayed, distributed, and weakly attributable consequences.[14, 15]

A key implication of wicked and commons-based problem is that they are constrained by Pareto conditions. In complex socio-technical systems, improvements for one actor often impose costs, risks, or compliance burdens on others, making unilateral improvement impossible without trade-offs. Research on circular economy transitions therefore highlights the importance of strategies following the satisficing principle—the search for solutions that are acceptable and legitimate across stakeholders rather than optimal for any single one.[16] In the life sciences, competitive pressure and intellectual property regimes further reinforce these constraints and make system innovation inherently coordination-intensive.

Under such conditions, firm-centric innovation approaches and local optimization logics are structurally inadequate and system approach is required. System

[11] Takács, F., Brunner, D., & Frankenberger, K. (2022). Barriers to a circular economy in small- and medium-sized enterprises and their integration in a sustainable strategic management framework. *Journal of Cleaner Production*, 362. https://doi.org/10.1016/j.jclepro.2022.132227

[12] Rittel, H. W. J., & Webber, M. M. (1973). Dilemmas in a general theory of planning. *Policy Sciences*, 4(2), 155–169.

[13] Schön, D. A. (1983). *The reflective practitioner: How professionals think in action.* Basic Books.

[14] Hardin, G. (1968). The tragedy of the commons. *Science*, 162(3859), 1243–1248.

[15] Ostrom, E. (1990). *Governing the commons: The evolution of institutions for collective action.* Cambridge University Press

[16] Simon, H. A. (1955). A behavioral model of rational choice. *Quarterly Journal of Economics*, 69(1), 99–118.

innovation reconfigures interdependent actors, practices, technologies, and governance structures at the level of socio-technical systems rather than individual products or organizations.[17,18]

Purpose-Driven Innovation

Purpose-driven innovation serves as the central explanatory concept that addresses the questions of why and how alignment and cooperation become possible. It denotes innovation processes where a shared societal purpose aligns heterogeneous actors, legitimizes trade-offs, and sustains collaborative engagement under uncertainty.[19,20] This understanding builds on work that conceptualizes organizational purpose as an overarching reason for existence that defines a societal problem, respectively, solution space and in this way guides strategic choices, stakeholder relationships, and long-term value creation.[21,22,23]

Recent research emphasizes the relevance of purpose in contexts characterized by competing institutional logics, regulatory constraints, and collective action problems. In such settings, purpose provides direction and legitimacy for action under ambiguity and enables coordination where neither market incentives nor hierarchical authority is sufficient.[24] Purpose orients exploration, legitimizes compromise, and stabilizes collaboration across organizational, disciplinary, and institutional boundaries.[25]

Design-Oriented Innovation Learning Processes

While system innovation and purpose-driven innovation clarify why coordination is necessary and what kind of change is required, they leave open the question of how actors learn to enact these principles in practice. Design-oriented innovation

---

[17] Geels, F. W. (2002). Technological transitions as evolutionary reconfiguration processes. *Research Policy*, 31(8–9), 1257–1274.

[18] Markard, J., Raven, R., & Truffer, B. (2012). Sustainability transitions: An emerging field of research and its prospects. *Research Policy*, 41(6), 955–967.

[19] Henderson, R. (2021). Innovation in the twenty-first Century: Architectural Change, Purpose, and the Challenges of Our Time. *Management Science*, 67(9), 5479–5488.

[20] Björck, A., Pregmark, J., Brandin, K., & Schoch, D. (2025). From purpose to circularity: Unpacking the strategic and systemic role of corporate purpose. *Swiss Journal of Business*, 25/3. doi.org: 10.5771/2944-3741-2025-3-262.

[21] Mayer, C. (2021). The Future of the Corporation and the Economics of Purpose. *Journal of Management Studies*, 58(3), 887–901.

[22] Gulati, R. & Wohlgezogen F. (2023) "Can Purpose Foster Stakeholder Trust in Corporations." *Strategy Science* 8, no. 2. 270–287

[23] Durand R. & Huynh C. (2024). Corporate Purpose research: Streams and promises. *Journal of Management Scientific Reports*. Vol. 2, Issue 3–4.

[24] Bushe, G. R. (2021). *Clear leadership: Sustaining real collaboration and partnership at work.* Berrett-Koehler.

[25] Tushman M., Volberda H., Di Minin A., Bushe G., Björck A., Pregmark J., Gionfriddo G. / Fredberg T. (2024). Purpose-driven Startegic Renewal, Open Innovation & Generative Change: Models, Governance, Practice. *Academy of Management Proceedings*. https://doi.org/10.5465/AMPROC.2024.12103symposium

provides a learning and process logic suited to uncertainty, ambiguity, and competing stakeholder demands.

Design-oriented innovation builds on the design thinking tradition, which emphasizes iterative problem framing, human-centered inquiry, prototyping, and reflection in situations where neither problems nor solutions are fully known in advance.[26,27] In this chapter, design thinking is not treated as a toolbox of methods, but as a learning architecture that enables collective sensemaking, exploration of trade-offs, and development of satisficing solutions. Crucially, design-oriented innovation operationalizes purpose. It reframes abstract societal objectives—such as environmental sustainability or patient safety—into actionable problem framings positioned in a user context.

Circular Innovation Outcomes and Applications

Circular innovation refers to innovations that reduce resource extraction, waste, and emissions by embedding reuse, recovery, transparency, and regenerative logic into products, services, business models, and governance arrangements.[28,29] In the life sciences, circular innovation targets material flows, waste streams, traceability, and environmental impact.

From a system innovation perspective, circularity cannot be achieved through isolated product redesign. Instead, it requires coordinated action across ecosystems involving manufacturers, healthcare providers, regulators, waste handlers, and recyclers. In this chapter, circular innovation is therefore treated as the applied manifestation and outcome of purpose-driven, design-oriented system innovation.

Although the literatures on purpose-driven innovation, system innovation, circular economy, and design thinking provide valuable insights into sustainability solutions, they remain weakly connected when it comes to explaining how actors develop the capabilities required to enact these principles over time. This chapter addresses this gap by arguing that design-oriented learning architectures provide a critical space in which coordination, sensemaking, and boundary-spanning capabilities can be developed.

6.3 Educational Design and Didactics for Purpose-Driven Innovation

Wicked, commons-based challenges require learning environments that actively cultivate the capabilities needed to navigate uncertainty, trade-offs, and multi-actor coordination. Design-oriented learning—exemplified by the Stanford Design

[26] Brown, T. (2008). Design thinking. *Harvard Business Review*, 86(6), 84–92.

[27] Dorst, K. (2015). Frame innovation: Create new thinking by design. MIT Press.

[28] Geissdoerfer, M., Savaget, P., Bocken, N. M. P., & Hultink, E. J. (2017). The circular economy—A new sustainability paradigm? *Journal of Cleaner Production*, 143, 757–768.

[29] Kirchherr, J., Reike, D., & Hekkert, M. (2018). Conceptualizing the circular economy: An analysis of 114 definitions. *Resources, Conservation and Recycling*, 127, 221–232.

tradition and the ME310 Global Engineering Design Innovation course—provides such an architecture. The section first introduces Stanford Design and the ME310 course as a didactic blueprint and then shows how this blueprint was extended to address international business and ecosystem coordination challenges in the life sciences.

6.3.1 Stanford Design Thinking as a Learning Architecture

The educational approach analyzed in this chapter builds explicitly on the design thinking tradition developed at Stanford University, particularly through the long-standing ME310 Global Engineering Design Innovation course.[30, 31] Within this tradition, design thinking is a learning architecture for navigating ill-defined problems characterized by uncertainty, multiple stakeholders, and competing constraints. Despite similarity with problem-based learning approaches, a key characteristic of Stanford design pedagogy is that it departs from linear problem-solving models by emphasizing the co-evolution of problem and solution. Empirical studies of ME310 show that teams rarely progress through neat stages; instead, problem framing, technical exploration, and stakeholder alignment develop iteratively through cycles of action and reflection.[32]

Three principles are central to this dynamic and open learning architecture:

First, learning through making. Prototyping is used early and repeatedly to surface assumptions, test hypotheses, and create shared understanding among heterogeneous actors.[33] Prototypes function as epistemic artifacts that make system dynamics tangible and discussable.

Second, *productive engagement with ambiguity.* Ambiguity is treated as a deliberately designed condition of learning. Students are expected to tolerate uncertainty, delay premature convergence, and revisit assumptions as new insights emerge.[34]

Third, *human-centered and stakeholder-embedded inquiry*. Design thinking at Stanford emphasizes deep immersion in user and stakeholder contexts through ethnographic methods, improvisation, and observation.

---

[30] Leifer, L., & Steinert, M. (2011). Dancing with ambiguity: Causality behavior, design thinking, and triple-loop learning. *Information Knowledge Systems Management*, 10(1–4), 151–173.

[31] Plattner, H., Meinel, C., & Leifer, L. (Eds.). (2011). *Design thinking: Understand—improve—apply.* Springer.

[32] Koppenhagen, F., Blümel, T., Held, T., et al. (2021). The challenging combination of agility and convergence in hybrid product development processes: An empirical analysis of Stanford's ME310 process model. *Proceedings of the Design Society*, 1, 2993–3002.

[33] Domingo, L., Moore, D., Sirkin, D., Toye, G., Leifer, L. and Cutkosky, M. (2020). "Strategic Prototyping to Learn in Stanford University's ME310 Design Innovation Course", *Proceedings of the Design Society: DESIGN Conference*. Cambridge University Press, 1, pp. 1687–1696. https://doi.org/10.1017/dsd.2020.176

[34] Leifer, L., & Steinert, M. (2011). Dancing with ambiguity: Causality behavior, design thinking, and triple-loop learning. *Information Knowledge Systems Management*, 10(1–4), 151–173.

6.3.2 Stanford ME310 as a Blueprint for Innovation Learning

The ME310 course operationalizes these design thinking logic and principles through a year-long, industry-sponsored, studio-based project structure (Fig. 6.1). Student teams work on open-ended challenges provided by corporate partners without predefined problem statements or solution paths.[35] The extended duration allows teams to experience the full spectrum of innovation process dynamic, including setbacks, reframing, and iterative convergence.

A distinctive contribution of ME310 is its explicit cultivation of ambidextrous innovation capability—the ability to alternate between exploratory, divergent modes of inquiry and analytical, convergent modes of engineering reasoning.[36] Students are expected to move fluidly between:

- divergent exploration and disciplined convergence,
- creative reframing and technical validation,
- open-ended sensemaking and structured decision-making.

This ambidexterity is particularly relevant for purpose-driven innovation, where actors must simultaneously explore systemic alternatives and respect stringent constraints related to safety, regulation, and feasibility.

The course further institutionalizes strategic prototyping as a learning strategy. Different prototype types—such as need-finding prototypes, critical function

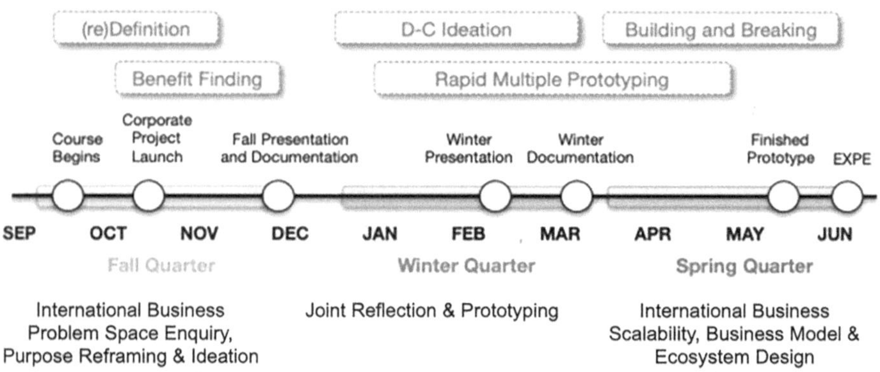

Fig. 6.1 Stanford ME310 structure and international business extensions (Source: adapted from Stanford GEDI). Developed by the authors, used with permission, all rights reserved

[35] Domingo, L., Moore, D., Sirkin, D., Toye, G., Leifer, L. and Cutkosky, M. (2020). "Strategic Prototyping to Learn in Stanford University's ME310 Design Innovation Course", *Proceedings of the Design Society: DESIGN Conference*. Cambridge University Press, 1, pp. 1687–1696. https://doi.org/10.1017/dsd.2020.176

[36] Koppenhagen, F., Blümel, T., Held, T., et al. (2021). The challenging combination of agility and convergence in hybrid product development processes: An empirical analysis of Stanford's ME310 process model. *Proceedings of the Design Society*, 1, 2993–3002.

prototypes, and "dark-horse" prototypes—serve distinct epistemic purposes at different stages of the process. Teams use prototypes to learn what matters, where trade-offs lie, and which assumptions require re-examination.

Instruction in ME310 follows a coaching model rather than traditional lecturing. Faculty and experts enable learning by coaching teams navigate uncertainty, access relevant expertise when needed, and reflect on their process.[37] Knowledge is introduced just-in-time, reinforcing learning as situated and emergent.

6.3.3 Extending the Blueprint: International Business and Ecosystem Coordination

The ZHAW–Stanford collaboration extends the ME310 design pedagogy into international business contexts in order to address innovation challenges that are embedded in global value chains, regulatory regimes, and multi-actor ecosystems (Fig. 6.1). While the Stanford ME310 blueprint was originally developed to support user- and product-centered innovation under technical uncertainty, the life sciences context introduces additional layers of complexity related to governance, incentive alignment, regulatory legitimacy, and system-level coordination. The purpose of this extension was therefore not to modify ME310's core design logic, but to test and adapt it for purpose-driven system innovation.

Throughout each of the projects, international business students from ZHAW contributed analytical perspectives on strategy, finance, ecosystem governance, internationalization, and scaling, complementing Stanford students' strengths in physical prototyping, mechanical engineering, materials knowledge, and early-stage entrepreneurship. This integration deliberately shifted the focus from product-centric problem-solving toward a system-level coordination. As a result, teams were required to engage with questions that lie beyond traditional design scopes: who orchestrates the ecosystem, how value and risk are distributed, how regulatory compliance is achieved, and how sustainability ambition can be aligned with economic viability.

Crucially, this extension supported and selectively broadened the Stanford design blueprint. Prototyping expanded from testing user interactions and technical feasibility to exploring ecosystem configurations, and business models. Human-centered inquiry was extended from individual users to include organizational and institutional stakeholders such as hospitals, regulators, waste handlers, and manufacturers as well as adopting a multicultural perspective. Ambidexterity, as a core capability, also evolved: teams had to navigate not only the tension between divergent exploration and convergent engineering rigor, but also tensions between economic performance, regulatory constraints, and societal purpose.

[37] Jung, M., Mabogunje, A., Sonalkar, N., & Leifer, L. (2007). Design knowledge coaching: A conceptual framework to guide practice and research. *Proceedings of the International Conference on Engineering Design (ICED).*

This extension enriched the nature of interdisciplinary collaboration. Rather than operating in parallel disciplinary streams, engineering and international business perspectives were integrated through joint system design activities. Across project phases, this integration followed a dynamic, bi-directional learning path. During early insighting and ethnographic research, international business students contextualized user observations within national regulatory systems, market structures, and stakeholder power relations broadening engineering perspectives. In ideation and early prototyping phases, both teams were encouraged to consider not only localized problems but their embedding within broader ecosystems. The deliberate "dark-horse" phases emerged as eye-opening experiences for business students how to endure and tackle provoked conceptual breaks and productive conflict, forcing teams to surface and negotiate assumptions across disciplinary and institutional logics.

As projects progressed toward system design, physical prototypes increasingly functioned as a core around which ecosystem architectures, and circular business logics were developed. In later phases, international business perspectives became particularly salient as teams addressed questions of orchestration, incentive alignment, regulatory feasibility, and global scalability—issues that could not be resolved through product or technology design alone but were deemed central for industry partners.

Across these extensions, purpose played a critical integrative role. Shared societal purpose—such as reducing environmental harm while safeguarding patient safety—functioned as a higher-order reference point that enabled teams to negotiate trade-offs, and sustain engagement across disciplinary, institutional, and cultural boundaries. In this sense, the extension of the Stanford design blueprint positioned purpose not only as a source of motivation, but as a coordination and legitimization mechanism within the learning process itself.

6.4 Empirical Evidence: Circular Innovation Use Cases in Life Sciences

This section presents use cases drawn from three interdisciplinary student projects conducted between Stanford University, ZHAW, and industry partners in the global life sciences sector. The empirical material comprises 30 semi-structured interviews with students, faculty, coaches, and industry partners, complemented by rich project documentation, design artifacts, and participant observation.

The cases address distinct problem domains—circular pharmaceutical products, circular medical devices in hospital operating rooms, and circular single-use delivery systems. They are examined using a common lens of educational design and purpose-driven innovation. The cases are not treated here as best practices, but they have been analyzed as empirical evidence from which recurring patterns of purpose-driven innovation can be induced.

6.4.1 Case A: Circular Medication Take-Back Systems

The first use case addresses circularity challenges in the pharmaceutical sector, focusing on the disposal of expired or unused medication and the environmental impact of active pharmaceutical ingredients (APIs). Despite growing evidence of pharmaceutical residues contaminating soil and water systems, end-of-life handling of medicines remains fragmented and weakly governed. Existing take-back schemes vary widely across regions and depend heavily on consumer awareness and voluntary participation. Participants described this problem space as particularly complex due to strict regulatory requirements, limited transparency regarding post-consumption pathways, and diffuse responsibility across actors.

Purpose-driven reframing positioned pharmaceutical waste as a public health and environmental integrity challenge extending beyond the point of patient use. This reframing enabled teams to question dominant assumptions about pharmaceutical ownership, responsibility, and disposal, and to legitimize exploration of end-of-life models traditionally considered outside the remit of pharmaceutical firms.

The circular solution required coordination among manufacturers, pharmacies, regulators, waste handlers, and consumers—actors with limited direct interaction under existing regimes. Participants consistently emphasized that no single organization could credibly orchestrate this system alone. Governance challenges centered on aligning regulatory feasibility with behavioral change and public trust, underscoring the institutional nature of pharmaceutical circularity.

A distinctive feature of this case was the pivotal role of iterative "dark-horse" exercise. Both teams initially converged on incremental improvements—such as packaging redesigns or enhanced logistics—before becoming conceptually stuck. The "dark-horse" exercise forced teams to abandon incremental fixes and interrogate assumptions about consumption, return behavior, and institutional responsibility. Interview data indicate that this moment of deliberate disruption was critical for shifting attention toward systemic interventions addressing return behavior, incentives, and institutional arrangements.

The resulting solution combined medication take-back schemes with a redesigned distribution and packaging concept. The case demonstrates that purpose-driven innovation in pharmaceutical medication relies less on behavioral change, regulatory alignment, and ecosystem coordination, supported by business models that integrate compliance and consumer incentives.

6.4.2 Case B: Circular Medical Devices in Hospital Operating Rooms

The second use case examines circularity in hospital operating rooms, focusing on single-use medical devices. Although these products contain high-value materials such as medical-grade plastics, metals, and glass, they are typically incinerated due to biohazard risks, regulatory requirements, and the absence of sorting infrastructure. Empirical observations revealed that waste handling practices in operating rooms are shaped by extreme time pressure, spatial constraints, and the prioritization of patient safety, leaving little room for environmental considerations in daily routines.

Purpose-driven reframing centered on the challenge of maintaining patient safety while reducing environmental harm. Purpose functioned as a legitimizing narrative that allowed sustainability to be discussed within clinical environments without being perceived as a competing priority. Participants reported that this reframing was essential for engaging clinical staff and hospital administrators in exploratory need-finding and problem framing.

Responsibility for operating-room waste is fragmented across hospitals, waste handlers, manufacturers, and recyclers. Firm-centric approaches quickly failed due to missing incentives and high-risk aversion. The project therefore proposed an ecosystem-based solution coordinated by a neutral orchestrator. Interview data indicate that this orchestration role was perceived as critical for managing interfaces, ensuring compliance, and building trust among actors operating under distinct regulatory and operational logics.

The final prototype consisted of an AI-enabled, contact-free sorting solution that enabled material recovery without exposing staff to biohazard risks. Digital traceability platforms made material and data flows, emissions savings, and compliance visible. Circular value creation relied on combining recovered material streams with service-based revenues from traceability, compliance documentation, ESG reporting, and a cross-industry marketplace for secondary materials. The case illustrates that circularity in operating rooms becomes viable only when safety and regulatory constraints are internalized within ecosystem governance and business model design.

6.4.3 Case C: Circular Single-Use Delivery Systems

The third use case focuses on single-use delivery systems such as prefilled syringes and safety injection devices. These products contribute significantly to healthcare waste and downstream emissions, yet their environmental impact remains poorly measured and weakly governed. Participants identified injectables as a boundary-spanning problem situated between pharmaceuticals and medical devices, subject to overlapping regulatory regimes and fragmented responsibility across manufacturers, healthcare providers, and waste management systems.

Purpose-driven innovation in this case centered on reducing environmental impact without compromising sterility, usability, or patient safety. Purpose provided a shared reference point that enabled dialogue across pharmaceutical and MedTech logics, which are often treated as distinct despite their interdependence in injectable products.

As in the previous cases, circular solutions required coordination across manufacturers, hospitals, logistics providers, decontamination services, and recyclers. Data asymmetries and competitive concerns initially inhibited collaboration. An orchestrator-led ecosystem was therefore proposed to manage data standards, interfaces, and material flows across the injectable lifecycle.

The team identified multiple automation opportunities, including robotic sorting, decomposition, and material recovery. Material composition databases and downstream traceability applications formed the core enabling infrastructure. Interviewees reported that making emissions and material flows visible transformed sustainability discussions from abstract aspirations into concrete, evidence-based decision-making. However, the case also revealed that standardization and governance were prerequisites for data to become actionable.

The circular business model depended on platform-based coordination rather than product redesign alone. Value was created through shared data services, material marketplaces, and certification mechanisms that aligned incentives across actors. Circularity thus emerged as a collective outcome rather than a firm-level achievement.

6.4.4 Cross-Case Learning Mechanisms Enabling Purpose-Driven Innovation

Across the three cases, four learning mechanisms—purpose anchoring, productive ambiguity, interdisciplinary interaction, and prototypical boundary objects—enabled purpose-driven system innovation. Importantly, they were not prescribed upfront as learning objectives but evolved through repeated engagement with the constraints and tensions of the projects.

First, purpose anchoring functioned as a learning mechanism for orientation.

In all cases, teams struggled with prolonged phases of ambiguity, slow progress, and competing demands from different stakeholders. The project purpose—defined as a shared societal challenge such as reducing environmental harm while safeguarding patient safety—helped participants maintain orientation when neither problem definitions nor solution pathways were stable. Interview data show that purpose enabled students to legitimize continued exploration in the absence of short-term results, reinterpret setbacks as necessary learning steps, and negotiate trade-offs. In this sense, purpose operated as an ecosystem-level coordination device and a cognitive and motivational anchor within the learning process itself.

Second, working with ambiguity emerged as a core learning condition.

Early project phases were consistently characterized by unclear system boundaries, incomplete data, and evolving interpretations of responsibility and feasibility. Rather than resolving this ambiguity prematurely, teams—supported by coaching and the ME310-inspired process logic—were encouraged to hold and revisit uncertainty through iterative inquiry. This included repeated reframing of the problem space, alternating divergence and convergence, and deliberate postponement of solution commitment. Over time, participants developed a greater tolerance for ambiguity and learned to treat uncertainty as a resource for system-level insight rather than a failure of analysis.

Third, interdisciplinary interaction functioned as experiential training for ecosystem coordination.

Friction between engineering, international business, and human-centered design perspectives surfaced divergent assumptions about feasibility, value, risk, and responsibility. These tensions were not resolved through disciplinary dominance but through repeated negotiation, supported by shared artifacts and structured reflection. The cases show that learning occurred not simply through knowledge exchange but through practicing coordination across persistent differences.

Fourth, physical and digital artifacts served as pedagogical boundary objects stabilizing collaboration.

Across cases, prototypes, system maps, and data models, externalized assumptions about the system and made abstract dynamics tangible. These artifacts enabled teams to align interpretations across disciplines, test hypotheses without full commitment, and engage stakeholders in concrete discussion. Artifacts thus played a dual role: enabling innovation progress and structuring collective learning.

6.5 Structural Challenges and Design Responses

This section synthesizes structural challenges that consistently shaped project dynamics and outlines design responses that enabled progress.

6.5.1 Structural and Institutional Misalignment

A central structural challenge across all projects was persistent misalignment between academic and industry logics. Universities and corporate partners operated under different temporal rhythms, evaluation criteria, and tolerances for ambiguity. Academic calendars and assessment regimes privileged exploration, reflection, and learning, while industry partners emphasized feasibility, credibility, and time-bound progress. Additional divergence existed between the academic year planning of both universities—sprint-like project durations in international business and 9-month elective course in mechanical engineering. Furthermore, different time zones and predominantly online, hybrid collaboration, challenged the development of a rhythm and collective learning. Students experienced these tensions directly, navigating between open-ended inquiry and expectations for concrete outcomes, while faculty balanced pedagogical integrity with the need to sustain long-term industry collaboration.

6.5.2 Regulatory Constraints

All three cases were embedded in highly regulated environments governing pharmaceuticals, medical devices, injectables, and healthcare waste. Participants initially perceived regulation as an external constraint that limited creativity and slowed progress, particularly in early project phases when students lacked familiarity with regulatory approval pathways, liability considerations, and risk management logics.

Over time, purpose-driven collaboration enabled a reframing of regulation from an obstacle into a design parameter. Anchoring innovation efforts in shared societal goals—most notably patient safety and environmental protection—legitimized early engagement with regulatory expertise and shifted the orientation from circumvention to accommodation. Rather than attempting to bypass regulation, teams learned to design within its boundaries, exploring solutions compatible with existing frameworks and approval logics. Purpose did not reduce regulatory complexity, but it enabled more constructive interaction with it by aligning experimentation with public value and long-term system integrity.

6.5.3 Data, Trust, and Competitive Tensions

A further structural challenge concerned limited access to reliable downstream data and widespread reluctance to share it because of competitive and reputational concerns by industry partners. Across cases, data transparency did not emerge automatically through technical solutions. Instead, the empirical evidence indicates that trust was a prerequisite rather than a consequence of transparency. Neutral or semi-neutral orchestrators, clearly defined data governance rules, and explicit agreements on data use were necessary to stabilize collaboration. This finding underscores that digital infrastructures alone are insufficient without accompanying social and governance arrangements that legitimate cooperation under competitive conditions.

6.5.4 Interdisciplinarity as a Persistent Tension

Interdisciplinary collaboration constituted both a core resource and a persistent source of tension. Engineering, business, healthcare, and design participants brought divergent assumptions about feasibility, risk, value creation, and responsibility. Early project phases were frequently marked by frustration, particularly when teams attempted to converge prematurely on solutions without resolving underlying differences in problem framing.

The cases show that productive collaboration did not depend on eliminating disciplinary tension, but on learning to work with it. Open-ended project design, iterative prototyping, and structured reflection transformed disciplinary differences into learning resources. Prototypes functioned as boundary objects that enabled dialogue across epistemic perspectives, while reflection practices surfaced implicit assumptions and made trade-offs discussable. Over time, participants developed the capacity to operate within unresolved tension—an essential competence for purpose-driven system innovation in complex, multi-actor environments.

6.5.5 Education as a Design Response and Enabling Condition

Across all projects, the educational architecture within which collaboration was embedded proved critical for navigating these structural. Education thus functioned as an active design response to structural challenges. It provided protected spaces in which participants could rehearse ecosystem coordination, negotiate competing logics, and develop shared understanding without immediate performance pressure. In this sense, education operated as a micro-foundation through which purpose anchoring, ambidextrous thinking, and system-level coordination could be learned, enacted, and stabilized over time.

Taken together, these findings reveal a consistent pattern: purpose-driven innovation in the life sciences unfolds under challenging conditions that are enduring features of system innovation in regulated sustainability contexts. Progress depends on how these conditions are navigated through deliberate design of educational, organizational, and governance arrangements.

6.6 A Framework of Purpose-Driven Innovation: Synthesis and Guidance for Teaching, Business, and Policy

Our framework synthesizes seven design principles that emerged inductively from the cases and address the structural conditions identified in Sect. 6.5. These principles show how purpose and design-oriented processes jointly enable system innovation: (1) purpose anchored in systemic problems; (2) reframing waste as a strategic resource; (3) ecosystem design anchored in artifacts; (4) orchestrator-centered governance; (5) digital enablement embedded in governance; (6) circular business model integration; and (7) open-ended, interdisciplinary processes. Together they explain how educational architectures, governance mechanisms, and technical infrastructures cohere to produce scalable circular outcomes.

(1) **Purpose anchored in a systemic problem**

Across all cases, purpose proved effective only when grounded in clearly articulated systemic challenges—such as linear waste regimes, downstream emissions, or environmental contamination—rather than narrowly defined organizational inefficiencies. Systemic problem framing enabled ecosystem-level coordination, while purpose anchoring sustained engagement under uncertainty.

(2) **Reframing waste as a strategic resource**

All cases required a cognitive shift from disposal toward regeneration. Waste was reframed not only as a material stream, but also as a source of data, learning, and coordination. This reframing did not occur automatically; it emerged through iterative sensemaking enabled by educational settings that tolerated ambiguity and experimentation.

(3) **Ecosystem-based solution design anchored in physical artifacts**

Ecosystem-based solution design emerged through iterative collaboration anchored around physical prototypes or system artifacts. These artifacts stabilized coordination across heterogeneous actors and functioned as boundary objects for interdisciplinary learning.

(4) **Orchestrator-centered governance**

Neutral or semi-neutral orchestrators played a critical role in aligning actors, managing interfaces, and governing sensitive data and material flows. From a learning perspective, orchestrators also provided continuity and structure in otherwise open-ended collaborations, reinforcing the link between governance design and educational architecture.

(5) **Digital tools for transparency**

Digital tools—such as traceability platforms, dashboards, and automated sorting technologies—were important for transparency and accountability. However, the cases demonstrate that digital enablement created value only when embedded in shared governance arrangements and collective interpretation.

(6) **Circular business model integration**

Purpose-driven innovation remained fragile unless translated into viable business models. Platform-based services, shared infrastructures, marketplaces, and certification mechanisms aligned incentives across actors and stabilized collaboration over time.

(7) **Open-ended and interdisciplinary innovation processes.**

Finally, the cases show that purpose-driven innovation depends on open-ended, interdisciplinary collaboration as a core capability. Iterative prototyping, reflection, and negotiated trade-offs enabled teams to work productively with uncertainty, regulation, and competing logics.

In a next step, we synthesized the capabilities necessary for purpose-driven system innovation, and the roles of and the interplay between purpose and design pedagogy necessary for educating for circularity (Table 6.1). Across the cases, a central capability enabling purpose-driven system innovation was ambidexterity: the learned ability to navigate persistent tensions between exploration and exploitation, creativity and regulation, and societal purpose and economic viability.

Table 6.1 Pedagogy for purpose-driven system innovation to achieve circularity. Developed by the authors, used with permission. All rights reserved

Problem characteristic	Required capability	Design thinking contribution	Purpose contribution
Wickedness (ill-defined, evolving problems)	Iterative sensemaking and reframing	Continuous problem framing, prototyping, learning loops, perspective variety	Provides orientation under uncertainty by anchoring exploration to a shared societal challenge
Tragedy of the commons	Collective legitimacy and engagement	User-centered inquiry, co-creation	Creates moral and normative justification for collective action beyond individual incentives
Pareto constraints	Satisficing through negotiated trade-offs	Exploration and visualization of alternatives	Legitimizes compromise over individual optimization
Simultaneous exploration and constraint pressure	Ambidexterity (exploration–exploitation balance)	Alternation between divergent exploration and convergent validation; rapid prototyping under constraint	Enables actors to hold economic viability, regulatory rigor, and societal purpose simultaneously without premature closure
Behavioral change requirements	Motivation and sustained commitment	Empathy work, iterative testing	Sustains intrinsic motivation beyond compliance
Fragmented ecosystems	Cross-boundary coordination	Boundary objects, collaborative processes	Aligns heterogeneous actors without hierarchy
Regulatory and risk constraints	Legitimated experimentation	Safe-to-fail prototypes, staged learning	Frames experimentation as public-value oriented
Long time horizons	Persistence and intergenerational orientation	Iterative progress tracking	Connects short-term action to long-term societal outcomes

6.7 Conclusion

This chapter set out to examine how purpose-driven innovation can be learned and enacted in the life sciences to enable circular outcomes. Drawing on three interdisciplinary projects implemented by Stanford University, ZHAW, and industry partners, the chapter demonstrates that the primary constraints to circular transformation are coordinating heterogeneous actors, governing shared resources, and learning to navigate uncertainty, trade-offs, and competing disciplinary and institutional logics. Across the cases, the project purpose emerged as a practical coordination and legitimization mechanism that enabled systemic problem framing and ecosystem-level collaboration.

The chapter makes three related theoretical contributions. First, it advances a learning-centered perspective on purpose-driven system innovation by showing how a shared purpose operates in practice as a stabilizing reference that legitimizes exploration and compromise. Second, it extends research on circular economy by demonstrating that circular innovation materializes only through ecosystem-level coordination anchored in physical and digital artifacts, orchestrated governance arrangements, and economically viable business models. Third, it contributes to system innovation theory by positioning education as a micro-foundation: a protected enactment space in which actors develop ambidextrous capabilities to move between exploration and exploitation, design and analysis, creativity and compliance.

For practitioners, the findings underscore that purpose-driven circular innovation cannot be achieved through firm-centric optimization or isolated sustainability initiatives. Managers in the life sciences must invest in ecosystem orchestration capabilities, shared data infrastructures, and business models that align incentives across manufacturers, healthcare providers, regulators, and downstream actors. Purpose, when jointly articulated and continuously revisited, can be operationalized as a governance mechanism that legitimizes data sharing, negotiated trade-offs, and experimentation within regulatory constraints. For policymakers, the results highlight the importance of enabling conditions for ecosystem collaboration, including neutral coordinating bodies, interoperable data standards, and regulatory spaces that allow learning without compromising safety and accountability.

The chapter also has clear implications for teaching and educational design. Preparing future leaders for sustainability-critical domains requires learning environments that mirror the ambiguity, interdisciplinarity, and coordination challenges of real-world system innovation. Open-ended, design-oriented, and purpose-anchored educational architectures such as those building on the Stanford ME310 tradition and extended through international business perspectives, provide protected spaces in which students can rehearse ecosystem coordination, and develop ambidextrous thinking. Education thus emerges as a central lever for accelerating purpose-driven system transformation.

Finally, the chapter points to several avenues for future research. Longitudinal studies are needed to examine how purpose-driven innovation capabilities developed in educational settings translate into organizational practice over time. Comparative research across industries, regulatory regimes, and national contexts could further test and refine the proposed framework. Future work should also examine power asymmetries, exclusion dynamics, and the conditions under which purpose loses—or regains—its coordinating force in complex ecosystems.

Together, these directions reinforce the chapter's central insight: advancing circular transformation in the life sciences requires sustained attention to the intertwined dynamics of learning, innovation, and governance dynamics that are likely to become increasingly salient across sectors facing systemic sustainability challenges.

Albena Björck is Associate Professor at the School of Management and Law, Zurich University of Applied Sciences (ZHAW). Her research activities lie in the areas of purpose-driven organizations and their strategic management in times of organizational transformation, crisis, and renewal, and draw from cross-disciplinary and human-centered perspectives. She is expert of the Swiss Innovation Agency and Chair of the Strategic Management Division at the European Academy of Management. Prior to academia, Björck spent more than 25 years in business practice and is currently also a consultant and executive coach in the mentioned areas of competence. Albena has been initiating and managing the industry-academia collaboration with Stanford University, ME Design Group, and ZHAW School of Management and Law from the academia side. She holds a PhD from the University of St. Gallen (Switzerland).

George Toye, Ph.D., P.E., is adjunct professor in Mechanical Engineering at Stanford University. While teaching advanced project-based engineering design thinking and STEM-based innovations at the graduate level as part of ME310, he also contributes to research in varied topics in engineering education, and effective globally distributed team collaborations. As well, he remains active in entrepreneurship and varied advising/consulting work. George earned his B.S. and M.S. degrees in Mechanical Engineering from U.C. Berkeley, and Ph.D. in Mechanical Engineering with minor in Electrical Engineering from Stanford University.

Cedric Gysel is a co-founder and executive director of Be Circular, a non-governmental organization for circular life sciences. In his role, he empowers organizations to embrace circular economy principles, facilitating a shift from traditional linear models. Cedric has a long-year experience as a leader in Sustainable Innovation in the Life Science industry across new and established products working with teams in Supply Chain, Research & Development and Commercial. He has been initiating and managing the industry-academia collaboration with Stanford University, ME Design Group, and ZHAW School of Management and Law from the industry side. He holds a master's degree from the Berne University of Applied Sciences.

7 Improving Corporate Environmental Responsibility: The Impact of Perceived Stakeholder Pressure, Benefits, and Barriers in Switzerland and Germany

Carlo Pugnetti, Nadine Berchtold, and Stefan Hunziker

Abstract

Organizations are subjected to a complex set of stakeholder pressures as they develop policies to drive their Corporate Environmental Responsibility (CER). These pressure points interact in complex ways to shape perceived priorities, benefits, and barriers associated with these policies. We leverage a survey of 296 Swiss and German companies to understand this interaction and how it varies by company size, geography, business model, and sector. We find that companies are susceptible to external pressure from legislative action but not from the local community, and to internal pressure from management. Image improvement rather than profit is the most significant perceived benefit, while cost is the largest perceived barrier. In addition, companies believe their activities are impactful. Some differences are notable, with large or German companies perceiving higher pressure and reporting higher implementation rates and benefits from these policies than small or Swiss companies. Based on these results, activists should pursue a two-pronged strategy. On the one hand, pushing for national environmental legislation, while on the other, interfacing with companies at the local level to support their actions through expertise and image-building. Conversely, executives should work to avoid becoming the single point of contact and interaction for a broad set of stakeholders. They otherwise risk being overwhelmed and adopting an excessively narrow approach.

C. Pugnetti (✉) · N. Berchtold · S. Hunziker
Lucerne School of Business, Rotkreuz, Switzerland
e-mail: carlo.pugnetti@hslu.ch; nadine.berchtold@hslu.ch; stefan.hunziker@hslu.ch

© The Author(s), under exclusive license to Springer Nature Switzerland AG 2026

P. Wollmann et al. (eds.), *The Sustainable Organization – Volume 3*, Future of Business and Finance, https://doi.org/10.1007/978-3-032-28949-0_7

7.1 Introduction

Since the 2015 Paris Climate Agreement, attention on how businesses address their corporate environmental responsibilities (CER) has intensified. Firms are increasingly expected to contribute to climate goals by mitigating their environmental impact and integrating sustainability into strategic decision-making. CER is considered one dimension of Corporate Social Responsibility (CSR), as reflected in definitions by the Swiss government,[1] the European Commission,[2] and the World Bank.[3] In this context, CER refers to corporate actions and policies aimed at mitigating adverse environmental impacts. In the double materiality concept, it is part of impact materiality. CER can be measured through practices related to emission reduction, energy efficiency, waste and water management, and product responsibility, and sometimes, external auditing is also included as an indicator of CER.[4]

A variety of factors may influence CER adoption. Several theories have been developed to explain and interpret the pressure exerted on companies to adopt CER policies. Stakeholder theory argues that organizations should account for the interests of various stakeholders, which can exert pressure on the organization. Legitimacy theory stresses acting responsibly to maintain social acceptance and the "license to operate," while institutional theory explains how organizations adapt to social norms under pressure. Under rational choice theory, additional perceived benefits drive behavior, while firm behavioral theory explains how pressure shapes these perceived benefits through obligation, social learning, and legitimacy. Lastly, resource-based theory highlights that limited resources, such as a lack of expertise or high costs, may hinder the environmental actions of a firm. To investigate which of the potential influencing factors play a role for German and Swiss companies in the case of CER policies, we investigate three distinct dimensions: (i) internal and external pressure perceived by the companies, (ii) benefits perceived by the companies, and (iii) barriers perceived by the companies. The focus is not on an objective measure of each factor investigated, but on the company's subjective perception of each. This is because perception is the ultimate driver and determinant of behavior. These determinants and their channels, however, may differ depending on firm

[1] SECO. (2022). *Corporate Social Responsibility (CSR).* https://www.seco.admin.ch/seco/en/home/Aussenwirtschaftspolitik_Wirtschaftliche_Zusammenarbeit/Wirtschaftsbeziehungen/nachhaltigkeit_unternehmen/gesellschaftliche_verantwortung_der_unternehmen.html

[2] European Commission. (2001). *GREEN PAPER.* Promoting a European framework for Corporate Social Responsibility. COM (2001) 366. Brussels.

[3] Mazurkiewicz, P. (2004). *Corporate environmental responsibility: Is a common CSR framework possible?* World Bank, 2 (1), 1–18.

[4] Cantele S., and Zardini A. (2020). What drives small and medium enterprises towards sustainability? Role of interactions between pressures, barriers, and benefits. *Corporate Social Responsibility and Environmental Management*, 27(1), 126–136. https://doi.org/10.1002/csr.1778; Eccles, R. G., Ioannou, I., & Serafeim, G. (2014). The Impact of Corporate Sustainability on Organizational Processes and Performance. *Management Science*, 60(11), 2835–2857. https://doi.org/10.1287/mnsc.2014.1984

characteristics. While large firms are typically more subject to public scrutiny, small and medium-sized enterprises (SMEs) play an equally vital role in the transition toward sustainability, due to their prevalence in most economies.[5] Yet SMEs and large corporations differ substantially in their structures and operations, necessitating a more nuanced understanding of potential differences in their CER strategies and implementation.[6] Beyond firm size, different business models can affect CER engagement. Business-to-Consumer (B2C) firms often face stronger public visibility and consumer expectations, whereas Business-to-Business (B2B) firms may experience more indirect environmental pressures through supply chains and partners. Country-specific contexts may also shape CER practices due to variations in regulatory frameworks, cultural norms, and market pressures. Differences in environmental regulation, stakeholder activism, and institutional environments can lead firms across countries to prioritize distinct aspects of CER. These differences may also extend to companies operating in other sectors, driven by both market and regulatory pressures.

The development of greater corporate environmental responsibility is part of a broader movement to integrate environmental, social, and governance (ESG) targets into a company's operations. Both regulatory guidelines and corporate activities have increased in number and complexity in recent years, as evidenced by the growing number of ESG reports. While greenwashing remains a threat to the credibility of these efforts, the progress made is real and tangible. A renewed effort is underway to integrate these efforts more closely into the company's strategy and culture under the heading of purpose.[7] Customers are also becoming more sophisticated in their expectations toward commercial enterprises. While they generally support all activities aimed at achieving the 17 United Nations Sustainable Development Goals (SDGs), they are particularly interested in those that align with a company's accepted business model. Thus, customers expect companies in the energy and utility sector to pursue activities to mitigate climate impact, while those in the healthcare sector should pursue health and well-being.[8]

[5] e.g., Reuter S., Lackner P., and Brandl G. (2021). Mapping SMEs in Europe. Data collection, analysis and methodologies for estimating energy consumptions at Country levels. https://leap4sme.eu/wp-content/uploads/2021/07/LEAP4SME-D2.1-SME-energy-and-economic-mapping-in-Europe.pdf

[6] Russo A., and Tencati A. (2009). Formal vs. Informal CSR Strategies: Evidence from Italian Micro, Small, Medium-sized, and Large Firms. *Journal of Business Ethics*, 85(S2), 339–353. https://doi.org/10.1007/s10551-008-9736-x

[7] Björck A., Pugnetti C., Steller N., Vogel B. and Aebischer A. (2025). Purpose in Business Models: Strategic Principles and Key Decisions. In Wollman, Pemler and Ndrevataj (Eds.), *The Sustainable Organization: How Organizations Address the 17 UN SDGs using the 3-P Model*, 145–170. Springer Nature. ISBN 978–3–031-89,548-7.

[8] Pugnetti C., Stricker L. and Egli S. (2025). The Role of Insurance in Sustainability. In Wollman, Pemler and Ndrevataj (Eds.), *The Sustainable Organization: How Organizations Address the 17 UN SDGs using the 3-P Model*, 301–316. Springer Nature ISBN 978–3–031-89,548-7.

This chapter explores key drivers of different CER policies, focusing on perceived stakeholder pressure, benefits, and barriers, and builds on Berchtold (2024).[9] While the previous research focused on the SME subgroup, this analysis also includes large corporations and emphasizes additional firm characteristics. It aims to provide differentiated insights for large companies versus SMEs, for companies operating in Switzerland versus Germany, for B2B versus B2C business models, and by sector. These differences in how environmental responsibility is perceived and enacted provide a valuable baseline for managers, public policy wonks, activists, and politicians interested in corporate environmental responsibility.

7.2 Methodology and Results

The analysis is based on a survey of companies in Switzerland and Germany conducted in the Summer of 2022 as part of the Enterprise Risk Management Report authored by the Lucerne and Kiel Universities of Applied Science.[10] A total of 296 companies participated in the survey, representing a broad cross-section of Swiss and German companies across company size and economic sector. The sample composition is summarized in Table 7.1.

Respondents rated their company's adoption of CER policies across six dimensions on a 5-point Likert scale (5 = very high relevance; 1 = very low relevance). Further, they are asked to score perceived barriers, perceived benefits, perceived external pressure, and perceived internal pressure for the implementation of these policies along different dimensions on a similar Likert scale. The setup and average results for the entire sample are summarized in Table 7.2 and ranked by score within each category.

The dataset has been analyzed to identify first-order effects and preferences within individual categories, and second-order correlations among CER policies and perceived carriers, benefits, and pressure routes. In addition, differences between companies in different geographies, of varying sizes, using different business models, or active in different sectors have been investigated for statistical

Table 7.1 Sample size across differentiating characteristics. Developed by the authors. All rights reserved. Used with permission

Size	Geography	Business model	Sector
Up to 250 FTEs: 192 More than 250 FTEs: 102	Switzerland: 243 Germany: 53	B2C: 167 B2B: 196	7 sectors represented by at least 25 companies each, excluding the finance and agriculture sectors

[9] Berchtold N. (2024). Stakeholder Pressure on Corporate Environmental Responsibility in SMEs. *Swiss Journal of Business*, 78(1), 5–29. https://doi.org/10.5771/0042-059X-2024-1-5

[10] Hunziker S., Vanini U., Durrer M., Unruh A., and Berchtold N. (2022). *Climate Risk Management in Schweizer und deutschen Unternehmen*. Hochschule Luzern. ISBN 978–3–906,877-53-2.

Table 7.2 Overall survey results by category (Likert scale: 5 = very high relevance, 1 = very low relevance). Developed by the authors. All rights reserved. Used with permission

CER policies	Benefits	Barriers	External pressure	Internal pressure
4.42 Waste 4.24 Energy 3.98 Emissions 3.93 Product 3.60 Water 3.10 Audit	4.21 Image Improvement 3.67 Labor Attractiveness 3.57 Competitive Advantage 3.22 Profitability	3.78 Cost 3.48 Lack of Time 3.08 Lack of Expertise 2.91 Lack of impact	3.80 Legislation 3.61 Clients 3.33 Media 3.11 Local Community 3.02 Competitors 2.71 Financial Institutions 2.39 Suppliers 2.10 External audit	3.82 Management 3.51 Owners 3.49 Employees

significance using single-factor ANOVA. The authors discuss the results to develop actionable suggestions for company executives, activists, and political leaders to enhance the continued adoption and development of CER policies.

The most common CER policies according to the dataset are waste and energy, with emissions and products, and to some extent water, also frequently represented, whereas audit is considerably less common. The perceived benefits of these policies strongly favor image improvements, followed by labor attractiveness and competitive advantage, with profitability a distant fourth. Barriers are felt particularly with respect to costs and insufficient time, whereas expertise and impact are less significant. External pressure is most often felt through legislation and clients, whereas the media, the local community, and competitors are less often felt. Financial institutions as lenders, suppliers, and audit are felt least. Internal pressure is felt most strongly by management and somewhat less by owners and employees. Overall, companies tend to follow a relatively broad set of CER policies, but do not necessarily focus on audit requirements to do so. Image is an essential motivator, while other market considerations, such as attracting and retaining qualified personnel, are less critical. Profit is perceived as a relatively weak benefit. Resource constraints, but not expertise, hamper the successful implementation of policies perceived as impactful. The most effective pressure points are through legislation and own management, and to some extent, clients and owners. Other stakeholders, such as local communities or suppliers, are perceived as less relevant.

The correlation coefficients among the factors roughly mirror the overall perception described above. However, some correlations deviate from this and are therefore particularly interesting. Among the barriers, the perceived lack of impact is most clearly negatively correlated with the CER policies. That is, if a company does not consider its own activities to (positively or negatively) impact the environment, this may slow down or stop the adoption of CER policies. The other barriers, especially cost, have a weaker impact on the policies. Among the CER policies, waste policies correlate least with perceived benefits, whereas audit policies correlate most strongly across all benefit dimensions. That is, policies already implemented across the board are not affected by perceived benefits, whereas emerging policies are more closely associated with them. This observation also holds true for pressure factors: in general, they affect the least implemented policies more visibly than those already widely implemented.

7.2.1 Differentiation by Company Characteristics

The overall results described in the previous section can be affected by several firm characteristics. In the dataset, we can distinguish companies by size (large vs. small/ medium), geography (Switzerland vs. Germany), business model (B2B vs. B2C), and sector. Large firms are defined as companies with 250 or more full-time equivalents (FTEs). Figure 7.1 presents a comparative analysis of the aggregate constructs[11] for Benefits, Barriers, External Pressure, and Internal Pressure across four primary grouping dimensions: the aggregated sample, company size, geographic location, and business model. Data are visualized through a series of boxplots, with the central horizontal line indicating the median, and an "X" indicating the mean. The box itself represents the interquartile range, spanning from the 25th to the 75th percentile. Whiskers extend to the minimum and maximum non-outlier values. Collectively, the relative vertical positions of these boxplots facilitate a rapid visual assessment of differences among the defined groups.

Company Size (Large vs. SMEs)

Large firms report stronger performance across all CER policies, particularly regarding emissions, energy efficiency, and environmental auditing. The difference in auditing is particularly significant, with large companies, on average, scoring 4.46 (high relevance to very high relevance), whereas SMEs score 2.36 (low relevance to neutral). The ranking of policies differs, with energy and emissions policies more readily implemented among large firms. Linked to higher implementation scores, benefits are consistently perceived as higher for large companies across all dimensions. The barriers to implementation are roughly consistent across the two cohorts, with the sole exception of the lack of impact, which was perceived as more significant for SMEs. Large companies also face greater stakeholder pressure, both internally from management and employees and externally from clients, legislation, and the media. Thus, smaller firms tend to perceive fewer benefits from environmental actions and experience lower external and internal pressure to engage in sustainability. Perhaps unexpectedly, perceived barriers among large firms show a stronger negative correlation with environmental policies, perceived benefits, and perceived pressure. The interpretation of this result may be linked to the dynamics of decisions to prioritize low-hanging fruit. Large firms have already progressed further than smaller ones, presumably prioritizing measures easier to implement. Thus, the remaining challenges are more complex and a more differentiated driver of future decisions. For both large and small firms, a perceived lack of impact appears particularly relevant in limiting further action, whereas for large firms, a lack of expertise also plays a significant role.

[11] The constructs are calculated as the equally weighted average of their respective components.

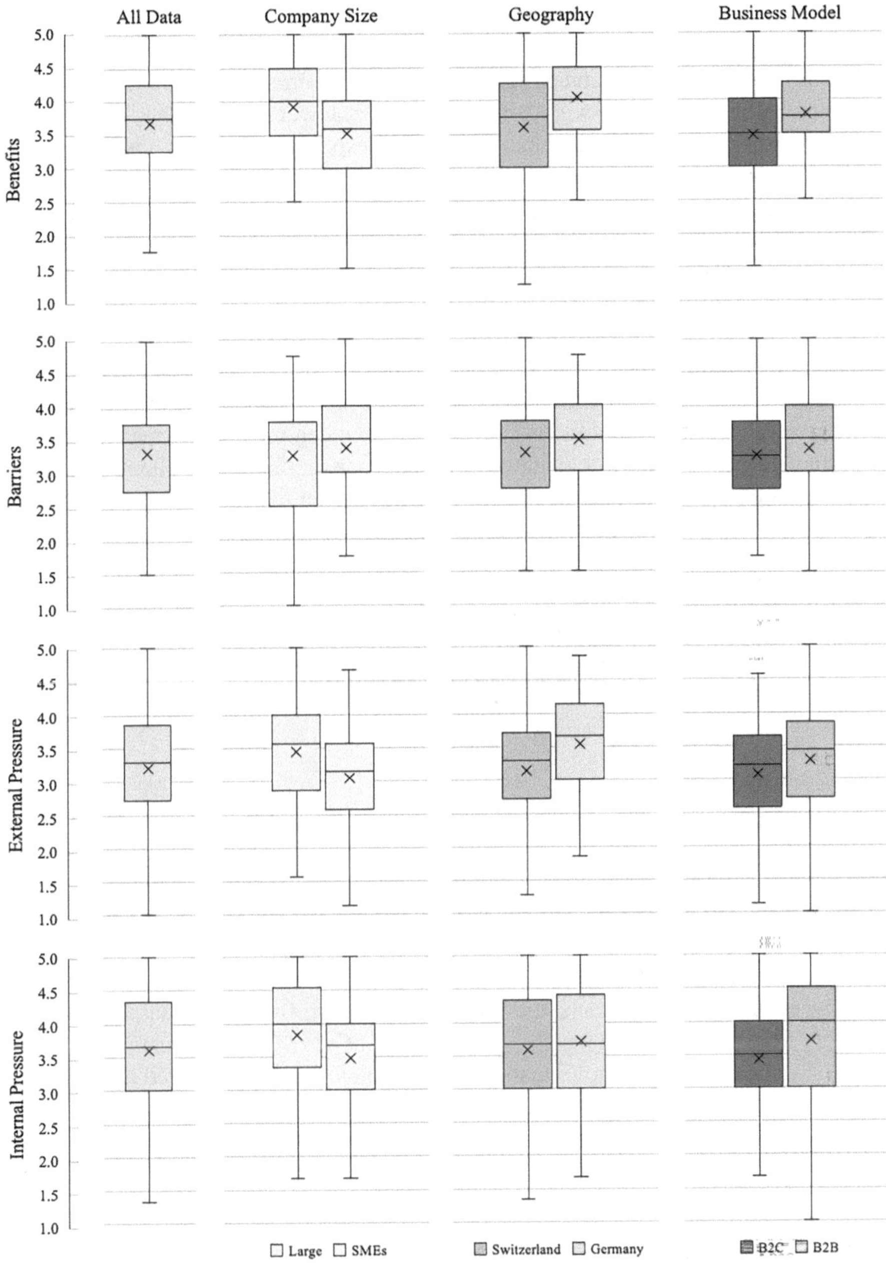

Fig. 7.1 Survey results by differentiating characteristics. Developed by the authors. All rights reserved. Used with permission

Geography (Switzerland vs. Germany)

The differences between Swiss and German companies are smaller than those between larger companies and SMEs; however, they remain significant across several dimensions. Overall, Swiss companies report lower implementation rates for environmental policies than German companies, especially in audits and emissions. The difference in environmental audit is particularly significant and is likely related to regulatory requirements. On the other hand, Swiss companies appear to be more advanced in their waste management policies. The higher implementation rates in Germany are linked to consistently higher perceived benefits, especially those related to attracting and retaining employees and achieving competitive advantage. Counterintuitively, German companies overall perceive higher barriers to implementing environmental policies. This may be linked to cultural factors in responses to questionnaires or to political and public environmental debates. Swiss companies also perceive lower stakeholder pressure across all categories. In Germany, pressure is particularly high due to legislation, internal management, and clients, possibly reflecting a tighter labor market and strong ESG-driven recruitment dynamics.

Although Swiss companies also rate those stakeholders with the highest pressure, the client and legislation score is still statistically significantly higher for German companies. Pressure from financial institutions is also higher in Germany, even when first controlling for firm size. Perceived barriers more directly affect environmental policies in Germany, especially when linked to expertise. Barriers are perceived as more relevant for German companies than for Swiss ones along several other factors.

Business Model (B2B vs. B2C)

Companies pursuing a business-to-business model report higher implementation of CER policies, particularly for audit, emissions, and product policies, compared with companies dealing directly with consumers. This correlates with higher perceived benefits across all categories. Perceived barriers, however, do not differ statistically between the two business models. B2B companies perceive greater pressure from competitors, perhaps due to the greater sophistication of their customers and employees. On the other hand, they feel less pressure from the local community. Pressure from the client and legislation is consistent across the two business models. Barriers impact both the implementation of policies and the perceived benefits to a greater extent for B2B companies. The exception is the lack of impact, which more strongly influences policy implementation in B2C companies. Despite these aspects, the overall differences between the two models are modest.

Sector

The differences among companies from different sectors are even more modest. Manufacturing firms are particularly engaged in emissions-related policies, while machinery manufacturers have implemented environmental audit processes at a significantly higher pace than other companies. In all other categories of barriers, benefits, and pressure, differences among sectors are very small. Thus, although there may be some different regulatory priorities and market dynamics, sectors tend to behave similarly, at least as far as the groups that could be identified and differentiated in the sample collected. It should be noted that the survey did not cover firms in the financial or agricultural industries.

7.3 Discussion

The dataset analyzed provides practical and powerful insights for several stakeholders. We will discuss insights for three such groups: legislators and policymakers, environmental activists, and company executives. While some recommendations overlap, these groups represent distinct interests pursuing specific goals in the context of CER.

Implications for Policy and Regulatory Developments

One potential challenge for policy, especially at the local level, is that CER policies appear already to be in an advanced stage of implementation in Switzerland and Germany, and that the perceived benefits are primarily image-related rather than profitability or competitive advantage. The opposite is observed for increased costs, which are perceived as an important barrier. Thus, there is a danger that CER is and will increasingly be seen as ancillary to core business rather than integral to it. This would clearly weaken the impact of any efforts toward cooperative evolution of guidelines. Local considerations are not perceived as a significant pressure point, and integration into local activities may therefore not be an effective means of improving a company's activities in a local setting. This result is somewhat counterintuitive, as several companies have supported and integrated with local efforts to better align with local priorities and improve their profiles as good corporate citizens. The mediating mechanism is potentially through company management. Thus, local regulators should aim to leverage connections with local management to align activities rather than emphasize a formal organizational relationship. Especially as cost is the predominant barrier, regulators should actively include practical knowledge on how to set effective regulations that result in a minimum of additional effort and cost, or which are even economically beneficial. The situation is different for national or supranational legislation. There, the pressure is felt and is very effective at guiding behavior. The drawback of this approach, of course, is that it takes time to develop legislation; it is by definition lagging behind the curve of needs and the frontier of activities. Thus, a combination of top-down national guidelines that codify behavior, linked to a more collaborative local approach that brings together local communities and company management, seems the most promising course of action. The focus should be on limiting costs, communicating the benefits, especially in image improvements, and having larger companies lead the development due to their greater susceptibility.

Implications for Environmental Activists

Activists, both as individuals and as organizations, face, to a large extent, the same constraints as local and national policymakers. Thus, activating a local community or the media to exert pressure does not appear to be an effective strategy, especially when the target is a small company. It is more conducive to improvement to pursue a differentiated local/national strategy. At the national level, activists should engage in the political process to establish guidelines. In this context, it is essential to both set maximalist requests to counter political inertia, but also to provide real expertise to the process and to escalate a number of local considerations as examples both of the needs and of the effectiveness of the proposed policies. At the local level, a personal connection with company management seems to be the most productive way to influence development. This is especially important for engagement investors, which could increase the relevance and influence of financial institutions on a firm's CER policies. Combining the two could foster a "good cop, bad cop" dynamic, with local activists adopting a more conciliatory, collaborative tone than their national counterparts. Providing cost relief is not within the remit or financial means of activists, but they can support these legislative measures. More directly in scope for activists is mitigating the lack of expertise by connecting companies with experts and case studies from other regions or industries. They can also be a highly effective multiplier of a company's image, given their credibility and independent evaluation. This is a critical component of activist leverage and should be deployed with great deliberation.

Implications for Company Executives

A key mechanism in connecting CER policies and their development with business activities and the company's organizational environment is the management. In this context, it is essential to note that the very high score given to management as a channel for pressure has been produced by the managers, who were typically the respondents to the survey. Thus, executives see themselves not only as important connectors between the business and organizational environments, but also as having significant agency in guiding this interaction. In some respects, this observation is understandable: executives can trade off among multiple targets and stakeholders, launch and guide initiatives, and interface with the external environment. On the other hand, they are a scarce resource, both in terms of time and expertise, and especially in their bandwidth to attend to multiple topics. Coordinating activities through management, therefore, potentially incurs a heavy opportunity cost. However, initial efforts to strengthen CER activities can enhance the firm's appeal to sustainability-oriented talent, thereby easing skill shortages and, in turn, freeing managerial capacity for more strategic rather than operational issues. In other respects, however, focusing on acting through management risks presents a too simplistic view of interactions between an organization and its environment. In a contingency-based view of organizations, firms manage a complex and changing environment by matching this complexity internally.[12] Thus, multiple parts of the organization interact with this complex external environment, and represent the insights they gain internally in order to develop structured organizational responses. All of these inter-

[12] Lawrence, P. R., and Lorsch, J. W. (1973). A reply to Tosi, Aldag, and Storey. *Administrative Science Quarterly*, 397–398.

actions and insights are substantial; reducing them to the role of management is not potentially putting too much pressure on a scarce resource, but also potentially short-sighted and leading to ineffectual responses. From a contingency-theoretical perspective, company executives should therefore encourage and endorse a broad set of points of interaction. This will partly reduce demands on their time and, more importantly, provide a more effective and differentiated pipeline for capturing and interpreting opportunities, constraints, and stakeholder needs. Important in this construct, of course, is an effective mechanism to capture this complexity in a structured organizational decision process.

7.4 The Three-Pillars Model

The Three-Pillars Model (3-P-Model)[13] provides a helpful structure to interpret sustainability efforts along three critical dimensions: 1. Sustainable Purpose as the raison d'être for an organization; 2. Traveling Organization, meaning an organization accepting to be in a state of flux and continuously evolving to incorporate new insights in its culture, business model, and process; and 3. Connectivity, i.e., a critical component of an interconnected whole, sharing insights and resources to achieve systemic improvements. The insights generated by this analysis can be placed in a broader context using the 3-P-Model.

Sustainable Purpose

The study shows how firms without an explicitly defined sustainability purpose can still advance an environmentally responsible economy, offering guidance relevant to the majority of active businesses. More than half of the respondents to the questionnaire hold C-level positions or serve on the Supervisory Board. Among all respondents, only about a quarter indicate that climate risks have a low or very low impact on the company's model. The aggregate picture, therefore, is that companies perceive significant effects on their core business from climate risks and that the topic and the solutions it requires are often positioned at the highest levels of the organization. It follows logically that the topic of environmental responsibility is linked to a company's strategy. Whether the company has an articulated sustainable purpose is secondary in this context. The reality of senior management attention and the perceived high impact of CER policies on business create a powerful de facto connection between sustainability and strategy.

Traveling Organization

One of the key messages in the dataset we analyzed is the extent to which several companies have implemented policies to honor their environmental responsibilities. This has been a direct result of companies evolving their stance from an initially

[13] First introduced in: Wollmann, P.; Kühn, F.; Kempf, M. (Eds.): Three Pillars of Organization and Leadership in Disruptive Times—Navigating Your Company Successfully through the twenty-first Century Business World. Cham: Springer Nature, © 2020

voluntary, unstructured approach to a structured approach. Thus, the past of these companies is a testament to the principle of organizational travel, with their activities evolving to meet the moment. The traveling organization pillar, however, requires companies to continue to grow. This is by no means guaranteed. Fortunately, there are several signs indicating that this may in fact be the case. First, several stakeholders impact the policies. Second, and more critically, company executives believe they have the agency to react to these pressures. Since markets, clients, regulators, public opinion, and other interested parties will continue to evolve, we expect executives to respond to these changes and, consequently, for organizations to continue to evolve.

Connectivity

Organizations are part of a complex network, connected to several other entities and groups in the context of environmental responsibility. At the receiving end of the relationship, companies are impacted by several stakeholders. As part of their regular business activities, they are impacted by commercial partners, especially their clients. In a broader social context, companies are affected by a wide variety of stakeholders, from legislators to local communities. Except for financial institutions and suppliers, all of these stakeholders exert pressure on companies beyond the midpoint on the scale of possible answers, indicating that the connections are not only broad but also impactful. Conversely, companies exert pressure on their partners. First, they, of course, function as clients and suppliers for other companies. Second, and perhaps more interestingly, respondents broadly believe their activities are impactful. This observation is based on the perceived barriers to CER policies: the lack of impact scores lowest among the possible responses. Thus, companies are part of a complex network where they influence and are influenced by other members.

7.5 Conclusions and Takeaways

This chapter has examined how companies in Switzerland and Germany perceive and implement Corporate Environmental Responsibility (CER) and how these activities are shaped by stakeholder pressure, perceived benefits and barriers, and firm characteristics. Several consistent insights emerge from the data.

First, CER is already widely established across firms, particularly in the areas of waste, energy efficiency, and emissions. While external audits remain less common, most companies report meaningful activity along core environmental dimensions. These activities are driven less by short-term profitability and more by reputational considerations, talent attractiveness, and competitive positioning. Image improvement clearly dominates other perceived benefits, suggesting that CER is primarily understood as a strategic and legitimacy-enhancing tool rather than a direct contributor to financial performance.

Second, cost and time constraints remain the most salient barriers to expanding CER efforts. Interestingly, the perceived lack of impact shows the strongest negative correlation with CER policies, underscoring that companies only invest in environmental activities when they believe these activities matter. Expertise gaps, by contrast, play a surprisingly limited role overall. However, they become more critical in large firms and in Germany, where implementation is already more advanced, and the "remaining" challenges are more complex.

Third, stakeholder pressure is highly asymmetric. Legislative pressure and managerial expectations are the most influential forces shaping CER behavior, while local communities, competitors, suppliers, financial institutions, and audit exert markedly weaker influence. This suggests that regulation and corporate leadership, not community activism or market peer dynamics, remain the primary levers for steering environmental responsibility. At the same time, the high perceived importance of management underscores both the influence and the vulnerability of executives, who serve as central coordinators of sustainability efforts.

Fourth, firm characteristics meaningfully modulate CER engagement. Large companies report substantially higher implementation levels, more substantial perceived benefits, and stronger stakeholder pressure. German companies also exhibit higher implementation, particularly in auditing and emissions, alongside higher perceived barriers. B2B firms show somewhat stronger CER adoption than B2C firms, although the differences are modest. Sector-level differences, with a few exceptions, are minimal. Overall, CER appears less driven by industry dynamics than by organizational size, regulatory context, and business model sophistication.

Taken together, these findings highlight three overarching takeaways:

- CER is strategically relevant and widely internalized, not only because of regulatory pressure but because firms perceive environmental responsibility as central to their legitimacy, talent strategy, and long-term positioning.
- The most effective levers for accelerating CER are regulatory clarity and managerial agency, supported by credible communication of reputational and societal benefits.
- Future progress will depend on strengthening companies' perceived impact, ensuring that environmental actions are not viewed as symbolic but as meaningful contributions to climate and sustainability goals.

Framed through the lens of the Three-Pillars Model, the results indicate that many firms are already connected to their environment, perceive CER as aligned with their strategic purpose, and have demonstrated the ability to evolve their practices over time. Sustaining this momentum will require continued learning, expanding points of interaction beyond management, and aligning regulatory frameworks with organizational realities. Strengthening these three pillars, purpose, organizational evolution, and connectivity, can help companies translate environmental responsibility into durable, effective, and systemically embedded practice.

Carlo Pugnetti is Lecturer at the Institute of Financial Services Zug IFZ. His research focuses on the evolution of customer behavior in insurance, with a particular focus on cybersecurity and on changes triggered by technology adoption. Before joining the IFZ, Carlo served as the CEO of Allianz Global Assistance in Switzerland and in several other functions within the Allianz Group. He began his career as a consultant for Oliver Wyman. Carlo Pugnetti holds a Ph.D. in Risk Analysis and a Master's degree in Electrical Engineering, both from Stanford University.

Nadine Berchtold is Senior Research Associate and has been working at the Institute of Financial Services Zug IFZ since 207. She has professional experience from Swiss retail banks and a microfinance firm in Kenya in the credit environment. Before joining the IFZ, she worked as a business consultant. She received her Master's degree in Banking and Finance from the Lucerne University of Applied Sciences and Arts and is currently pursuing her Ph.D. in Finance at the University of Neuchâtel.

Stefan Hunziker is Professor of Risk Management with extensive expertise in Enterprise Risk Management, Internal Control, and Compliance. He holds a Habilitation in Management and Organization Sciences (University of Sopron) and a Ph.D. from the University of St. Gallen. At the Lucerne University of Applied Sciences and Arts, he has led major programs, including the MSc in International Financial Management and several postgraduate certificates in Governance, Risk, and Compliance. His research and teaching focus is on Enterprise and Strategic Risk Management. He leads the Competence Center for Risk & Compliance Management at the Institute of Financial Services Zug (IFZ).

From Metrics to Meaning: Fostering Sustainability by Reclaiming Leadership as a Conscious and Archetypal Way of Being

8

Jacqueline Hofste

Abstract

In today's sustainability movement, many organizations fall into the trap of performative action: focusing on metrics, ticking boxes, and implementing isolated initiatives that may solve one issue but fail to address the deeper, systemic causes of the environmental crisis. This chapter challenges the reduction of sustainability to corporate performance indicators and argues for a redefinition of leadership, not as title or function, but as a conscious way of being that reflects a deeper connection to our shared humanity and purpose.

At its core, leadership is an archetypal expression, a universal pattern of human potential that can manifest through either constructive or destructive qualities of the psyche. When unconscious and shaped by scarcity-driven mindsets, it often takes the form of control, domination, or manipulation, reflecting the very systems that contribute to ecological degradation and social injustice. However, when approached with conscious intent, leadership becomes an act of stewardship, grounded in inner qualities such as integrity, wisdom, courage, and compassion.

8.1 The Illusion of Progress as the Driver for Well-Being

In the 1970s, a common fear was that we were hurdling toward a new ice age. Growing up in Germany, where winters were fierce, and I remember our house doors buried in snow, forcing us to find other ways out, this was a fear that felt so real. Then, in the 1980s, during my student years, the narrative shifted dramatically with *An Inconvenient Truth*, Al Gore's groundbreaking film that reframed our

J. Hofste (✉)
Enheduana, Mosman, NSW, Australia
e-mail: hello@enheduana.com.au

© The Author(s), under exclusive license to Springer Nature Switzerland AG 2026
P. Wollmann et al. (eds.), *The Sustainable Organization – Volume 3*, Future of Business and Finance, https://doi.org/10.1007/978-3-032-28949-0_8

collective anxiety toward the opposite extreme: global warming. Eventually, that too evolved into a new term, climate change, a phrase that aimed to capture the complexity of what we are truly facing.

As I have a background in science and can appreciate the attempt to decode such a vast, entangled system as earth's climate, and it is not a judgment on the difficulty of defining climate change. The work of experts and scientists is monumental, and their conclusions continue to evolve, shaped by new data and deeper understandings, which is a reflection on the shift we have to endeavor in the leadership realm to manage these uncertainties with unimaginable impact on all of humanity while not having the privilege to fully grasp the extent of the challenges we are faced with. History also shows that our responses to overwhelming challenges are often shaped by fear, uncertainty, and competing narratives, leaving many disoriented.

Despite the confusion—we are no longer dealing with abstract theories or distant warnings. The measurable signs of climate change are here. We are witnessing mass extinction, the degradation of ecosystems, and a deep moral reckoning with how our choices, personal as well as societal, while often made in the name of progress, have poisoned air, water, and soil. Our children are inheriting the consequences of a culture that has prioritized profit over principle, convenience over conscience. This has placed us in what can only be described as a double-bind: we are both the creators of the crisis and the ones who must resolve it. Yet the very mindset that brought us here cannot lead us out. It is this illusion of progress as the goal that we have succumbed to for so long that calls for a deeper shift: one that cannot be addressed through metrics or KPIs alone and it starts with another definition of leadership.

In this chapter, we will explore why our path forward requires more than just innovation or compliance. It requires a transformation in consciousness, a reawakening of what it means to lead, to act, and to be part of the world. Through the lens of archetypes and the ontological qualities of the Being Framework we will examine how leadership must move beyond performative sustainability toward something more integrated, more courageous, and more deeply aligned with life itself. The urgency of the ecological and societal crises requires more than incremental improvement which means leadership can no longer be just about performance; it has to be about authenticity, higher purpose, and a new paradigm. Drawing on a new lens for viewing organizational structures as described in the Three Pillars of Leadership by Peter Wollmann,[1] this chapter invites a shift from "doing" leadership to "being" leadership. It calls for a new level of awareness, one that recognizes our role in the global ecosystem not as passive participants or status-driven actors, but as conscious collaborators and responsible stewards of the Earth and all its inhabitants.

[1] Wollmann, P.: Kühn, F.: Kempf, M. (Eds.): "Three Pillars of Organization and Leadership in Disruptive Times, Future of Business and Finance". Cham: Springer Nature. ©2020. Following books on the Three-Pillar Model see: https://link.springer.com/search?query=&content-type=Book&dateFrom=&dateTo=&contributor=%22Peter+Wollmann%22&sortBy=relevance

8.2 The Limits of the Current Leadership Paradigm

We live in a world where productivity is valued over authentic engagement, and where doing more is mistaken for progress; leadership becomes reactive rather than regenerative. The result is a landscape marked by widespread burnout, moral disengagement, and accelerating environmental degradation, all clear signs that the current model is no longer fit for the complexity and urgency of our times. Within this context, leaders are systematically trained to function as efficient managers of fundamentally extractive and unsustainable systems. Individuals in leadership roles often excel at operational execution yet increasingly lack the capacity for deeper qualities of being such as authenticity, responsibility, and commitment to a purpose greater than personal advancement or relentless competition. Immersed in the overwhelming demands of their roles, many have lost touch with the deeper meaning of their work and lives.

Today's leaders have largely been shaped by the legacy of industrialization, shareholder capitalism, and mechanistic worldviews that prioritize output over insight, performance over authentic service, and growth over the imperative to consider adverse consequences. These forces have shaped organizational cultures and systems that once appeared effective in driving economic growth, yet were rooted in exploitative and short-sighted targets, methods that never truly served humanity or the planet's living systems. At its core, this paradigm is misaligned with human nature—the innate drive to serve, to connect to something greater, and to care for the environments that sustain life. Although we may not yet fully comprehend what it would entail to live in true harmony with our ecosystems and with each other, it is increasingly evident that the ongoing path of ruthless commodification and short-termism is both unsustainable and profoundly damaging to the environment and human well-being. The time has come to acknowledge that the very structures we have built for progress are now undermining the quality of life and the future we hope to protect.

The symptoms of this collapse are everywhere around us: widespread burnout, moral disengagement in teams, collapsing trust in institutions, and environmental degradation that now threatens the very systems on which human life depends. These are not isolated issues but rather interconnected crises revealing a systemic failure of our current conception of leadership, not just in operational capacity, but in vision, ethics, and foundational assumptions about what it means to lead.

8.2.1 The Mechanistic Mindset: Leading from Separation

Prevailing leadership models remain deeply rooted in a mechanistic worldview that regards people as resources, nature as a mere backdrop, and success as a linear accumulation of measurable outcomes. This perspective, shaped by centuries of industrial organization, fosters the belief that the world is predictable, controllable, and reducible to data points.

In such a paradigm, the role of a leader is reduced to one of control and optimization. Complexity is simplified to key performance indicators (KPIs), relationships become subordinated to outcomes, and creativity is stifled by risk aversion and the need to maintain image. The notion of human "being" has been replaced by human "doing."

8.2.2 The Culture of Performance: Image Over Purpose

The rise of personal branding, public messaging, and digital platforms has placed an enormous emphasis on how leadership appears rather than what it is. Leaders are encouraged to project confidence, clarity, and decisiveness even when these qualities are absent or inauthentic. This has severed the link between leadership and authenticity. Leaders are not required to engage in the deeper dimensions of their being, the doubts, fears, values, and personal conflicts that are part of what make them human. Instead, leadership is frequently reduced to a curated persona,[2] often disconnected from the leader's actual experience or capacity to relate to others in a meaningful way. This leads to what many are now calling the "integrity gap," the difference between who leaders present themselves to be and who they are. Over time, this fragmentation leads to burnout, moral fatigue, and quiet disengagement, the feeling that even one's best efforts are not truly changing anything.

8.2.3 The Ethical Collapse: Leadership Without Responsibility

Within the current paradigm, many leaders unwittingly maintain systems that harm life while claiming neutrality. Decisions are depersonalized and justified by adherence to policy and the pursuit of short-term gains, thereby avoiding difficult questions: What are the unintended consequences of my leadership? Whom does it ultimately serve? What legacy am I leaving?

This ethical detachment enables the rebranding of deforestation as resource management, or the rationalization of exploitative labor practices as efficiency. And because the system rewards those who play by its rules, it is possible to rise to the highest levels of leadership while never having cultivated the inner maturity to hold this responsibility with depth and care. Instead, leaders may retreat into the abstraction of their duties, projecting an image of responsibility while reducing their role to mere platitudes.

[2] The topic of a curated persona was from another perspective stressed in the third Three-Pillar Book under the headline "Development of the Personal Ability to Transform by Means of Actor Coaching and Training"; see: https://link.springer.com/chapter/10.1007/978-3-031-06904-8_11

8.2.4 Burnout as a Symptom of Systemic Disconnection

To function in such roles, individuals often find themselves compelled to suppress aspects of their own humanity, empathy, intuition, and even a sense of justice. This internal fragmentation steadily depletes their vitality; sustaining a façade or upholding a false narrative requires considerable energy. Over time, this leads to burnout, not just as a personal struggle, but a broader collective crisis, marked by an erosion of connection both to others and to one's own sense of meaning.

This, in part, explains why resilience programs and stress management interventions frequently prove insufficient. They tend to address the surface symptoms, rather than the underlying causes. What is genuinely called for is a foundational shift in the way leadership is conceptualized, a model that recenters our full humanity, and reimagines leadership as a collaborative, sustainable practice. Such an approach must be grounded in a meaningful integration of technology with the broader environment, ensuring that leadership remains fundamentally aligned with the flourishing of both individuals and the collective.

8.2.5 A New Urgency: Reframing the Role of the Leader

In today's context, where prevailing norms often promote disconnection from sustainable values and an excessive focus on performance, the meaning of leadership demands thorough reconsideration. Leadership, in this sense, should not be reduced to a title or a set of tasks, but rather approached as a collective responsibility rooted in the ongoing development of self-awareness, relational presence, and a profound integration of one's purpose, values, and worldview.

Importantly, this perspective does not advocate discarding strategy or efficiency. Instead, it calls for these elements to be approached with heightened consciousness and intentionality. It means seeing leadership not as a climb to the top, but as a journey inward and outward, continual practice of aligning inner being with outer impact. Within this context, the integration of archetypal intelligence and ontological awareness, together with the Three Pillars Model,[3] emerges as a transformative approach. These concepts, which will be examined in greater detail in the following section, offer a high-level lens through which leadership can be reimagined in both theory and practice.

[3] Ibid. The 3-P-Model is based upon the interacting concepts of (1) Sustainable Purpose—raison d'être of an organization, bringing new orientation and certainty to the people for their joint endeavor and success, (2) Travelling Organization—the mindset of an organization in a permanent state of flux, interacting with the markets' & customers' journey, with rapid adaptivity, (3) Connected Resources—Interconnecting all needed resources inside and outside the silos, creating consistency between goals and concepts, strategies and processes, competencies and roles.

8.3 Archetypes as Maps of Human Potential

If we are to truly understand and reevaluate the concept of authentic leadership, it is helpful to view it through an archetypal lens. Archetypes as described in my book *The Free Spirit of your Life Purpose*[4] reveal the deeper, often unconscious motivations that shape how individuals lead, respond to power, navigate conflict, and relate to others. Beyond learned skills or surface-level behaviors, archetypes provide insight into the inner architecture of the psyche, the behavioral patterns that guide our choices, fears, and aspirations. By exploring leadership archetypally, we begin to see that authenticity is not merely about personal integrity or transparency; it is about alignment with deeper truths and the ability to navigate both light and shadow aspects of one's role using archetypal language.

The light side is referred to as the embodiment of empowerment, when the true motivation of a person is aligned with their actions, and the shadow or dark side is the part that is suppressed or unknown and therefore out of alignment with a person's true intentions. The perspective of archetypal motivation and personal imprints opens the door to empowered, truth-based leadership that refers to the reality of matters and is therefore responsive to the needs of both the individual and the collective.

History shows that people have long tried to understand the archetypes influencing our thoughts, actions, and shared experiences. At the heart of this inquiry lies the concept of archetypes, universal patterns that transcend time, culture, and personality. Whether seen as Plato's eternal Forms, Carl G. Jung's psychological archetypes, or John Campbell's mythic archetypal figures, referring to some of the most renowned experts in archetypal understanding, these symbolic structures point to something deeper: the blueprint of the soul and the path of human evolution.

Plato, the ancient Greek philosopher, introduced one of the earliest frameworks through his theory of Forms.[5] While the term "archetype" may not have been developed at this time, his idea of existing forms in the invisible world relates to the view of archetypes, which Carl G. Jung, the Swiss psychologist, refers to in his body of work, which is detailed later in this chapter. In Plato's view, all things in the physical world are imperfect reflections of perfect, unchanging ideals that exist in a metaphysical realm, accessible not through the senses, but through contemplation and the soul's inner knowing. These Forms, such as truth, beauty, and justice, are not human inventions, but what he refers to as timeless realities, ultimate truth or principle that holds true beyond the boundaries of time.

As individuals progress through life and align with their principles, they begin to express deeper truths in their experiences; truths that go beyond appearances and are not simply accepted at face value. Integrating these deeper truths into our lives is a source for sustainable growth as a person and sustainable living. In the language

[4] Hofste, J.: "The Free Spirit of your Life Purpose". Authority Connect, 2021, 3rd Edition

[5] Plato. Edited: G.R.F. Ferrari University of California, Berkley, Translated: Tom Griffith: "The Republic: Book 3 and Book 5" University Printing House, Cambridge CBs 8BS, United Kingdom, 2018, 3rd Edition

of archetypes, this movement toward ideal expression is what we call living from the light side, whereas distortion or misalignment expresses the shadow as mentioned earlier. Plato's perspective adds a spiritual dimension to his concept of Forms, suggesting that human potential is metaphysical in nature. To lead consciously, then, is to align with eternal principles and express them in action. This is something that can also be transferred into the leadership realm.

Carl G. Jung,[6] the Swiss psychiatrist, brought this concept into modern psychology by identifying archetypes as innate, universal patterns in the collective unconscious, the shared psychic inheritance of humanity. Archetypes such as the Hero, the Sage, the Shadow, or the Mother are not learned but inherited, also referred to as instinctual structures of the psyche that influence how we think, relate, and mature as a person. They are symbolic energies that emerge during life transitions, challenges, or moments of deep reflection, guiding us toward greater wholeness. In this context, the term *symbolic* means that life's lessons are not to be taken personally but understood as a neutral feedback loop for learning; much like tripping over a rock is not a personal humiliation, but simply the law of gravity at work.

Jung's core contribution is the understanding that each archetype carries both light and shadow. The shadow represents the disowned or unconscious aspects of ourselves, not necessarily negative, but unintegrated into a person's life.

The human psyche as seen by Jung can be broken down into three main components: the Conscious, the Personal Unconscious, and the Collective Unconscious, as shown in Fig. 8.1. These main components are made up of the Ego, the Persona, the Shadow, Archetypes, and the Anima/Animus. According to Jung, the Anima is an archetype associated with the feminine aspect of the psyche, while the Animus is linked to masculine aspects. Both aspects are present in women as well as men to different degrees.

- The Ego and the Persona are part of the Conscious.
- The Archetypes and the Anima/Animus are part of the Collective Unconscious.
- The Shadow is part of the Personal and Collective Unconscious, which includes Archetypes and the Anima/Animus (Fig. 8.1).

The psyche of a person or a person in general becomes more conscious by moving parts of the Unconscious into the Conscious and this happens through our life experiences that confront us with the unknown. The tension between the Ego (our conscious identity) and the Self (the archetype of wholeness) lies at the heart of this journey. While the ego seeks control and certainty, the Self calls for depth, truth, and soul alignment. Archetypes aren't meant to be controlled but related to consciously. When leaders partner with these deeper forces, they can for example move from manipulation or a fake persona to embodied authentic leadership.

By mapping these patterns, leaders cultivate symbolic self-awareness and make decisions rooted not in fear or ego, but in clarity and purpose. Embracing this way

[6] Jung, C. G.: 1981, 2nd Edition. "Collected Works of C.G. Jung, Volume 9 (Part 1): Archetypes and the Collective Unconscious" Princeton University Press.

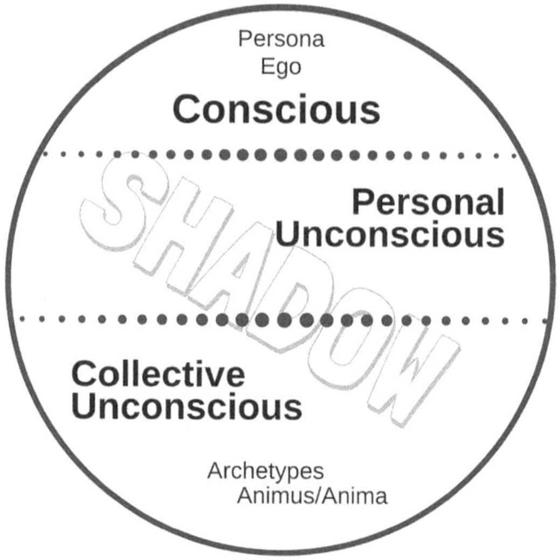

Fig. 8.1 Diagrammatic view of the aspect of the psyche—diagram by J. Hofste, used with permission, all rights reserved

of being is empowering, while operating from fear undermines one's energy and effectiveness, contributing to the burnout and disengagement described earlier. Archetypes, then, are not just tools for understanding behavior; they are at the root of true transformation. They challenge, guide, and awaken us. True growth and leadership, in Jung's view, require acknowledging, accepting, and transforming the shadow into the conscious light aspect. This inner work moves us toward what Jung referred to as individuation, the process of becoming our whole authentic self.

According to the book *Sacred Contracts* by Caroline Myss,[7] a renowned contemporary expert on archetypes, the leadership archetype represents a core energetic pattern that governs how individuals relate to power, influence, and responsibility. In its light aspect, this archetype embodies qualities such as stewardship, vision, courage, and service. A leader in this mode acts from a deep sense of inner authority and alignment with purpose, seeing their role as a custodian of collective well-being rather than a position of personal gain. Leadership becomes an act of guidance and inspiration, rooted in integrity and the willingness to serve something greater than the self.

In contrast, the shadow side of the leadership archetype emerges when power is driven by the Ego rather than the Self. Here, leadership can manifest as dominance, manipulation, control, or image-consciousness. The leader acting from the shadow may seek influence not to serve, but to validate their identity or protect their insecurities. According to Caroline Myss, this can show up as using one's authority to exploit others, maintaining one's power through inducing fear in others, or performing the appearance of leadership without true inner alignment. Carl Jung states that

[7] Caroline Myss PhD. Sacred Contracts: Awakening Your Divine Potential.
Harmony Books: New York. New York Member of the Crown Publishing Group. 2001, 1st Edition.

this occurs when the ego tries to possess the archetypal energy rather than surrendering to its higher wisdom resulting in imbalance, projection of fear to the external world, or leadership burnout.

As a student of Caroline Myss', I align with her view that the leadership archetype is not a role one plays, but a profound inner motivation expressed in a pattern that must be consciously integrated to become the fully matured leader. The journey of authentic leadership requires awareness of both its light and shadow expressions, and the humility to engage in continuous self-inquiry, allowing the archetype to unfold in service of the collective rather than the service to self. This developed maturity forms the bedrock of a sustainable and authentic expression of leadership that is steady, and capable of evolving without collapsing under pressure or turning dysfunctional over time.

This level of leadership is not dependent on short bursts of charisma, control, or performance, but grounded in self-awareness, integrity, and inner alignment that endures through challenges and change. It doesn't drain the leader or those they lead, but nourishes growth, energy, trust, and long-term well-being within the system (team, organization, community). And in the context of this book it leads to leadership that is ethically consistent and rooted in values that support the flourishing of others and the environment, rather than exploiting them for short-term or self-serving gains.

Leaders who connect with a purpose greater than themselves and act from a place of genuine meaning and responsibility tend to serve causes that transcend personal ambition or the pursuit of external approval. Peter Wollmann, in his Three Pillars Model, refers to this as a "sustainable purpose," a foundational principle that bridges authentic leadership intentions with enduring impact. This sustainable purpose essentially sets the direction for the kind of leadership needed today: one that brings about a new, sustainable approach to business, an approach that values integrity, human flourishing, and long-term outcomes over fleeting, short-term results.

This caliber of leadership is not for the faint of heart. It demands substantial courage to question established norms, resilience to withstand ambiguity and opposition, and a steadfast commitment to leading from a place of alignment rather than ego or convenience. Ultimately, such leadership is not simply an idea; it is an embodied, ongoing commitment.

8.3.1 Embodying Archetypal Leadership in Practice

The shift toward a more conscious and regenerative leadership paradigm requires not only conceptual understanding but lived practice. Archetypes offer profound lenses through which leaders can examine, refine, and align their inner world with the external demands of leadership. This alignment becomes essential for building organizations that not only meet complex challenges and navigate the resistance to change but contribute meaningfully to the evolution of society.

Each leader or person for that matter carries a constellation of archetypes, universal patterns of expressing themselves through behavior, and deeper motivation through meaning, which shape how they relate to themselves, others, and the systems they lead. Identifying core leadership archetypes involves understanding the underlying influences that shape an individual's leadership style, rather than simply selecting preferred traits. This begins with reflective practices: journaling, guided self-inquiry, archetypal profiling, and observing recurring patterns in one's decision-making, conflicts, and aspirations.

For example, a leader expressing the Sovereign archetype may find themselves drawn to order, stewardship, and vision setting, yet must also be wary of lapsing into authoritarian control. The Magician archetype as another example brings transformation and innovation, but if ungrounded, may drift into detachment or manipulation. Integration involves not just recognizing these forces but consciously choosing how to express their light aspects while becoming aware of and working through the shadows. This ongoing inner work becomes the foundation for authentic, values-based leadership that resonates at both the personal and systemic levels.

8.4 Leveraging Archetypes and Ontology: Tools for Transformational Leadership

Ontology, as a branch of metaphysics, is concerned with the study of reality and the question of what it means to be human. Even though as a formal study emerged in the early seventeenth century, its roots can be traced back to philosophers such as Plato and Aristotle. In contemporary times, Martin Heidegger in his publication *Being and Time*[8] leaned strongly into Ontology and reframed the philosophical inquiry by asking not only "what exists" but also "what it means to exist." For Heidegger, ontology was inseparable from questions of authenticity, presence, and the human condition, approached not as abstract speculation but as a practical and lived inquiry. This perspective and orientation are typically less common among philosophers as they tend to follow a more abstract approach.

This chapter explores the intersection between ontology and archetypes, two powerful frameworks that, when integrated, provide a transformative lens for leadership. Archetypes offer a psychological and symbolic map of human experience, while ontology grounds leadership in the reality of being itself. Together, they create a foundation for a form of leadership that is both psychologically profound and existentially authentic, capable of meeting the complexity of our times. To address the challenges of our time such as social fragmentation and widespread institutional distrust, we need leaders who are not only efficient but also self-aware and courageous to bring the new forward. They are required to be more than visionaries and grounded in ontology: the reality about being and existence (Fig. 8.2).

This shift points toward what can be called sustainable leadership, a form of leadership anchored in who we truly are, capable of integrating technological advancements responsibly, while minimizing harm to our environment and social

[8] Heidegger, M.: "Sein und Zeit (Being and Time)". Max Niemeyer Verlag Tübingen, Germany. 2001, 18th Edition.

Fig. 8.2 Copyright owned by J. Hofste, used with permission, all rights reserved

structures. Unlike leadership models focused on control, strategy, or superficial change, this approach does not drain or exhaust as it is grounded in an ontological model that fosters relationships aligned with human needs and integrated with our broader environment. As Peter Wollmann emphasizes in the Three Pillars Model, this requires building meaningful relationships among peers and stakeholders, shifting the focus from competition to collaboration via connectivity.

8.4.1 The Ontological Approach to Leadership: Reclaiming the Ground of Being

The ontological approach to leadership begins with the premise that who we are being matters more than what we are doing. In business terms, this relates to skills, strategies, and knowledge; while important, they are secondary to the inner state from which action arises. Leadership is not just a function; it is an expression of one's relationship to self, others, and reality.

Ontology invites us to look beneath behavior into the qualities of Being that shape our leadership presence as described in the *Being* book by Ashkan Tashvir.[9] The term "Being" while vague at first glance can be deconstructed into many individual aspects of being such as:

- Authenticity—leading from truth, rather than image and intentionally seeking out the reality of things.
- Integrity—being whole and aligned in thought, word, and deed.
- Responsibility—owning one's impact and the systems one contributes to.
- Presence—being grounded, attentive, and fully available in the moment.

These qualities are not merely techniques to be acquired, but intrinsic aspects of humans, qualities of what it means to be human, which can be cultivated and refined in their expression. The more an individual develops awareness of these aspects of

[9] Tashvir, A.: "Being—The Source of Power". Engenesis Publications, 2021.

being and intentionally aligns them with reality, the more effective and grounded their presence becomes in both personal and professional life and leading them to live a life of integrity.

For instance, a person who authentically embodies their brand is more likely to cultivate trust among prospective clients and positively influence their decision to engage. Conversely, when a person projects a brand that does not reflect their genuine self, initial success may still be achieved. However, such success is rarely sustainable. Over time, as clients recognize the dissonance between promise and experience, trust is eroded, and loyalty is lost.

These qualities of being form the roots from which ethical, effective, and human-centered leadership can authentically emerge. In their absence, even the most sophisticated strategies risk becoming performative, superficial, or ultimately unsustainable. This deficit is increasingly evident in our prevailing economic, business, and political systems, which too often prioritize short-term gains at the expense of long-term well-being while pretending "all is good." By straying from what aligns with the fundamental realities of human nature and the living environment, we have constructed systems that exploit rather than sustain life and the very systems they operate from. The consequences of this misalignment are starkly visible: the degradation of ecosystems, the depletion of natural resources, and a growing toll on human well-being, ranging from physical and mental health crises to social fragmentation and widespread disillusionment. And the root causes are not immediately apparent, as dysfunctional ways of being have become so normalized and we are so entrenched in it that we can't see it anymore for what it is. As a result, we often overlook the link between our state of *being* and the outcomes we create and experience in the external world.

To genuinely alter our current trajectory, it is essential to reconsider our fundamental relationship with life itself. This necessitates a model of leadership that moves away from superficial performance and instead embraces authentic integration, reconnecting with what is tangible, ethical, and sustainable. Ontological leadership, in this sense, redirects the focus of power inward. It moves beyond external authority or the mechanics of task management and instead invites us to consider deeper questions: Who am I becoming as a leader? What underlying worldview informs my actions? Am I truly embodying the values I promise to stand for?

8.4.2 From Personality-Driven to Purpose-Driven Leadership

Most conventional leadership models emphasize personality, individual style, behavioral traits, or adaptive strategies shaped by upbringing, culture, and ego needs. While personality is undeniably part of the leadership equation, it often confines us to a narrow set of familiar responses. We lead from habit rather than awareness, becoming reactive instead of intentional. Although the psychology behind these theories does not assert that personality and behavioral preferences are fixed, only over time has the popular interpretation evolved into the belief that such traits are largely unchangeable.

Purpose-driven leadership, grounded in the ontology of being human, challenges this assumption that personality style is what it is and can't be changed. It is rooted in essence rather than surface traits, and as such it can transcend ingrained preferences, behaviors, and perceptions. This form of leadership aligns with a deeper reason for being, one that reaches beyond self-interest and connects the leader to a wider field of meaning, service, and contribution. The shift from personality to purpose is not about negating individuality but about freeing it from the narrow confines of ego or limiting belief systems. In doing so, leaders expand into a more mature, authentic, and integrated version of themselves, capable of embodying a form of leadership that is both deeply human and profoundly transformative.

The question, then, is where to begin the journey of transformation. Archetypes provide a powerful point of entry. They illuminate the deeper patterns and forces shaping our leadership, some of which strengthen and guide us, while others distort or constrain. By bringing these archetypal dynamics into awareness, we gain the capacity to move beyond the surface layers of personality. Ultimately, archetypes guide us into an encounter with our power or our perceived lack of it. In this context, *power* refers to our natural ability and strength to be effective in life, not, as it is often perceived, as a means of manipulation or control.

Only when we learn to recognize and transform the shadow aspects of our archetypal nature can we act from our motivations in a way that serves others. This is the essence of genuine empowerment, which is far more than a fashionable buzzword. True empowerment arises from the willingness to face uncomfortable truths about us, truths that can be challenging, even confronting. Yet it is precisely by moving through these difficult aspects of our being that we cultivate real empowerment, confidence, and the possibility of change that goes beyond perceived limitation. This is the starting point for developing leadership capable of pushing us beyond our comfort zones and beyond limiting perceptions of reality, giving us a chance to create new solutions that break free from the double-bind of our deteriorating and failing systems. This requires courageous leadership willing to take the first step, leaders who draw on the deeper aspects of their being, allowing the mirror to reflect their inauthenticity in this double-bind situation.

Archetypes are not roles we assume such as project manager or team leader; they are universal templates of human experience and psychological potential that we can tap into. The Mediator, the Mentor, the Sage, the Rebel, the Detective—these are just a few examples of archetypes that shape how we think, feel, relate, and act based on their altruistic motivations which are unique to each of the archetypes. For example: the Rebel is motivated by challenging a stifling and undermining status quo that leads people into suffering, or the Mediator is motivated by the alignment and strength that comes from unity rather than competing parties. As previously mentioned, archetypes can be expressed in a light or shadow form, which means either from an empowered state or a distortion of true motivation based on fear, rejection, or scarcity. The archetypal journey, so to speak, is realizing the power from being authentic and acting out of conviction and courage rather than fear or avoidance.

In leadership, archetypes reveal the inner architecture behind our external behavior. For example:

- The Mediator archetype, in its light expression, carries the gift of fostering fairness, facilitating collaboration, and guiding strategy. In its shadow form, however, it may distort these same capacities, resorting to manipulation or subtle coercion to advance personal interests.
- The Mentor archetype can inspire trust and loyalty through empathy, guidance, and the sharing of wisdom. In its shadow expression, however, this same role may foster unhealthy dependency, where the mentor's presence inhibits the growth and autonomy of the one being guided.
- The Sage or Crone archetype embodies insight, reflection, and truth-telling. In its shadow expression, however, it may withdraw into detachment or adopt an air of arrogance when disconnected from compassion and the heart.

The power of working with archetypes lies in their ability to guide us from unconscious enactment to conscious integration, unlocking the freedom to express ourselves fully and authentically. Understanding what drives a person with a specific archetypal imprint as well as what may be undermining their efforts is essential, including the deeper fears they may be compensating for. It is important to be reminded that a person rarely embodies just one archetype; rather, they often relate to a set of archetypes, with one or two being most active at a given point in time. The table below highlights a selection of archetypes commonly seen in the business context. While there are countless archetypes, this selection focuses on those most relevant to this environment (Table 8.1).

8.4.3 Archetypes and Leadership

Archetypes can provide leaders with clarity and direction, revealing the underlying patterns that influence their behavior, decisions, and interactions. By understanding

Table 8.1 Selection of archetypes frequently observed in organizations

Archetype	Motivation	Shadow	Related archetypes
Advocate	The advocate archetype champions humanitarian and environmental causes, giving a voice to those affected by injustice who can't speak for themselves and standing up for what is right.	Putting their agenda above the cause, undermining objectivity and harming dissenters. **Fears:** being silenced, misunderstood, or betrayed	Humanitarian, avenger

(continued)

Table 8.1 (continued)

Archetype	Motivation	Shadow	Related archetypes
Creator	The creator archetype is driven by the desire to bring ideas into reality, crafting original works that express their imagination and inspire others	Hoarding insights for personal recognition, or the fear of public judgment. **Fears:** stagnation, indifference, mundaneness	Artist, artisan, performer
Entrepreneur	The entrepreneur is a visionary who sparks creativity in business and pursues opportunities for personal and collective growth	Compromising their vision for personal gain or selling their insights to the highest bidder **Fears:** loss, public failure, chaos	Philanthropist, visionary
Hero/heroine	The hero/heroine archetype faces formidable challenges to overcome obstacles and champion the rescue of others or a worthy cause	Engaging in escapism and false heroism, pursuing a cause that harms others or pretending heroic acts **Fears:** weakness, helplessness	Warrior
Mediator	The mediator archetype values unity, honors all perspectives, and restores harmony by rebuilding trust and open communication	Driven by self-interest they lose balance in disputes and evade direct confrontation **Fears:** confrontation, disharmony, loss of influence	Motivator
Mentor	The mentor archetype nurtures others' growth through wisdom and experience, guiding development across . personal, professional, and community realms	Undermining others by withholding wisdom to impede their growth and safeguard their own position **Fears:** irrelevance, loss of importance and status	Guide, teacher
Networker	The networker archetype thrives on building alliances, fostering connection, and sharing information to create unity and progress	Spreading misinformation to advance personal agendas and leveraging influential connections for self-gain **Fears:** disconnection, betrayal, lack of support	Messenger, evangelist

(continued)

Table 8.1 (continued)

Archetype	Motivation	Shadow	Related archetypes
Rebel	The rebel archetype challenges outdated systems to ignite innovation, reform, and renewal	Rejecting rightful authority out of anger or self-interest and valuing revenge over justice **Fears:** conformity, stagnation, complacency	Outcast, maverick
Ruler	The ruler archetype embodies power and authority guided by a commitment to serve and uplift their people	Driven by a desire to dominate, seeking control without restraint and mistaking it for power **Fears:** loss of control, insignificance, weakness	King, queen
Storyteller	The storyteller archetype imparts wisdom through captivating narratives that inspire imagination, ease fears, and spark vision and action	Distorting the truth to mislead or harm others **Fears:** rejection, being found out, invisibility	Spellcaster, magician
Sage/crone	The sage/crone archetype embodies early wisdom and a generous willingness to share it, earning trust and guidance from others while pursuing wisdom as a lifelong calling	Abusing trust for personal gain and manipulation **Fears:** ignorance, insanity, misinformation	Thinker, researcher, detective
Pioneer	The pioneer archetype is driven to explore uncharted territories in pursuit of discovery and the promise of new horizons	Chasing an unattainable ideal, imposing their vision of freedom to dominate others **Fears:** confinement, immobility, awakening to illusion	Explorer, adventurer
Provocateur	The provocateur thrives on challenging ideas and beliefs through sharp intellect and debate, embodying the Devil's Advocate to provoke deeper reflection	Thriving on advantage, using wit and intelligence to confuse others **Fears:** accountability, being ostracized	Free spirit, troublemaker

which archetypal expressions are active within them, leaders can consciously cultivate empowerment. This can look like the Mediator, building alignment and unity within teams, the Crone or Sage, sharing wisdom to support others, or the Entrepreneur aligning resources in the most beneficial way while mitigating disempowered or shadow tendencies like arrogance, control, or dependency. An awareness of the archetypes allows leaders to navigate complex organizational dynamics with intentionality, authenticity, and psychological insight, fostering cultures that are resilient, adaptive, and human-centered.

8.4.4 Ontology and Leadership

Ontology provides leaders with ethical clarity, helping them discern whether their actions and decisions are aligned with reality and their true nature of being. By grounding leadership in ontological awareness, authenticity, responsibility, integrity, and presence, for example, leaders can move beyond reactive or performative approaches. This alignment enables them to make choices that serve the broader system, sustain human and environmental well-being, and model a conscious way of being that resonates throughout their teams and organizations.

Together, these two frameworks enable leaders to connect inner awareness with tangible impact and effectiveness. They cultivate both the symbolic understanding of archetypal imprints and ontological maturity, allowing leaders to express themselves at their highest potential, rather than falling back into limiting behavior patterns based on fear and scarcity thinking.

8.4.5 From Archetypal Awareness to Ontological Responsive Leadership

The Being Framework as developed by Ashkan Tashvir, with its emphasis on ontological qualities like authenticity, responsibility, and integrity, also provides a structure for working with archetypes. Where archetypes offer symbolic psychic forces and narrative, making them very relatable, the Being Framework provides a compass on how to anchor that psychic energy in integrous, aware, and authentic leadership.

The integration of ontology and archetypes is not just a conceptual exercise; it is a practice of embodiment. It asks the leader to bring their full self, including mind, heart, body, and purpose into their leadership. This includes:

- Self-reflection—regularly noticing which archetypes are active and how they show up in different contexts and identifying the root causes for that.
- Ontological inquiry—deepening awareness of one's inner state and intention before acting. This can be developed through a step-by-step process of analyzing how misleading beliefs and perceptions of a situation can trigger emotional or blindsided reactions, rather than calm, centered responses.

- Conscious choice—by cultivating a high level of awareness, actions can be aligned with the qualities of Being and the fullest expression of one's archetypal potential.

This approach leads to leadership that is dynamic and grounded, visionary and responsible, powerful and compassionate. It fosters a leadership culture that is not merely reactive to external pressures, but capable of responding to complexity with creativity, depth, and care. In ontological terms, this authentic level of leadership or being in general is referred to as ontological responsiveness. A leader demonstrating ontological responsiveness can recognize when they are triggered and prone to an emotional reaction, versus when they have sufficient emotional distance to respond more effectively in an interaction.

8.4.6 Sustainable Transformation

As described by A. Tashvir in the *Human Being*[10] book the Being Framework offers a methodology and practices for transformation that are both sustainable and repeatable. It is grounded not in quick behavioral fixes, but in the gradual expansion of perception and provides a systematic approach with concise practices to achieve real transformation. At its core, this work is about shifting how we see ourselves, others, and the systems we inhabit. When perception expands, new choices become possible: choices that are more aligned with truth, integrity, and conscious responsibility.

This transformation does not necessarily occur all at once. It unfolds step by step, as we bring awareness to the unconscious patterns driving our behavior, our habitual reactions (ways of being), protective mechanisms, and blind spots. Through ontological inquiry, we begin to see where we are acting out of fear, control, or conditioning, and instead open to more empowered ways of being. With each insight, there is an opportunity to integrate this new understanding into daily life, through communication, decisions, and intentionally directed awareness.

Over time, these incremental shifts create lasting transformation, not because we are trying harder, but because we are seeing more clearly and adapt our perception, thus creating a more realistic perspective of situations. True transformation is not an effort to become something different, but a return to **who** we already are beneath the layers of our experiences and perceptions that we have accumulated from life and our understanding of past events.

This mirrors the process of working with archetypal shadow. As we identify the limiting patterns of certain archetypes, such as the Rebel, the Mediator, or the Ruler, we begin to witness how these distortions have shaped our leadership. With awareness, we are no longer trapped in their grip. We can choose to express the light side of the archetype instead as it is that brings about enduring and fulfilling results.

[10] Tashvir, A.: "Human Being—Illuminating the reality beneath the façade". Engenesis Publications: Sydney. 2022.

In both frameworks, archetypal and ontological, transformation becomes sustainable when it is integrated, not rushed, not performed, but experienced and consciously executed. One shift at a time, a leader becomes more whole, more conscious, and more capable of meeting the organizations with wisdom, depth, and presence. By weaving together archetypal wisdom and ontological presence, we move beyond performative leadership into the realm of transformational service, leadership that awakens, aligns, and heals. This is the kind of leadership our time demands, not because it is trendy or idealistic, but because it is what will allow us to navigate complexity with integrity and co-create a future worth living for. It also reminds us that leadership is not just about strategy, charisma, or control; it is about consciousness, character, and contribution. It is about who we are and how we choose to show up.

8.4.7 Navigating Resistance and Staying True to the Path

Leading from this place of inner clarity will be undermined if the systems around us are still operating from outdated paradigms. Organizational design must evolve alongside leadership consciousness. Culture, structure, and strategy must be recalibrated to support life-affirming principles, such as transparency, distributed power, ecological responsibility, and human development. This path, however, is not without resistance. Whether internal (fear, doubt, ego) or external (legacy systems, performance metrics, stakeholder pressure), the forces that seek to maintain the status quo can be persistent. The leader committed to transformation must develop resilience, not as over-the-top stoicism, but as the ability to stay centered in the face of dissonance.

It is here that both archetypal and ontological work converge: the inner Sovereign offers vision and direction; the Warrior brings courage to confront pushback; the Lover keeps the work connected to meaning and relationship; and the Sage brings discernment and long-term perspective. Ontologically, presence becomes the anchor, allowing the leader to respond rather than react, to remain in integrity rather than collapse under pressure. Ultimately, leading from this space is not about achieving perfection, it is an ongoing iterative process which is guided by practicing alignment to a deeper truth rather than mitigating symptoms. The leader becomes both the instrument and the invitation: someone who models a different way of being and doing, thereby catalyzing transformation not just in themselves, but in the systems they touch.

While this approach may seem daunting at first, it becomes real with patience and endurance. Step by step, leaders can influence others, reshape systems, and inspire their communities to embrace this transition. The path is by no means easy; it may be seen as the odyssey of our modern times. Like the epic journeys of old, today's leaders must navigate uncertainty, face trials that test their resolve, and resist the temptation to retreat to the familiar shores of comfort and conformity. Yet, it is precisely through these challenges that strength, wisdom, and clarity emerge (Fig. 8.3).

Fig. 8.3 Copyright owned by J. Hofste, used with permission, all rights reserved

This modern odyssey is not about conquest or heroic individualism, but about awakening a deeper, collective courage. It calls for leaders to embody authenticity, awareness, and responsibility in a world that often rewards speed, performance, and surface appearances.

Initiating change often falls to a single individual, and it's true the initial steps can feel quite isolating. Still, over time, such efforts tend to build momentum, almost like a ripple turning into a wave. When a movement reaches its critical point, even those who were previously hesitant or disengaged may find themselves compelled to participate. They begin to recognize that this new approach to leadership, one focused on service, interconnectedness, and a higher purpose, is not merely a vague idea, but a real possibility for our collective future. The journey starts with one person willing to act, despite the solitude that may accompany those early steps. With continued effort, momentum gathers. Eventually the movement grows, transforming what was once a solitary initiative into a shared endeavor.

8.5 The Travelling Organization[11]: From Scarcity to Innovation

As outlined previously, at the heart of today's ecological, social, and leadership crises lies a pervasive operating assumption: scarcity. It is the belief that there is not enough time, money, resources, talent, opportunity, and so we must compete,

[11] Numerous use cases can be found in the Three-Pillars Books (https://link.springer.com/search?query=&content-type=Book&dateFrom=&dateTo=&contributor=%22Peter+Wollmann%22&sortBy=relevance) and a striking metaphor for the Travelling Organization at https://www.peterwollmann.com/post/about-travelling-in-the-unknown-in-the-19th-century-and-today-a-pattern-for-leadership-and-managem

extract, and optimize at all costs. This mindset drives urgency over clarity, efficiency over ethics, and silos over shared intelligence. Yet it is precisely this scarcity orientation that blocks the innovative systemic solutions we so urgently need.

This reframe also requires leaders to be profoundly intentional in how they show up and how they shape their organizations. Drawing on the work of Peter Wollmann, this means cultivating the three essential pillars of evolved leadership: being flexible to emerging demands by building "travelling" organizations that are adaptive and mobile in spirit, developing deeper connectivity across people, systems, and meaning, and being purposeful in our intentions, with clarity about the "why" behind what we build and lead. Leadership in this new paradigm is not reactive or performative; it is regenerative and rooted in authenticity.

In our reality, our actions do not occur in isolation. Leadership becomes an ecosystemic responsibility, not merely a professional role. This shift demands more than new techniques or cosmetic sustainability initiatives; it asks for a fundamental transformation in how we see ourselves, others, and our place in the living web of existence. This shift isn't about "doing more with less"; it's about seeing different. It will affect all areas of science shaped by this limited perspective and self-serving interests, which often results in inadequate and outdated theories, concepts, and solutions. It means moving from the question "How do I get ahead?" to "How do we move forward together?" and "What truly serves both the individual and the collective?"—questions that cannot be answered outside the context of our environment. How can we live in abundance and prosperity without harming the very branch we sit on? Is it possible to create enough for all without depleting the planet? If we seek meaningful answers to these seemingly impossible questions, we must first confront and rethink our values and ways of thinking: a shift that can be profoundly disruptive and fear provoking.

In his book *Think Again*, Adam Grant[12] highlights the importance of questioning even our most trusted best practices, a process that can challenge the very core of our beliefs and disrupt our familiar world. To show what this entails in detail, I will draw on a personal experience, as the process is far from simple. During my time as a Software Development Manager our team was tasked with reviewing our existing procedures to identify areas for improvement in an area of quality that was considered the "holy grail" of safety procedures that worked and were deemed perfect and failproof. While these procedures were considered best practice, we had to set aside our preconceived notions and question everything, even the standards we had long considered untouchable.

This wasn't easy. The software we developed was used in high-stakes automation systems such as chemical plants, traffic control, and mining operations, where strict adherence to safety protocols was critical, and failure could cost lives. Our quality standards were treated as the best practice of safety and performance in their field with awards to show for, and many of our engineers were understandably very protective of them. The idea of challenging these procedures, even in the name of greater effectiveness, was met with resistance and genuine concern. But we chose to

[12] Grant, A.: "Think Again". Viking Publishing: US. 2021.

lean into discomfort. We dared to look more closely, and in doing so, discovered opportunities to improve our processes without compromising safety or quality.

It was a powerful lesson: even the most rigorous systems can evolve, and progress often starts with the courage to question what we think is set in stone. This shift from separation to interconnectedness lays the foundation for a new kind of innovation, one that moves beyond scarcity-driven problem-solving and embraces possibility-based systems thinking. It calls for challenging questioning of our assumptions, including how we perceive science and technology. Instead of accepting things as they are, how we generate energy, build machines, or structure systems, we must open ourselves to entirely new paradigms, concepts, and methods.

In *The Five Literacies of Global Leadership*[13] Richard David Hames argues that one of the greatest challenges to leadership today is not just external turbulence, but our capacity to perceive and engage with the future. He contends that our deep-seated mental, emotional, and cultural frameworks constrain our ability to reflect on alternative and unknown futures. He refers to five root causes which we will briefly describe.

First, current leadership paradigms remain anchored in outdated models, focused on hierarchy, heroic individuals, and predictable outcomes. These paradigms limit imagination and restrict our ability to envision truly novel futures.

Second, the human tendency toward resisting cognitive change makes new and in particular novel ideas inherently difficult to hold. Cognitive change refers to a shift in how we think, perceive, and make sense of reality. It is a transformation in the underlying mental models, assumptions, and frameworks that shape our interpretation of the world. It's not just "learning new information" but restructuring thought patterns, how we process experience, what we consider possible, and how we connect ideas. In his book *Metacontent* Ashkan Tashvir[14] presents a comprehensive framework for understanding complexity. It helps us examine how we develop our mental models, the multilayering of their origins and meaning which can make them so deceptive to analyze in their complexity as we have the tendency to oversimplify. He then guides us through the process of interpreting events whose reality is obscured by complexity and deception, which can easily mislead us.

Third, Hames emphasizes that future orientations require more than conventional leadership skills. We must cultivate advanced literacies to envision multiple possible futures, practice what he called "Deep Design" that goes beneath surface-level aesthetics or functionality and engages with underlying structures, values, and long-term impact to shape systems that can realize them.

Fourth, fostering these literacies also necessitates expanding our collective awareness. Techniques like transformational narratives that reshape paradigms and systems-based foresight will enable us to shift beyond dominant frames of meaning.

---

[13] Hames, R. D.: "The Five Literacies of Global Leadership". John Wiley & Sons Ltd. The Atrium, Southern Gate, Chichester, West Sussex, England. 2007.

[14] Tashvir, A.: "Metacontent. The Intellectual substrates for sense-making". Engenesis Publications. 2024.

This can guide us to move from default reactions into generative, future-informed choices.

Fifth, the emotional landscape is equally crucial. Modern life, for many, is marked by alienation, fatigue, and despair, to imagine, or to commit to futures that require moral and psychological courage. As noted earlier in this chapter, this state is so deeply ingrained in our society that we often fail to see it, leaving us trapped without a clear way out.

Ultimately, Hames suggests that meaningful engagement with alternative and unknown futures requires a simultaneous transformation in leadership cognition, emotion, and structure. It entails letting go of limiting paradigms, cultivating new mental literacies, expanding our imaginative and ethical faculties, and healing the emotional wounds of disillusionment. Only then can we shift from reactive modes of leadership to truly generative practices, and begin shaping better futures not merely for ourselves, but for the systems and societies we serve. Metaphorically, it means having the courage to jump into ice-cold water and discover what unfolds.

To summarize this leadership evolution, we need to:

- **Reframe mental models**: moving from old, habitual ways of thinking to new perspectives, e.g., from "leadership is about control" to "leadership is about enabling collective intelligence."
- **Shift our perception:** seeing the same situation differently because your interpretive lens has changed, e.g., viewing conflict not as a threat but as a catalyst for innovation.
- **Update assumptions**: challenging ingrained beliefs that unconsciously guide behavior, e.g., letting go of the assumption "I must have all the answers" to embracing "I can move through the new territory while collecting some answers as I go."
- **Expand awareness**—becoming capable of holding greater complexity, ambiguity, or paradox without defaulting to simplistic conclusions.

In many ways, it requires us to let go of well-understood frameworks and be willing to reimagine from the ground up. Rather than dismissing unconventional ideas, we must be willing to test them, explore their potential, and see where they lead. This demands leaders who are courageous and vulnerable enough to question the status quo, even when it invites resistance or rejection. It also requires a level of moral clarity in leaders who are intentional in exposing and challenging corrupt systems, rather than becoming complicit in them.

This is where the Three Pillars Model—Travelling Organizations, Sustainable Purpose, and Connectivity—becomes more than just a glossy buzzword or theoretical model. In truly interconnected teams and networks, authenticity and compassion cannot be performative; they must be lived. Reconnecting to a higher purpose requires leaders to engage in deep self-reflection, to confront their own blind spots, and to cultivate a leadership presence that is aligned not just with performance outcomes, but with what genuinely serves both people and the planet.

This includes practices such as:

- Designing with the whole system in mind, not just the immediate problem.
- Cross-pollinating diverse perspectives to dissolve echo chambers and invite fresh insight.
- Slowing down to sense what wants to emerge, rather than forcing outcomes from old assumptions.
- Embracing the unknown as a space of creative potential rather than a threat to be controlled.
- Rooting innovation in meaning, so that it serves a higher collective purpose rather than fragmented market trends.

In this way, effective innovation becomes inseparable from ontological leadership and archetypal integration. When leaders are attuned to their own inner motivation, their archetypal drives, shadow patterns, and capacity for presence, they are far more equipped to lead from responsiveness rather than reactivity. Their presence can then become a catalyst for cultures of trust, creativity, and long-term value, not just productivity.

This is the kind of innovation the world needs, not more cleverness, but greater consciousness, not more information, but deeper integration. It is not technology alone that will change our trajectory, but awareness guiding how and why we use it. I recognize this is a tall order, and to be honest, I don't have the answers, but I suggest we can begin by asking the right questions and taking the first steps in this direction.

8.5.1 Business as Reflective Practice, Leadership as a Sovereign Stewardship

If we are to move beyond survival-mode capitalism, we must begin to see business as a reflective practice, a place where values are lived, relationships are honored, and the soul of humanity is expressed in tangible ways. Reflective practice requires presence, intention, and integration. It is the art of showing up whole and aligned, choosing not only what we do but who we are "being" while "doing" it. In this frame, meetings become spaces for connection, strategy becomes an act of service, and leadership becomes a form of devotion to the possible.

Leadership, then, is not a performance. It is not a title. It is a revered responsibility to steward the conditions for life to flourish within individuals, organizations, and the broader systems we touch. It asks us to align with truth even when it's uncomfortable, to make decisions that ripple forward with integrity, and to walk our talk not for reputation, but for the sake of coherence.

8.6 Conclusion: The Leader as Re-Weaver of the World

We are not simply experiencing a period of disruption; we are undergoing collective initiation into a whole new way of living. Established models and paradigms are disintegrating, making way for an emergent reality that has yet to fully take form. What is required now is not an improved iteration of the past, but a fundamentally different mode of innovation and way of being that is based on the full acknowledgment of what is serving humanity and their integration into the environment. By shifting from scarcity to interdependence, from separation to sacredness, from mechanistic to life-affirming systems, we reclaim the essence of leadership, not as power over [others], but as service [for others], not as conquest, but as contribution.

When leaders take this responsibility to heart, they bring humanity back into the systems they guide. They shift from simple transactions to meaningful, transformational exchanges. They create room for pause, curiosity, creativity, and care. And they recognize that the greatest resource in any organization isn't money or technology, but people who are aware, authentic, and courageous.

Leading in this way means becoming a re-weaver of the world, someone who recalls what has been forgotten, helps repair what is broken, and dares to imagine what could be. It's a path that is neither easy nor conventional, but absolutely necessary. And it's not a journey to be undertaken alone, but one joined with the wider flow of life and community.

Jacqueline Hofste is a team culture architect and leadership mentor who helps leaders and their teams unlock authentic, purpose-driven performance. Drawing on her background in senior corporate leadership, a master's degree in physics, and certifications in coaching and archetypal personality profiling, she bridges scientific insight with intuitive and ontological approaches to leadership. Her work empowers individuals to embody confident, authentic, and intuitive leadership while cultivating high-performing team cultures that thrive on clarity, trust, and shared purpose.

Part III

Cross-Sectoral Approaches to Sustainability from Industry to Fashion

Case Study from Swiss Post: Sustainability Matters—How to Transform an Organization

9

Vivien Büchler-Bezzola and Christian Plüss

Abstract

This chapter examines how Swiss Post reconfigures its purpose, governance, and business model to pursue a sustainability-oriented transformation under public-mandate conditions. It demonstrates that credible change requires understanding institutional foundations and path dependencies, while deliberately shifting organizational trajectories where they constrain long-term relevance. The chapter shows how integrated thinking provides the analytical architecture and managerial mindset for linking financial and nonfinancial performance, enabling a multi-capital approach and aligning the organization with national and global sustainability objectives, including the United Nations Sustainable Development Goals. The Climate & Energy case study illustrates how strategic ambition is translated into organizational practice—showing in concrete terms how targets, decarbonization pathways, and operational levers interact to deliver sustainability transformation at scale.

9.1 Introduction

Swiss Post stands at a crossroads. Few organizations embody continuity as strongly as a national postal service with 175 years of history—yet few face such a profound break with their past. Letter volumes decline as customers use communication and financial services increasingly in digital form, and core services that define Swiss Post—logistics, passenger mobility—require transformation on an unprecedented scale due to electrification and decarbonization. The organization must continue to guarantee a reliable universal service for 9 million people while redesigning its

V. Büchler-Bezzola (✉) · C. Plüss
Swiss Post Ltd, Berne, Switzerland
e-mail: vivien.buechler-bezzola@post.ch; christian.pluess@post.ch

© The Author(s), under exclusive license to Springer Nature Switzerland AG 2026

P. Wollmann et al. (eds.), *The Sustainable Organization – Volume 3*, Future of Business and Finance, https://doi.org/10.1007/978-3-032-28949-0_9

business model in real time. Swiss Post is therefore not simply modernizing; it is navigating through a disruptive change.

Sustainability becomes both compass and catalyst in this environment: a long-term orientation toward societal value and a practical framework for redesigning the business model under uncertainty. In this sense, *Sustainability Matters—How to Transform an Organization* is not a slogan but the strategic task Swiss Post must master as it evolves into a forward looking, climate aligned, digitally enabled organization.

To understand Swiss Post's trajectory, the Three-Pillar Model provides an instructive lens. Designed as a mindset model for navigating complex, nonlinear contexts, it emphasizes meta-capabilities that become essential in transformation.[1]

- **Sustainable Purpose** is anchored in Swiss Post's identity as a state-owned universal service provider. It must guarantee essential services for the country while contributing to Switzerland's sustainability strategy. This dual mandate becomes tangible in the Climate & Energy transformation, with a net-zero goal in 2040, and the decarbonization of fleets, buildings, and value chains turns purpose into measurable action.
- **Travelling Organization** reflects what Swiss Post must be to navigate declining letter volumes, rapid digitalization, and technological changes with large uncertainty accompanied by fundamental operational transformation. Linear planning is insufficient; the organization must experiment, learn, and recalibrate continuously, moving through disruptive times while ensuring service reliability.
- **Connectivity** is essential because Swiss Post operates in a dense societal and operational ecosystem involving diverse stakeholder groups. Decarbonizing fleets, redesigning infrastructure, and delivering trustworthy digital services all depend on collaboration across organizational and societal boundaries.

Building on these Three Pillars, this chapter is structured as follows: Sect. 9.2 establishes the institutional foundations that shape Swiss Post's sustainability trajectory; Sect. 9.3 maps the business model through an integrated thinking lens and aligns it with the UN SDGs; and Sect. 9.4 presents the Climate & Energy case study, demonstrating how Sustainable Purpose, Travelling Organization, and Connectivity converge in a measurable, strategically anchored transformation. Taken together, the chapter shows how Swiss Post's sustainability transformation is not a marginal add-on, but a redefinition of how the organization creates value and designs its business model—using sustainability as the organizing logic through which it continues to serve Switzerland.

[1] Wollmann, P.; Pemler, D.; Ndrevataj, M. (Eds.). (2025). *The Sustainable Organization—How Organizations Address the 17 UN SDGs Using the 3-P-Model.* Cham Springer Nature.

Wollmann, P.: Püringer, R. (Eds.) (2022). *Transforming Public and Private Sector Organizations—Implementing Sustainable Purpose, Travelling Organization and Connectivity for Resilience.* Cham: Springer Nature.

Wollmann, P.; Kühn, F.; Kempf, M. (Eds.). (2020). *Three Pillars of Organization and Leadership in Disruptive Times—Navigating Your Company Successfully through the 21st Century Business World.* Cham: Springer Nature.

9.2 Establishing the Baseline for Swiss Post's Sustainability Transformation

Sustainability transformation cannot be designed in a green-field approach; it must start from an organization's history and governance. Research on organizational path dependence shows how self-reinforcing mechanisms and institutional obligations bound the feasible space for change, making it necessary to surface legacy decision rules and stakeholder expectations before any redesign. This lens emphasizes that history matters: prior choices create trajectories that can lead to persistence or lock-in, while meaningful change typically requires critical junctures and deliberate interventions to break or reroute existing paths.[2]

This section applies this path-dependence perspective to establish a structured baseline for Swiss Post's sustainability transformation. Sect. 9.2.1 identifies three institutional foundations as products of historically imprinted routines and expectations that shape today's option space and the credibility of potential interventions. Sect. 9.2.2 reconstructs the 2010–2020 phase, showing how bottom-up initiatives emerged within the inherited constraints—signaling cultural readiness while illustrating how incremental moves can reinforce or gently reroute existing paths unless a critical juncture is intentionally created. Sect. 9.2.3 traces the strategic anchoring of sustainability in governance as a path-shaping move—consistent with path-dependence arguments about the need to institutionalize change after a critical juncture.

9.2.1 Institutional Foundations

The section examines first how stakeholder dialogue has become an organizational practice; second how the balancing of economic, social, and environmental responsibilities functions as a historically sedimented norm set guiding tradeo-ffs; and third how resilience reflects accumulated routines and governance practices that have proved viable across shocks.

9.2.1.1 Stakeholder Dialogue as Organizational Practice

Since its founding in 1849, Swiss Post has been far more than an ordinary company. It perceives itself as an institution with a special responsibility toward society and Switzerland. Its original mandate was to ensure reliable transportation of goods,

---

[2] Schreyögg, G., & Sydow, J. (2010). Understanding institutional and organizational path dependencies. In *The hidden dynamics of path dependence: Institutions and organizations* (pp. 3–12). London: Palgrave Macmillan UK.

Sydow, J.; Schreyögg, G. & Koch, J. (2009). Organizational path dependence: Opening the black box. *Academy of Management Review*, 34(4), 689–709. https://doi.org/10.5465/amr.34.4.zok689

Sydow, J., Schreyögg, G. & Koch, J. (2020). On the theory of organizational path dependence: Clarifications, replies to objections, and extensions. *Academy of Management Review*, 45(4), 717–734. https://doi.org/10.5465/amr.2020.0163

information, money, and people for the Swiss population and economy. These services have always been—and will remain—essential for fostering social cohesion and enabling a connected, functioning Switzerland.[3]

Swiss Post's identity is rooted in a democratic principle: it serves everyone. As an organization operating in the public interest and owned by the state, it designs its offerings to meet the needs of Switzerland and its people. Consequently, the expectations and requirements of diverse stakeholder groups—from citizens and businesses to political decision-makers—continuously influence its strategic direction as well as the development of services.[4]

Swiss Post's role as a public service enterprise requires ongoing dialogue with a variety of external stakeholders. The regulatory framework, as defined by the Postal Services Act[5] and the Federal Act on the Organization of Swiss Post,[6] is not established unilaterally; it results from democratic legislative processes. Accordingly, the scope and quality of the universal service, as well as the purpose of Swiss Post, are statutorily defined. Therefore, interaction with various stakeholder groups is not merely an obligation but a central organizational practice throughout Swiss Post's history of over 175 years, which has been shaped by continuous changes to its business model. Navigating between different interests and aligning activities with external requirements are deeply embedded in Swiss Post's governance and management practices.

9.2.1.2 Balancing the Sustainability Dimensions as Organizational Culture

Until 1997, Swiss Post was part of the Swiss Federal Administration. Following the liberalization of the national parcel market, it was transformed into an independent public enterprise in 1998 and, in 2013, into a special-law public limited company wholly owned by the Swiss Confederation.[7] As such, it remains subject to the strategic goals set by the Federal Council every 4 years, which define what the Swiss Confederation seeks to achieve as the owner of Swiss Post. These goals, currently defined for the period 2025–2028, require Swiss Post to pursue a strategy that is sustainable and guided by ethical principles, while remaining within its economic capabilities.[8] The strategic goals go beyond purely financial targets: Swiss Post must secure long-term corporate value (economic responsibility), reduce

---

[3] Swiss Post. (n.d.). *Swiss Post is founded—how it all began.* Retrieved January 31, 2026, from https://www.post.ch/en/about-us/profile/the-history-of-swiss-post/1849-swiss-post-is-founded

Swiss Post. (n.d.). *The history of Swiss Post.* Retrieved January 31, 2026, from https://www.post.ch/en/about-us/profile/the-history-of-swiss-post

[4] Swiss Post. (n.d.). *Swiss Post & politics.* Retrieved January 31, 2026, from https://www.post.ch/en/about-us/profile/swiss-post-and-politics

[5] Swiss Confederation. (2010). *Postgesetz (PG) / Postal Services Act.*

[6] Swiss Confederation. (2012). *Postorganisationsgesetz (POG) / Federal Act on the Organisation of Swiss Post.*

[7] Swiss Confederation. (n.d.). *TERMDAT—The terminology database of the Federal Administration.* Retrieved January 31, 2026, from https://www.termdat.bk.admin.ch

[8] Swiss Confederation. (2025). *Strategische Ziele für die Post 2025–2028.*

greenhouse gas (GHG) emissions (environmental responsibility), and address regional needs while maintaining a progressive, socially responsible human resources policy (social responsibility).

These owner expectations have led the organization to operate within the interplay of economic, social, and environmental responsibilities, embedding the three dimensions of sustainability into governance and management practices. The integrative approach—considering economic, social, and environmental criteria jointly and balancing them in decision-making—has shaped Swiss Post's organizational culture, even if not always explicitly labeled as "sustainability."

9.2.1.3 Resilience as an Organizational Capability

Operating at the intersection of economic, social, and environmental dynamics, Swiss Post positions itself as an organization in continuous transformation. Its ability to adapt to evolving customer needs and broader economic, environmental, societal, and technological developments has been fundamental to its sustained viability. Historically, this adaptability is evident in the transition from horse-drawn mail coaches to motorized vehicles and, more recently, in the shift toward fossil-free propulsion systems and the development of digital business models.[9]

Excellence in transformation processes constitutes a defining characteristic of Swiss Post and is reflected in its international standing. Since the introduction of the Universal Postal Union (UPU) ranking in 2017, Swiss Post has consistently been ranked as the world's leading postal operator.[10] The UPU ranking evaluates postal organizations globally based on four dimensions: reliability, reach, relevance, and resilience. Resilience, in this context, refers to the capacity to withstand external shocks—including environmental, societal, and technological disruptions—while developing business models that remain viable over the long term.[11] Sustainable governance is a critical driver of this resilience, as it enhances organizational adaptability and fosters innovation. Swiss Post's repeated distinction as the world's best postal operator thus underscores not only its service excellence but also its resilience, rooted in a consistently sustainability-oriented management approach.

Over recent decades, these institutional foundations have created the conditions for Swiss Post to pursue a systematic, strategy-anchored transformation toward

---

[9] Swiss Post. (n.d.). *1906: Swiss Post in motion—the first Postbus*. Retrieved January 31, 2026, from https://www3.post.ch/en/about-us/profile/the-history-of-swiss-post/1906-swiss-post-in-motion-the-first-postbus

Swiss Post. (2025). Annual Report 2024: Swiss Post is managing its transformation with seven strategic ambitions. Retrieved January 31, 2026, from https://geschaeftsbericht.post.ch/24/ar/en/swiss-post-is-managing-its-transformation-with-seven-strategic-ambitions

[10] Universal Postal Union. (n.d.). All publications. Retrieved January 31, 2026, from https://www.upu.int/en/universal-postal-union/activities/research-publications/all-publications: Postal Development Reports 2017–2022 and The State of the Postal Sector 2023–2025

[11] Universal Postal Union. (n.d.). *Integrated Index for Postal Development (2IPD)*. Retrieved January 31, 2026, from https://www.upu.int/en/Universal-Postal-Union/Activities/Research-Publications/Integrated-Index-for-Postal-Development

Universal Postal Union. (2025). *State of the Postal Sector 2024*.

sustainability. They have enabled the alignment of the business model and governance with internationally recognized frameworks, including the United Nations Sustainable Development Goals (UN SDGs) (Sect. 9.2.3). Building on this foundation, the following section examines the first phase of Swiss Post's sustainability transformation, focusing on 2010–2020.

9.2.2 The First Phase of Sustainability Transformation (2010–2020)

Between 2010 and 2020, Swiss Post's initial sustainability transformation efforts demonstrate how the institutional foundations described above materialized in practice. The transition toward sustainability did not begin as a centrally orchestrated program; it emerged through bottom-up initiatives that revealed a latent cultural readiness for change.[12] A salient illustration is the roll-out of three-wheeled electric delivery vehicles beginning in 2011—an innovation that signaled environmental awareness well before climate and energy became explicit strategic priorities.[13] These grassroots activities spanned business units and hierarchical levels, highlighting the commitment and creativity of employees across the organization.

Despite the momentum, the effort remained fragmented: initiatives proceeded without a common framework, defined accountabilities, or decision rules. As a result, impact could not be measured consistently, cross-unit coherence was limited, and long-term effectiveness was difficult to secure. While these early efforts underscored Swiss Post's adaptability and cultural predisposition toward sustainability, they simultaneously highlighted the need for a structured approach to embed sustainability within strategy, governance, and processes—an imperative that would shape the next phase of transformation.

9.2.3 The Push for a Strategically Anchored Sustainability Transformation

Throughout the 2010s, sustained digitization throughout society reshaped Swiss Post's business model: addressed letter volumes declined by around 30 percent over the decade, while parcel volumes rose with the growth of e-commerce.[14] The COVID-19 pandemic acted as a catalyst: in 2020, Swiss Post processed record parcel volumes (183 million; +23 percent year-on-year) as addressed letters fell almost 6 percent, with lockdowns and remote work accelerating digital interactions and

---

[12] Wolf, A., Heim, M., & Wyss, L. (2018). CSR as a Driver of Innovation at Swiss Post. In R. Altenburger (Ed.), Innovation Management and Corporate Social Responsibility (pp. 231–243). Springer. https://doi.org/10.1007/978-3-319-93629-1_13

[13] Swiss Post. (2013). *Die Lupe 2/2013*. Retrieved January 31, 2026, from https://www1.post.ch/-/media/post/ueber-uns/dokumente/post-unter-strom-lupe-2-2013.pdf?la=de

[14] Swiss Post. (2023, January 24). *Swiss Post carried 194 million parcels last year* [Press release].

further reducing on-site transactions.[15] These behavioral shifts materialized at physical access points: since 2010 over-the-counter payments fell by 70 percent and traditional counter transactions declined by 50 percent, reflecting the broad adoption of digital solutions.[16]

Swiss Post's financial services subsidiary, PostFinance, simultaneously faced pressure from a prolonged low or even negative interest rate environment that compressed the margins at the core of its business model. Moreover, legal restrictions prohibit lending and mortgages, constraining PostFinance's earnings potential and strategic flexibility.[17]

A decisive turning point came in 2018. Beyond the structural shifts, Swiss Post was shaken by the PostBus subsidies affair.[18] Following a review, the Federal Office of Transport found that from 2007 to 2016 PostBus had concealed profits in subsidized regional passenger transport by rebooking them to other business areas. This led to excessive subsidy payments and ultimately to a repayment of approximately CHF 200 million to the Swiss Confederation, cantons, and municipalities.[19] In February 2018, the Federal Council mandated the Federal Office of Police to open administrative criminal proceedings.[20] In June 2018, the then CEO of Swiss Post resigned, and further management changes followed.[21]

With the appointment of a new CEO in 2019, Swiss Post began to address the governance fallout of the PostBus affair and, in parallel, launched a holistic business review that led to a comprehensive strategic realignment. The overarching aim was to ensure that Swiss Post remains relevant and trusted for people and businesses in Switzerland in the face of digitalization and the predictable decline in letter volumes.[22] Under the banner "Swiss Post of tomorrow," the new Group strategy entered into force in January 2021 for an initial period to 2024, with a horizon that extends

---

[15] Swiss Post. (2021, January 19). *Paketmengen 2020: Siebter Rekord in Folge.* Retrieved January 31, 2026, from https://www.post.ch/de/weitere-angebote/briefmarken-und-philatelie/welt-der-briefmarken/2021/paketmengen-2020

[16] Swiss Post. (2024, October 29). *Swiss Post invests in the future: 2000 locations to remain staffed* [Press release].

[17] PostFinance Ltd. (2025). Annual report 2024.

[18] Swiss Post. (n.d.). 2018—The PostBus scandal. Retrieved January 31, 2026, from https://www.postauto.ch/en/about-us-and-news/organization/the-history-of-postbus/2018

[19] Federal Office of Transport. (2018, September 21). *Bund, Kantone und Post verständigen sich auf Rückerstattungen im «Fall PostAuto»* [Press release].

Federal Office of Transport. (2018, December 18). *Die Vereinbarung zu den Rückerstattungen im «Fall Postauto» tritt in Kraft* [Press release].

[20] Swiss Federal Council. (2018, February 27). *PostAuto: fedpol führt das Verwaltungsstrafverfahren* [Press release].

[21] Swiss Post. (2018, June 11). *External investigation and expert opinion confirm unacceptable misconduct—Swiss Post Board of Directors takes action* [Press release].

[22] Swiss Post. (2018, November 22). *Roberto Cirillo appointed new Swiss Post CEO* [Press release].

Swiss Post. (2019, April 3). *Roberto Cirillo: "I want Switzerland to be proud of Swiss Post"* [Press release].

Swiss Post. (2020, May 14). *"Swiss Post of tomorrow" to combine realignment with a focus on traditional strengths* [Press release].

across the decade until 2030.[23] It focused on expanding parcel logistics, developing trustworthy digital services, broadening PostFinance's offerings, and modernizing the branch network and customer touchpoints while opening branches to partners.

At the outset of the 2021 strategy period, sustainability was elevated to one of four strategic pillars—development, growth, efficiency, and sustainability—to futureproof the business model. This decision functioned as a critical juncture: it deliberately reset decision rules, roles, and oversight so that sustainability moved from dispersed, bottom-up initiatives to a path-shaping, Group-wide transformation anchored in governance. By institutionalizing sustainability as a pillar, Swiss Post reduced the risk of regression to prior routines and increased the credibility of subsequent sustainability interventions—an approach consistent with the management literature's argument that breaking, rerouting, and then stabilizing new paths requires formal reinforcement after the juncture.[24] Against this backdrop, the central question becomes how to integrate economic, social, and environmental dimensions into one coherent value-creation logic. The following section addresses this by examining Swiss Post's business model through an integrated lens.

9.3 Swiss Post's Business Model Through the Lens of Integrated Thinking

With sustainability formally anchored as strategic pillar, management must move from signaling intent to organizing the sustainable transformation of the business model, thereby positioning the company for long-term success. This requires a holistic, integrated value-creation logic that connects resources, activities, outputs, outcomes, and impacts across financial and nonfinancial domains—so sustainability is internalized in day-to-day business rather than treated as adjacent. Empirical evidence shows that firms embedding such practices develop stronger board oversight and incentive alignment, deeper stakeholder engagement, and a longer-term orientation—and tend to outperform peers in long-run accounting and stock-market performance.[25] In path-dependence terms, a consistent integrated value-creation approach locks in the post-juncture sustainability trajectory by embedding routines and control points that resist regression. Swiss Post converted the 2021 sustainability juncture into a path-shaping transformation by fostering integrated thinking as its managerial mindset (Sect. 9.3.1). Building on this logic, Sect. 9.3.2 maps Swiss Post's business model and value chain, and Sect. 9.3.3 aligns its business model to the UN SDGs.

[23] Swiss Post. (2021). *Annual Report 2020: The Swiss Post of tomorrow*. Retrieved January 31, 2026, from https://geschaeftsbericht.post.ch/20/ar/en/the-swiss-post-of-tomorrow/

Swiss Post. (2022). *Annual Report 2021: The Swiss Post of tomorrow*. Retrieved January 31, 2026, from https://geschaeftsbericht.post.ch/21/ar/en/the-swiss-post-of-tomorrow/

[24] See footnote 2.

[25] Eccles, R. G., Ioannou, I., & Serafeim, G. (2014). The impact of corporate sustainability on organizational processes and performance. *Management science*, *60*(11), 2835–2857. https://doi.org/10.1287/mnsc.2014.1984

9.3.1 Integrated Thinking as the Managerial Mindset for Sustainability-Oriented Business Model Transformation

To transform a business model toward sustainability, management should adopt a multi-capital perspective—extending beyond financial capital—across strategy, governance, planning, risk oversight, and decision-useful disclosure. The Integrated Reporting Framework codifies this perspective by emphasizing connectivity of information, accountability for all capitals, and support for integrated decision-making over the short, medium, and long term.[26] Integrated thinking operationalizes this logic as a management discipline: it assesses dependencies and trade-offs among the capitals, provides a structured basis for coherent decisions, and thereby enables renewal of the business model.[27] Conceptually, integrated reporting and the Integrated Thinking Principles provide the foundation; since 2022, the International Accounting Standards Board (IASB) and International Sustainability Standards Board (ISSB) jointly steward the Framework within the International Financial Reporting Standards (IFRS) Foundation.[28]

Building on a dynamic capabilities mechanism, integrated thinking organizations using a multi-capital view sense sustainability relevant opportunities and risks, seize them through resource allocation and process redesign, and then transform assets and routines—thereby reconfiguring and renewing the business model.[29] An international study finds that deeper embedding of integrated thinking is associated with stronger social and environmental performance without a long-term financial trade-off (while short-term financial performance can dip), underscoring integrated thinking as a practicable route to sustainability-oriented business model transformation.[30] In asset-intensive service systems, sectoral work in the port industry illustrates how governance and key performance indicator (KPI) structures link energy and emissions outcomes with sustainable and efficient operations—offering a blue-

[26] IFRS Foundation. (2021). International Integrated Reporting Framework.

IFRS Foundation. (n.d.). Integrated reporting: About. Retrieved January 31, 2026, from https://www.ifrs.org/issued-standards/integrated-reporting/

[27] IFRS Foundation. (2022). Transition to integrated thinking: A guide to getting started.

[28] IFRS Foundation. (2022, May 25). Integrated reporting—articulating a future path [Press release].

IFRS Foundation. (2022, August 1). IFRS Foundation completes consolidation with Value Reporting Foundation [Press release].

[29] Teece, D. J. (2010). Business models, business strategy and innovation. *Long Range Planning, 43*(2–3), 172–194. https://doi.org/10.1016/j.lrp.2009.07.003

Teece, D. J. (2018). Business models and dynamic capabilities. *Long Range Planning, 51*(1), 40–49. https://doi.org/10.1016/j.lrp.2017.06.007

Di Vaio, A., Syriopoulos, T., Alvino, F., & Palladino, R. (2021). "Integrated thinking and reporting" towards sustainable business models: A concise bibliometric analysis. *Meditari Accountancy Research, 29*(4), 691–719. https://doi.org/10.1108/MEDAR-12-2019-0641

[30] Reimsbach, D., & Braam, G. (2022). Creating social and environmental value through integrated thinking: International evidence. *Business Strategy and the Environment, 32*(1), 304–320. https://doi.org/10.1002/mbse.3131

print for fleets, and buildings with long asset lifecycle.[31] Field research also documents how integrated thinking is actually practiced inside firms and translated into purpose-driven strategy.[32] To assess an organization's integrated thinking maturity, a framework distills five principles—awareness, leadership, structures, performance management, and external communication—and produces maturity scores that align with integrated report quality and with ratings of governance sophistication.[33] As integration matures, firms shift from an inside-out efficiency view to an outside-in orientation in which societal challenges shape value propositions and control systems.[34]

In recent years, Swiss Post has advanced integrated thinking at corporate level by aligning governance and management processes with the Integrated Reporting Framework and the Integrated Thinking Principles. Concretely, this has led to the multi-capital view being operationalized in planning and investment decisions, strengthening decision-useful disclosure for monitoring and accountability, and embedding connectivity between financial and nonfinancial performance in steering practice. This integrated mindset materializes in two ways: first, in the evolution of the strategy architecture—from exclusively financial ambitions in the 2021–2024 strategy period to seven ambitions across financial and nonfinancial domains (Sect. 9.4.2),[35]; and second, in external disclosures. For the year 2023 Swiss Post issued its first Non-Financial Report complementing the financial statements[36]; for 2024 it added a dedicated Sustainability Report within the annual reporting suite[37]; and for 2025, it presented an integrated reporting package that brings financial and nonfinancial performance together.[38] As the external disclosures mature, the Swiss Post's business model is understood as a holistic value chain—linking capital and activities to outputs and outcomes across the areas of Economy, Society, Employees, and Environment (Fig. 9.1).

[31] Di Vaio, A., Varriale, L., & Alvino, F. (2018). Key performance indicators for developing environmentally sustainable and energy-efficient ports: Evidence from Italy. *Energy Policy, 122*, 229–240. https://doi.org/10.1016/j.enpol.2018.07.046

[32] See footnote 29.

[33] Maroun, W., Ecim, D., & Cerbone, D. (2023). Refining integrated thinking. *Sustainability Accounting, Management and Policy Journal, 14*(7), 1–25. https://doi.org/10.1108/SAMPJ-07-2021-0268

[34] Dyllick, T., & Muff, K. (2016). Clarifying the meaning of sustainable business: Introducing a typology from business-as-usual to true business sustainability. *Organization & Environment, 29*(2), 156–174. https://doi.org/10.1177/1086026615575176

[35] Swiss Post. (2025). Annual Report 2024: Swiss Post is managing its transformation with seven strategic ambitions. Retrieved January 31, 2026, from https://geschaeftsbericht.post.ch/24/ar/en/swiss-post-is-managing-its-transformation-with-seven-strategic-ambitions

[36] Swiss Post. (2024). *Non-Financial Report 2023.*

[37] Swiss Post. (2025). *Sustainability Report 2024*

[38] At the time this chapter was finalized in February 2026, the 2025 Integrated Report had not yet been published. Accordingly, all subsequent data pertain to the year 2024.

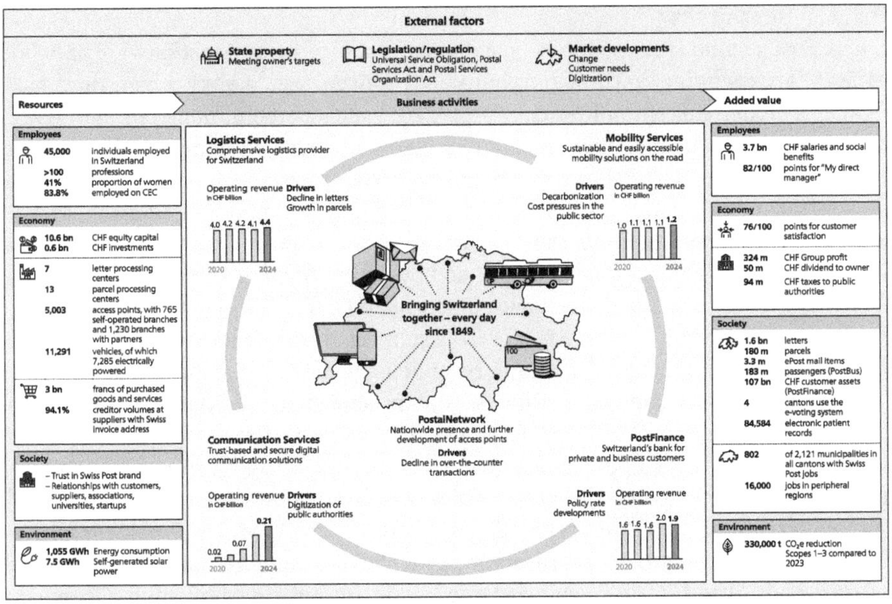

Fig. 9.1 Swiss Post's business model (Swiss Post Sustainability Report 2024, p. 10, used with permission). How to read the graphic: The figure flows from top to left to right: external factors shape the resources, which enable business activities in five business units; these generate outputs and, over time, added value for employees, economy, society, and environment

9.3.2 Mapping Value Creation: Swiss Post's Integrated Business Model

Swiss Post is a state-owned enterprise with a statutory public service mandate. Its remit is to guarantee reliable postal, logistics, communication, mobility, and financial service provision nationwide while financing operations from its own resources. The institutional and legal framework—above all the Postal Services Act and the Federal Act on the Organization of Swiss Post[39]—sets boundary conditions: the business model must remain service-driven, financially sound, and adaptive to digitization and the structural decline in letter volumes (Sect. 9.2.3). Like an ordinary shareholder, the Swiss Confederation expects the enterprise value of Swiss Post to grow.[40] The following sections describe how Swiss Post's business model creates value, based on 2024 figures.[41]

[39] See footnotes 5 and 6.

[40] See footnote 8.

[41] Swiss Post. (2025). *Sustainability Report 2024.*
Swiss Post. (2025). *Financial Report 2024.*
See footnote 38.

9.3.2.1 Resources

Swiss Post's value creation relies on a diverse resource base. A nationwide network of 5003 access points ensures proximity and convenience across urban, rural, and alpine regions; this footprint comprises 765 self-operated branches and 1230 branches with partners, complemented by parcel lockers and home service formats that safeguard reach as customer behavior digitizes. Twenty sorting and distribution centers form the logistics backbone, and a fleet of 11,291 vehicles operates logistics and public transportation. On the financial side, Swiss Post's capacity for modernization and decarbonization is underpinned by equity of CHF 10.6 billion, enabling CHF 0.6 billion in annual investments and around CHF 3 billion in annual procurement of goods and services each year. It is supported by a market-exposed revenue mix in which 88 percent of operating revenue is generated in competitive markets. The workforce is substantial and diverse, comprising some 37,000 full-time equivalents (FTE) across over 100 professions. In Switzerland, around 45,000 people work for Swiss Post; 84 percent are covered by the collective employment contract, and the workforce spans 139 nationalities with 41 percent women. Energy consumption within Swiss Post is 1055 GWh (gigawatt-hour), with a 17 percent renewable share. These tangible assets are complemented by intangible capital—a position of trust built over 175 years and sustained by stable relationships with customers, partners, and private and public sector organizations.[42]

9.3.2.2 Business Activities

Swiss Post operates in four markets—logistics, communication, mobility, and financial services—through an operating model organized into five business units: Logistics Services, Digital Services, Mobility Services (including PostBus), PostFinance, and PostalNetwork. As a universal service enterprise owned by the Swiss Confederation, this structure connects the company's extensive physical reach with secure digital interaction and enables coordinated steering of both universal service and market-exposed activities.[43]

- **Logistics Services** runs national letter, parcel, and press logistics at scale, handling 1.6 billion letters, 180 million parcels, and 667 million newspapers. The unit offsets letter decline (−5.5 percent) and parcel market dynamics with price and efficiency measures, added capacity, and route optimization, and it continues to electrify delivery. In addition, the Logistics Services unit operates goods logistics at Swiss Post Cargo, which links domestic and neighboring markets to cross-border flows.
- **Digital Services** extends Swiss Post's traditional position of trust into secure digital offerings—from hybrid letter solutions to the electronic patient record and e-voting—in line with public sector digitization needs.

[42] Swiss Post. (2025). *Sustainability Report 2024,* page 10–11, 81–88.
Swiss Post. (2025). *Financial Report 2024*, page 260.

[43] Swiss Post. (2025). *Sustainability Report 2024,* page 10–11, 81–88.
Swiss Post. (2025). *Financial Report 2024*, page 4–7, 12, 26–28.

- **Mobility Services** provides essential regional public transport, recording 183 million passenger journeys per year. The unit manages the Group's entire fleet and offers mobility solutions to third parties. In this market, it is progressing toward fossil-free propulsion and thereby contributing to the Group's net-zero climate ambition, coupling service reliability with a staged technology transition (Sect. 9.4).
- **PostFinance** supplies payments and fee-based financial services within a distinct regulatory perimeter that restricts traditional lending and mortgages. It holds CHF 107 billion in average customer assets and manages portfolio and risk appetite in a volatile interest rate environment.
- **PostalNetwork** orchestrates the hybrid access network—self-operated branches, partner-operated branches, parcel lockers, and home service—and adapts local formats as over-the-counter transactions decline, maintaining proximity without compromising service accessibility.

The five units generate roughly 90 percent of revenue in Switzerland, supported by a focused international footprint across 18 countries.[44]

9.3.2.3 Added Value

Swiss Post's business model is designed to create value across all dimensions of sustainable development in line with the strategic goals set by the Swiss Confederation (Sect. 9.2.1.2). These goals require Swiss Post to provide a high-quality universal service from its own resources while managing economic capacity responsibly and operating sustainably and ethically. Accordingly, the economic, social, and environmental contributions are treated not as separate aims, but as mutually reinforcing elements of an integrated value-creation logic.[45]

- **Economy**: Swiss Post provides the universal service without using taxpayers' money, financing it from its own resources. In 2024, the Group achieved CHF 7.6 billion in operating revenue and CHF 0.3 billion in Group profit. Swiss Post distributed a dividend of CHF 50 million to its owner, the Swiss Confederation, and contributed value to the public sector via taxes and levies. Serving around 9 million residents and hundreds of thousands of businesses, overall customer satisfaction stood at 76/100. This combination of economic performance and service quality underpins sustained investment in network modernization, fleet, and digital capabilities, and, in turn, supports Switzerland's market infrastructure.
- **Society**: Swiss Post provides a high-quality universal service, maintaining nationwide accessibility and strengthening social cohesion by combining local proximity with national standards. In 2024, 97 percent of the population could reach an access point within 20 minutes on foot or by public transport. Regulatory service quality targets are consistently met or exceeded: on-time delivery stood at 97 percent for Priority mail, 96 percent for Priority parcels, and 98 percent for

[44] Swiss Post. (2025). *Financial Report 2024*, page 12.

[45] Swiss Post. (2025). *Sustainability Report 2024,* page 10–11, 81–88.
Swiss Post. (2025). *Financial Report 2024*, page 14, 260

subscription newspapers. At the same time, secure digital channels—such as the digital letter, the electronic patient record, and electronic voting—expand access where physical presence is not required. PostBus supports mobility and social inclusion across urban, rural, and alpine regions, reaching even the most remote valleys, transporting 500,000 passengers daily.

- **Employees**: An attractive work environment is reflected in an employee net promoter score (eNPS) of 57. Swiss Post maintains a strong talent pipeline and a distinctive learning culture, making it a major provider of vocational training. In 2024, the company employed 1887 apprentices, an apprentice share of 6 percent, and invested CHF 9.3 million in external training, with CHF 7.5 million contributed by Swiss Post. Organizational indices confirm engagement and leadership quality: "My direct manager" reached 82 out of 100, and "Our corporate culture" reached 76 out of 100. CHF 3.7 billion was distributed to employees in the form of wages, salaries, and social benefits.
- **Environment**: In 2024, total GHG emissions stood at 1770 ktCO_2e (kilotonnes of carbon dioxide equivalents) across Scopes 1–3. Scopes 1 and 2 accounted for 240 ktCO_2e, while Scope 3 amounted to 1530 ktCO_2e. Operational levers are scaling visibly: city delivery in major cities such as Bern, Zurich, Geneva, and Basel is fully electric, and around 7300 electric vehicles were in operation, supported by charging infrastructure delivery depots. Swiss Post purchases 100 percent of its electricity from Swiss renewable sources, including ecologically certified power for vehicle charging and access points.

Swiss Post aligns resources and activities to generate economic performance, societal benefits, employee well-being, and environmental progress. Subsequent section will link this integrated value-creation logic to the UN SDG.

9.3.3 Aligning Value Creation with the United Nations Sustainable Development Goals

Swiss Post has been an Active Participant in the United Nations Global Compact (UNGC) since 2014—an engagement that entails a voluntary commitment to implement the Ten Principles on human rights, labor, environment, and anticorruption and to support the UN SDGs.[46] For a nationwide universal service institution operating in logistics, mobility, communication, and financial services, the UN SDGs provide a globally recognized reference to embed value creation in a common language of sustainable development.

- **SDG 3—Good Health and Well-being**: Swiss Post contributes by ensuring safe working conditions and promoting employee health, as well as by guaranteeing

[46] United Nations Global Compact. (n.d.). *Swiss Post Ltd—participant profile*. Retrieved January 31, 2026, from https://unglobalcompact.org/what-is-gc/participants/31971-Swiss-Post-Ltd-

United Nations Global Compact. (n.d.). Principles. Retrieved January 31, 2026, from https://unglobalcompact.org/what-is-gc/mission/principles

access to essential goods (targets 3.8 universal health coverage; 3.9 reduce deaths and illnesses from hazardous chemicals and air, water, and soil pollution and contamination).[47]

- **SDG 7—Affordable and Clean Energy**: Through renewable electricity sourcing and the expansion of fossil-free mobility and infrastructure, Swiss Post supports increasing the share of clean energy and improving energy efficiency (targets 7.2 renewable energy share; 7.3 improvement in energy efficiency).[48]
- **SDG 8—Decent Work and Economic Growth**: By providing fair working conditions, training opportunities, and job creation across all regions, Swiss Post promotes decent work and sustainable economic growth (targets 8.5 productive employment; 8.8 protection of labor rights).[49]
- **SDG 9—Industry, Innovation, and Infrastructure**: Swiss Post's investments in digital solutions, modern logistics, mobility infrastructure, and innovation contribute to sustainable and resilient systems, and support Switzerland's competitiveness (targets 9.1 sustainable, resilient infrastructure; 9.4 resource efficiency through modernization).[50]
- **SDG 11—Sustainable Cities and Communities**: With a nationwide universal service and moving toward low-emissions delivery and mobility solutions, Swiss Post fosters the development of inclusive, safe, and sustainable cities and communities (targets 11.2 accessible transport systems; 11.6 reduction of urban environmental impacts).[51]
- **SDG 12—Responsible Consumption and Production**: Swiss Post advances sustainable consumption and production patterns through efficiency improvements, recycling processes, and circular logistics approaches (targets 12.5 waste prevention and recycling; 12.6 sustainable corporate practices and reporting).[52]
- **SDG 13—Climate Action**: Swiss Post reduces GHG emissions along its decarbonization pathway—electrifying logistics and public transport fleets, sourcing certified renewable electricity for operations, replacing heating and cooling systems, improving building energy efficiency, and engaging suppliers and investments on Scope 3—while embedding climate considerations into governance and finance. Swiss Post strengthens resilience of the universal service, climate literacy of the workforce, management and partners, and the integration of early warning systems and operational contingency procedures (Sect. 9.4) (targets

[47] United Nations. (n.d.). *SDG 3 Targets & indicators*. Retrieved January 31, 2026, from https://sdgs.un.org/goals/goal3

[48] United Nations. (n.d.). *SDG 7 Targets & indicators*. Retrieved January 31, 2026, from https://sdgs.un.org/goals/goal7

[49] United Nations. (n.d.). *SDG 8 Targets & indicators*. Retrieved January 31, 2026, from https://sdgs.un.org/goals/goal8

[50] United Nations. (n.d.). *SDG 9 Targets & indicators*. Retrieved January 31, 2026, from https://sdgs.un.org/goals/goal9

[51] United Nations. (n.d.). *SDG 11 Targets & indicators*. Retrieved January 31, 2026, from https://sdgs.un.org/goals/goal11

[52] United Nations. (n.d.). *SDG 12 Targets & indicators*. Retrieved January 31, 2026, from https://sdgs.un.org/goals/goal12

13.1 strengthen resilience and adaptive capacity to climate-related hazards and natural disasters; 13.2 integrate climate change measures into national policies, strategies, and planning; 13.3 improve education, awareness-raising and human and institutional capacity on climate change mitigation, adaptation, impact reduction, and early warning).[53]

The UN SDGs set the global frame that situates Swiss Post's contribution to sustainable development. In Switzerland, the UN SDGs are translated into country-specific priorities through the 2030 Sustainable Development Strategy (SDS 2030) and the Agenda 2030 policy cycle, which embed sustainability in domestic steering and accountability.[54]

Swiss Post refines this orientation via a double materiality analysis in line with the European Sustainability Reporting Standards (ESRS), identifying topics that are material in impact and finance and aligning them with SDS 2030 and, consequently, the UN SDGs (Sect. 9.4.2).[55] In communicating effects, Swiss Post adopts a contribution over attribution stance, using theory of change-based evidence to make credible claims about the firm's role alongside public policy and market dynamics, with assumptions verified and key external factors accounted for.[56] This approach aligns with management research urging explicit mechanisms and contingencies when linking corporate actions to outcomes and helps to reduce the risk of greenwashing by grounding impact narratives in verifiable evidence, thereby sustaining a coherent chain from global goals to national priorities to enterprise execution.[57]

Highlighting the systematically and strategically anchored transformation toward a more sustainable organization since 2021, the next section presents a case study of Swiss Post's work in Climate & Energy. It shows how targets and measures were embedded in governance, strategy, and operations to contribute to UN SDG 7 Affordable and Clean Energy, UN SDG 9 Industry, Innovation, and Infrastructure, UN SDG 11 Sustainable Cities and Communities, and the overarching UN SDG 13 Climate Action.

---

[53] United Nations. (n.d.). *SDG 13 Targets & indicators*. Retrieved January 31, 2026, from https://sdgs.un.org/goals/goal13

[54] Swiss Federal Council. (2021). *2030 Sustainable Development Strategy*.

Federal Department of Foreign Affairs (FDFA), & Federal Department of the Environment, Transport, Energy and Communications (DETEC). (2025, January 17). *Implementing the 2030 Agenda for Sustainable Development*.

[55] EFRAG. (2023). ESRS Set 1.

EFRAG. (n.d.). ESRS Implementation Guidance—IG1 Materiality assessment. Retrieved January 31, 2026, from https://knowledgehub.efrag.org/eng/interactive/ig/ig1/09-2025

[56] Mayne, J. (2012). Contribution analysis: Coming of age?. *Evaluation, 18*(3), 270–280. https://doi.org/10.1177/1356389012451663

[57] Aguinis, H., & Glavas, A. (2012). What we know and don't know about corporate social responsibility: A review and research agenda. *Journal of management, 38*(4), 932–968. https://doi.org/10.1177/0149206311436079

Delmas, M. A., & Burbano, V. C. (2011). The drivers of greenwashing. *California management review, 54*(1), 64–87. https://doi.org/10.1525/cmr.2011.54.1.64

9.4 Case Study on Climate & Energy: Swiss Post's Transformation Toward a More Sustainable Organization

This section explains how Swiss Post has systematically and strategically anchored its sustainability transformation since 2021, using Climate & Energy as the focal case. It builds on the organizational and governance realignment outlined earlier (Sect. 9.2.3) as well as the managerial mindset of integrated thinking and UN SDGs orientation that frame Swiss Post's business model (Sects. 9.3.1–9.3.3). The emphasis is on how climate ambition has moved from declaration to execution and how it is embedded in governance, strategy, and day-to-day responsibilities.

Climate & Energy is selected for three reasons. First, it is the most material topic within the environmental dimension, which means it has decisive credibility with stakeholders. Second, it directly shapes the business model, as fleet operations, building heat, and electricity use are essential for the universal service and to Swiss Post's operating resilience. Third, it is the area where Swiss Post aims to take a clear leadership role in Switzerland, in step with market developments and evolving customer and stakeholder expectations. As a state-owned enterprise, Swiss Post operates under a public mandate that requires exemplary action in Climate & Energy. Following the nationwide vote of June 2023, the Climate and Innovation Act legally anchored Switzerland's net-zero 2050 goal (effective January 1st, 2025), establishing clear expectations for companies and public entities.[58] The bill specifies a 2040 target for the public administration and state-affiliated enterprises. In line with this expectation—and well before the popular vote—the Swiss Confederation launched the Exemplary Energy & Climate initiative in 2013 under the Energy Strategy 2050 to encourage federal bodies and state-affiliated enterprises to move toward efficiency gains, renewable energy expansion, and transparent progress reporting.[59] Swiss Post's net-zero 2040 ambition aligns with this 2040 public sector horizon and positions the company a full decade ahead of the economy-wide 2050 target consistent with the Paris-aligned pathway.

The transformation toward a more sustainable organization enfolds over five steps: strengthening executive commitment, connecting materiality and strategy, turning ambition into measurable actions, institutionalizing integrated steering, and fostering cultural change (Fig. 9.2).

[58] Swiss Confederation. (2023). *Federal Act on Climate Protection Targets, Innovation and Strengthening Energy Security (Climate and Innovation Act).*

[59] Swiss Confederation. (n.d.). Exemplary Energy & Climate—Basics. Retrieved January 31, 2026, from https://www.vorbild-energie-klima.admin.ch/en/basics

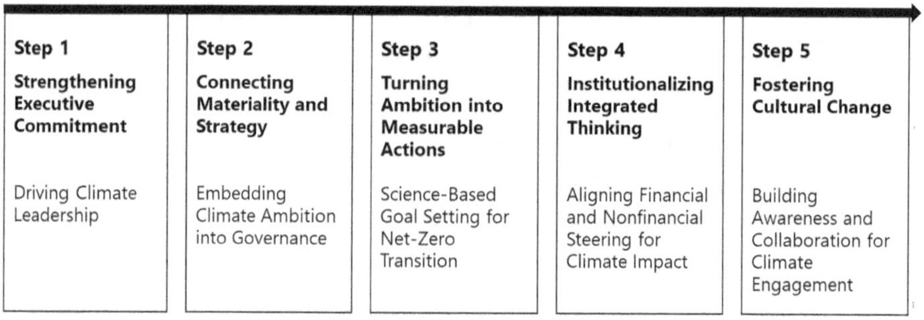

Fig. 9.2 Five-step model for transforming an organization toward sustainability, illustrated through the Climate & Energy case (own illustration)

9.4.1 Step 1: Strengthening Executive Commitment: Driving Climate Leadership

Executive commitment is recognized as a success factor in sustainability transformations. Without leadership endorsement, even well-intended initiatives remain fragmented and fail to gain strategic relevance. For Swiss Post, the decision to embed sustainability as one of four pillars in the "Swiss Post of tomorrow" strategy for 2021–2030 (Sect. 9.2.3) marked the turn toward sustainability leadership. To be effective, however, the sustainability agenda had to be more tangible. Given its high degree of materiality, Climate & Energy became a key enabler of Swiss Post's sustainable development. The business model needed to align with a net-zero trajectory and to address rising regulatory and societal expectations. A general strategy statement would not be enough; without strong governance and visible leadership, the climate ambition would risk remaining aspirational rather than transformational.

Swiss Post therefore elevated Climate & Energy to a top management priority with explicit CEO support. The process began with focused dialogue between the Board of Directors and Executive Management to build a shared understanding of climate risks, opportunities, and transition options. A two-day Board strategy workshop dedicated exclusively to Climate & Energy, informed by international practice, was held. Following an intensive decision-making phase in Executive Management, the workshop provided the basis for joint validation. This sequencing ensured that climate priorities were a key part of strategic deliberations in top management.

The commitment to climate leadership was formalized in governance. A newly formed Board Committee on People, Sustainability & Governance assumed responsibility for strategic sustainability topics, including Climate & Energy, while the Audit, Risk & Compliance Board Committee integrated sustainability-related audit and regulatory oversight. Within Executive Management, Swiss Post appointed a dedicated Delegate of Executive Management for Sustainability at Swiss Post with an explicit Climate & Energy remit, establishing a clear transformation mandate at the highest decision-making level. In addition, a focus group comprising Executive Management members from the major emitting business units—Logistics Services,

Mobility Services, and PostFinance—together with the CFO and the Chief of Staff, was tasked with developing science-based climate targets and a Group-wide climate strategy for executive approval and subsequent Board endorsement.

The effects were twofold. First, the Board of Directors began to consider climate-related risks and opportunities systematically in strategic decisions, including mergers and acquisitions (M&A). Second, Executive Management assumed clear responsibility for the implementation of Climate & Energy targets, assigning measures to specific members and linking accountabilities directly to markets and corporate functions. Beyond structures, the personal engagement and time investment of individual leaders strengthened organizational alignment.

This step is essential because it turns Climate & Energy into a strategy-critical mandate. Executive commitment embeds climate responsibility at the core of governance and ensures that resource allocation, decision rules, and accountability follow. By translating leadership at the top into clear structures and cascading roles down to line management, Swiss Post made the Climate & Energy target actionable across the organization. Responsibilities were linked to the business lines with the largest levers, and climate leadership became visible in how decisions are prepared, taken, and reviewed. Step 1 establishes the foundation for the steps that follow.

9.4.2 Step 2: Connecting Materiality and Strategy: Embedding Climate Ambition into Governance

Research indicates that organizations achieve greater sustainability impact and superior performance when they focus on material topics and connect them to strategy and governance.[60] Furthermore, materiality-aligned efforts are associated with higher value creation, whereas activity on immaterial issues shows no comparable benefits.[61] Step 2 converts materiality from a standalone sustainability exercise into a strategy-shaping input. For Swiss Post, this means closing the gap between external expectations and internal priorities by placing Climate & Energy—as a highly material topic—at the center of corporate strategy, planning, and decision-making, rather than confining it to sustainability reporting.

Swiss Post redefined the role of materiality within its strategy process to ensure that insights from the materiality assessment inform corporate strategy. Previously, materiality analysis had been conducted, but it was largely perceived as an isolated sustainability exercise. The redesign included a double materiality assessment, developed in line with the ESRS logic, which examined both financial materiality (effects on business success) and impact materiality (effects on society and the environment). Crucially, the assessment was integrated with risk management and corporate development, so that climate-related risks, opportunities, and impacts were evaluated alongside strategic, operational, and financial matters.

---

[60] See footnote 25.

[61] Khan, M., Serafeim, G., & Yoon, A. (2016). Corporate sustainability: First evidence on materiality. *The Accounting Review, 91*(6), 1697–1724. https://doi.org/10.2308/accr-51383

In line with the conducted double materiality assessment, Swiss Post reframed its strategic architecture for 2025–2028.[62] For the first time, the Group strategy combined financial and nonfinancial ambitions at the top level, creating a single, Group-wide set of priorities. Within this set, Climate & Energy was elevated from a transversal theme in the first strategy period of "Swiss Post of tomorrow" (2021–2024) to a formal strategic ambition under the label "On track for net-zero," which defines the 2030 Climate & Energy goal and links directly to the target system elaborated in Step 3. This elevation changed both status and process: climate ambition moved from advocacy to mandated strategy content.

Integration extended beyond the corporate level: business unit strategies translated the ambition into market-specific transformation plans. Corporate functions embedded climate-relevant criteria into their processes and aligned functional development accordingly. In finance, climate planning was linked to strategic financial planning and capital allocation, so that investment decisions, portfolio choices, and resource prioritization reflected the ambition "On track for net-zero." In parallel, technical climate content—such as GHG emissions trajectories, energy efficiency measures, and decarbonization levers—was translated into business language, enabling managers to evaluate options in terms of operational relevance, total cost of ownership (TCO), and risk-adjusted returns.

The result is a single strategy stack: materiality insights feed strategic priorities, which steer planning, execution focus, and performance management. Climate & Energy becomes a named, resourced strategic ambition at corporate level and an operational reality in business unit strategies and roadmaps. It is reviewed in the same steering cadence as financial ambitions. Decision papers routinely reference the "On track for net-zero" implications, and potential interdependencies or conflicts between ambitions are surfaced early for executive deliberation—preventing late-stage execution delays and ambiguity in resource allocation.

By embedding Climate & Energy in the strategic core and coupling it to planning and finance, Swiss Post replaces ad hoc coordination with governance-backed steering. This alignment reinforces the ability to deliver on the net-zero ambition while maintaining strategic coherence. Trade-offs between economic, social, and environmental dimensions become transparent at executive level: decisions follow a unified set of priorities; the sustainability transformation gains credibility and executability. In short, materiality functions as a management instrument rather than a compliance routine—providing the structured bridge from leadership intent (Sect. 9.4.1) to goal setting and its translation into measurable actions that accelerate execution (Sect. 9.4.3).

[62] Swiss Post. (2024, May 29). *Swiss Post looks to the future: Customer needs and digitization determine development steps* [Press release].

9.4.3 Step 3: Turning Ambition into Measurable Actions: Science-Based Goal Setting for Net-Zero Transition

Translating climate ambition into measurable actions requires targets that are science-based, time-bound, and governed with the same rigor as financial objectives. Between 2021 and 2023, Swiss Post moved from high-level climate intent to a science-based target system covering all material GHG emission sources across Scopes 1–3. Building on the intent to reach net-zero ahead of the standard 2050 horizon, and in line with the role model expectation of Swiss Post's owner, the Swiss Confederation,[63] the company adopted the Science Based Targets initiative (SBTi) framework,[64] set near-term targets for 2030 on the 2021 baseline, and committed to net-zero across the value chain by 2040.[65] In May 2023, SBTi validated Swiss Post's target set (near-term plus net-zero commitment), confirming alignment with a 1.5 °C pathway.[66] In 2025, the near-term targets for 2030 were anchored as a strategic ambition alongside six other financial and nonfinancial ambitions in the corporate strategy (Sect. 9.4.2).[67]

Based on the 2021 GHG inventory prepared under the GHG Protocol,[68] Swiss Post's SBTi validated near-term targets require, by 2030, a 42 percent reduction in Scopes 1 and 2, a 42 percent reduction for the Scope 3 category "use of sold products," and a 25 percent reduction across all other Scope 3 categories (Figs. 9.3 and 9.4).[69]

Long-term climate goals such as Swiss Post's net-zero 2040 are difficult to integrate into business planning because they exceed typical corporate horizons; Swiss Post plans in four-year strategy cycles. The company addressed this by establishing a Group-wide decarbonization pathway to 2040 that embeds the near-term 2030 targets and specifies them into annual reduction milestones. The corporate pathway is cascaded into business unit transition plans with defined owners, budgets, and KPI and is embedded in corporate and business unit management cycles. The Business Unit Mobility Services, with its public transport bus fleet, illustrates the logic. Because buses remain in service for more than 12 years, backward scheduling

[63] See footnotes 58 and 59.

[64] Science Based Targets initiative (SBTi). (2021). *SBTi Corporate Net-Zero Standard.*

[65] Swiss Post. (2023, June 22). Swiss Post awarded SBTi quality label for climate target [Press release].

[66] Science Based Targets initiative. (2023, May). Swiss Post Ltd.—approved science-based target [Certificate]. Retrieved January 31, 2026, from https://www3.post.ch/-/media/portal-opp/k/dokumente/corporate-responsibility/sbti-swiss-post-certificate.pdf?vs=2

[67] See footnote 62.

[68] World Resources Institute (WRI), & World Business Council for Sustainable Development (WBSC). (2004). *The Greenhouse Gas Protocol: A corporate accounting and reporting standard (Revised Edition).*

WRI, & WBSC. (2011). *Corporate Value Chain (Scope 3) accounting and reporting standard.*

WRI, & WBSC. (2015). *GHG Protocol Scope 2 guidance: An amendment to the GHG Protocol Corporate Standard.*

[69] Swiss Post. (2025). *Sustainability Report 2024,* page 28–33.

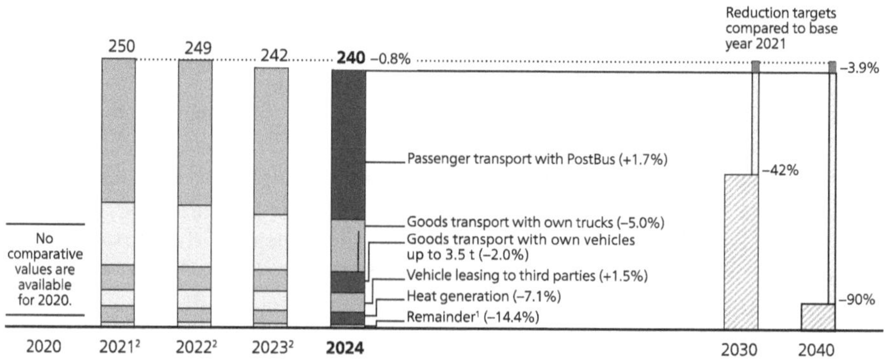

1 Includes emissions from cooling systems in properties, industrial vehicles, the company's own business vehicles and electricity consumption in buildings of subsidiaries entered retroactively.

2 Emissions with subsequent entry: for acquisitions/disposals, the emissions of acquired/sold companies in previous financial years are calculated retroactively.

Fig. 9.3 Scope 1 and 2 GHG emissions in 1000 t of CO2 equivalents (direct emissions from own vehicles and buildings) with 2021 as base year (Swiss Post Sustainability Report 2024, p. 29, used with permission)

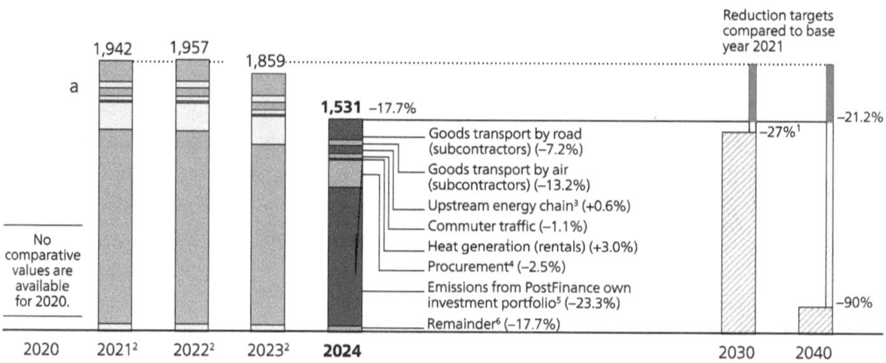

1 Volume-weighted value from the target for Scope 3 emissions from use of products sold (base year volume: 29,760 t CO_2 e) of at least 42 percent and the target for other Scope 3 emissions (base year volume: 1,912,131 t CO_2 e) of at least 25 percent compared to the base year 2021.

2 Emissions with subsequent entry: for acquisitions/disposals, the emissions of acquired/sold companies in previous financial years are calculated retroactively.

3 Includes emissions from all fuel and energy-related activities in fuel procurement.

4 The emissions from purchased goods and services were calculated using emission factors per goods group (secondary data, coverage at approx. 90 percent). See Methodology for more information.

5 The emissions values shown relate exclusively to the corporate bonds for which data is available in the bank's own investment portfolio, which corresponds to 89.0 percent of the corporate bonds and 44.2 percent of the entire own investment portfolio. Only the Scope 1 and Scope 2 emissions of all material greenhouse gases of the invested counterparties are considered. See Methodology for more information.

6 Includes emissions from leased property, plant and equipment (excluding heat generation), goods transport by rail and water, business travel, vehicle leasing (finance leasing), fuel sales and use and disposal of products sold.

Fig. 9.4 Scope 3 GHG emissions in 1000 t of CO_2 equivalents (indirect emissions from the value chain) with 2021 as base year (Swiss Post Sustainability Report 2024, p. 30, used with permission) from net-zero 2040 implies that the last fossil fuel buses would need to enter service by 2028. To meet the Swiss Post's near-term targets 2030 for Scopes 1 and 2, Executive Management accelerated bus fleet electrification to 2035 (Sect. 9.4.4). Planning and procurement were aligned to deploy electric buses wherever feasible, which required a rethinking of scheduling, depot design, routing, and sourcing.

Mobility is the pivotal lever for Scope 1 and 2 abatements. Around 80 percent of these GHG emissions come from combustion engines in public transport and logistics, so the shift to zero-GHG-emission mobility is the priority transition measure.[70] In 2020, Swiss Post, supported by a research institute and a consultancy, assessed battery electric, hydrogen, and synthetic fuel options and selected battery electric as the lead technology. On that basis, Executive Management launched a program to electrify the largest fleet in Switzerland: more than 5000 delivery vehicles (target 100 percent electrified by 2030), the three-wheeled delivery vehicles were electrified earlier (Sect. 9.2.2), roughly 2300 buses (target 100 percent electrified by 2035), and more than 500 trucks. Early charging rollouts revealed practical hurdles, including long installation and permitting lead times, hardware reliability issues, power capacity constraints, vehicle delivery delays, and routing adjustments due to range. To manage scale-up, Swiss Post created the E-Mobility Charging Solutions unit in 2025 to plan, build, and operate reliable, cost-effective charging and to optimize energy procurement and load management, recognizing electricity as a strategic resource on the decarbonization pathway.[71] In 2024, Swiss Post operates the largest electric fleet in Switzerland with around 7300 electric vehicles.[72]

Buildings are the second major lever for Scope 1 and 2 abatements. Swiss Post prioritizes fossil-free heating through heat pump deployment, with a target of around 80 percent of floor area heated without fossil fuels by 2030. This is complemented by targeted efficiency upgrades and expanded on-site photovoltaics (PV). PV at 52 sites generates 7.5 GWh per year; in 2024, the company added 1.7 MWp (megawatt peak) and is on track to reach 30 MWp by 2030, helping meet its electricity needs from fleet electrification. A flagship example is the logistics center in Villmergen, the first logistics building in Switzerland to obtain the DGNB Label Switzerland, the Swiss application of the German Sustainable Building Council (DGNB) system.[73]

Scope 3 quantification is challenging. Swiss Post is improving data quality to identify the main levers in each category and to steer toward the near-term targets 2030.[74] Managing these levers requires different instruments depending on the level of influence on the GHG emissions. Influence is high in subcontracted transport services for logistics, particularly in Switzerland. Here, Swiss Post imposes

---

[70] See footnote 69.

[71] Swiss Post. (2025). *Sustainability Report 2024,* page 33.

[72] See footnote 69.

[73] See footnote 71.

DGNB. (n.d.). DGNB's sustainability approach. DGNB. Retrieved January 31, 2026, from https://www.dgnb.de/en/sustainable-building/dgnbs-sustainability-approach

[74] See footnote 69.

low-carbon logistics standards and phases in contractual zero-GHG-emission service expectations. Financed GHG emissions in PostFinance's own investment portfolio follow a transition plan to reduce emissions by at least 25 percent by 2030 versus 2021, increase the share of financed counterparties with validated climate targets, and exclude issuers with more than 30 percent coal-based revenue, establishing a measurable contribution to Group Scope 3 reductions. Purchased goods and services are managed through a Supplier Code of Conduct, scorecards, and SBTi-aligned cascades. Swiss Post also supports and enhances low-carbon commuting. While commuting choices are personal, Swiss Post decided to shape outcomes through parking management, public transport offers, on-site charging, and supported carpooling. Swiss Post is piloting these incentives to reduce commuting-related GHG emissions while preserving accessibility and employee acceptance.

Beyond the decarbonization pathway, achieving net-zero requires a credible strategy for CO_2 removal and storage to address hard-to-abate residual GHG emissions. To uphold its role model position among Swiss corporations, Swiss Post is building a diversified portfolio that combines nature-based and technological approaches with clear permanence and integrity criteria. The company has set up a dedicated unit. As a first step, it invested in a forestry project that channels harvested timber into long-lived uses ("cascading use"), storing biogenic carbon over extended periods. As there are no established standards for this approach yet available, Swiss Post had to develop a methodology, accompanied by a research institute and validated by an independent certification institution. Technological solutions will be tracked and integrated when viable and scalable. CO_2 removal and storage complement rather than replace emission reductions along the pathway.[75]

A science-based and well-governed target system gives strategic coherence by aligning targets with global decarbonization trajectories. It strengthens capital discipline through quantified pathways that guide capital expenditures and investment decisions, improves regulatory readiness by defining metrics and targets based on sustainability standards, and builds stakeholder trust through transparent near-term targets for 2030 and credible long-term net-zero commitments supported by operational evidence. The next section shows how these targets are institutionalized through integrated financial and nonfinancial steering for Climate & Energy.

9.4.4 Step 4: Institutionalizing Integrated Thinking: Aligning Financial and Nonfinancial Steering for Climate Impact

Integrated thinking links financial and nonfinancial dimensions and enables organizations to steer holistically toward long-term value creation.[76] For Swiss Post, this principle is pivotal to translate the strategic ambition "On track for net-zero" into operational accountability. This section demonstrates how Climate & Energy

[75] See footnote **Fehler! Textmarke nicht definiert.**

[76] See footnote 27 and 29.

targets are embedded into the steering logic and how the financial performance model evolves into an integrated performance model.

Swiss Post institutionalized integrated performance steering by embedding Climate & Energy into planning, budgeting, investment, and reporting processes. Twice per year, aligned with the strategy cycle, business unit leadership teams review the progress on their transition plans. Based on these reviews, Executive Management assesses progress toward the strategic ambition "On track for net-zero." Actuals and plans from all business units are consolidated into a forecast of target attainment, enabling early identification of gaps. This forecasting approach also supports scenario thinking, allowing alternative pathways to be explored in response to emerging trends, risks, and opportunities. These scenarios are aligned with risk considerations and inform strategic decision-making. When deviations occur, Executive Management decides on corrective measures such as accelerating transition plans, resequencing activities, or reallocating capital to keep the ambition on track while balancing economic, social, and environmental priorities. An example is the acceleration of the bus fleet electrification plan from 2040 to 2035 (Sect. 9.4.3) to close the attainment gap for the near-term target for 2030 covering Scopes 1 and 2.

To enable consistent steering, Swiss Post introduced a nonfinancial KPI architecture that links transition plans and their measures to overall Group performance. Leading and lagging indicators are co-defined by business units and the central sustainability team for each transition measure, ensuring relevance at business unit and corporate level. It is essential that these KPI are business-oriented and reflect market-specific levers. A single source of truth for nonfinancial Climate & Energy data and KPI underpins internal steering and external disclosure. Data quality is enhanced through activity-based inputs, supported by clear data governance roles and audit trails. Improvements in coverage, granularity, and automation follow a cost–benefit logic and focus on data that is critical for steering operational transition measures and achieving the strategic ambition "On track for net-zero."

Financial integration has been a cornerstone for Swiss Post in institutionalizing integrated performance steering. The decarbonization pathway is embedded in the annual budget and the rolling four-year strategic financial plan. Capital and operational expenditures for transition measures are planned at business unit level and aggregated for Group steering, ensuring that the climate ambition is reflected in financial governance. To make the economics of decarbonization transparent and actionable, Swiss Post applies four complementary concepts. The "incremental view" isolates the additional cost of a measure compared to a fossil baseline, showing what is truly attributable to decarbonization beyond normal replacement cycles. "TCO analysis" captures lifetime economics, including operating expenses such as energy and maintenance, so decisions reflect full cost implications. "Cost parity" highlights where low-carbon technologies outperform fossil alternatives and where savings can be realized. And "abatement cost" expresses the cost per ton of CO_2e avoided by combining financial and nonfinancial data, enabling prioritization across measures and guiding portfolio decisions. Climate & Energy measures are assessed within the investment portfolio review to maintain enterprise-wide capital discipline. Investment decisions also consider long-term benchmarks such as assumed prices for CO_2 removal and storage, ensuring that capital allocation remains resilient under future regulatory and market conditions.

Integrated external disclosure complements performance steering and reinforces transparency and credibility. Reporting is aligned with sustainability and climate frameworks such as the Task Force on Climate-related Financial Disclosures (TCFD), ensuring comparability and regulatory readiness.[77] Climate data is prepared for assurance, which strengthens stakeholder trust and positions Swiss Post for evolving disclosure requirements.[78] The company's achievement of a CDP Climate score of A in two consecutive cycles underscores its commitment to transparency and accountability.[79]

Yet steering systems alone do not deliver transformation. Achieving impact at scale requires cultural change, climate literacy, and collaboration across the organization and its ecosystem. The last section explores how Swiss Post fosters awareness and partnerships to embed Climate & Energy into everyday practice.

9.4.5 Step 5: Fostering Cultural Change: Building Awareness and Collaboration for Climate Engagement

Awareness at all levels drives long-term success because climate impact is realized in daily decisions, from driving delivery vehicles to goods procurement, and customer interaction. For a universal service provider with a diverse value chain, partnerships inside the organization and with external stakeholders, combined with high climate literacy, are decisive enablers for implementing the decarbonization pathway. This requires not only technical innovation but also a deep cultural shift.

At the leadership level, the Board of Directors and Executive Management were engaged early to establish the relevance and urgency of sustainability, signaling top-level endorsement (Sect. 9.4.1). Their commitment created the basis for organization-wide alignment and catalyzed the cultural shift: high climate literacy at the top made Climate & Energy a part of leadership practice. Executive Management amplified this through the leadership cascade, addressing Climate & Energy with their direct reports and embedding accountability for progress into performance dialogues. To foster ownership and cross-functional alignment, information sessions were held across management tiers.

Swiss Post recognized that organizational change was essential at every level of the organization. It professionalized implementation structures and tailored collaboration to the complexity of the transition, from virtual teams and communities of practice to competence centers and dedicated units. The E-Mobility Charging

[77] TCFD. (2017). Recommendations of the Task Force on Climate-related Financial Disclosures.

[78] Swiss Post. (2025). *Sustainability Report 2024,* page 93.

[79] Keystone-SDA. (2025, May 3). Nine Swiss companies achieve top position in climate ranking. SWI swissinfo.ch. https://www.swissinfo.ch/eng/multinational-companies/nine-swiss-companies-achieve-top-position-in-climate-ranking/89256712

Solutions unit, for example, originated as a virtual team and was later transformed into a business unit with a dedicated investment budget to provide the charging infrastructure across the Group. Additionally, a specialized CO_2 removal and storage unit evolved from the central sustainability team, reflecting the professionalization of this capability (Sect. 9.4.3).

External pressure from stakeholders, especially clients who increasingly demanded sustainable practices such as low-emissions delivery, was leveraged to accelerate internal momentum and reinforce the business case for change. In parallel, the narrative was sharpened among leadership: climate-friendly investments are contributors to customer value, operational efficiency, and risk reduction, and are consistently underpinned by robust numbers and disciplined business case calculations.

Furthermore, communication played a pivotal role in the change process. External messaging addressed customers, the owner, and the wider Swiss public, underscoring Swiss Post's commitment to transparency and sustainability leadership. In parallel, internal communication ensured that employees understood the strategic direction and their role in delivering the ambition "On track for net-zero." While setting a positive long-term vision, Swiss Post reported progress and gaps and explained corrective actions both internally and externally. This transparency, reinforced by consistent Climate & Energy messaging, builds trust and credibility and is especially important for employees whose engagement amplifies climate action. Visible touchpoints such as electric vehicles deployed across Switzerland act as ambassadors, making the transition tangible for staff and the public and raising societal awareness for decarbonization (Sect. 9.4.3).

Consistent communication on Climate & Energy sharpened understanding, but differing starting points required targeted support. Swiss Post therefore strengthened climate literacy through a structured approach: central programs established a common baseline, while decentralized modules in business units addressed specific operational contexts and needs, translating learning into everyday decisions that influence energy performance and GHG outcomes. Experience from the last few years shows, however, that the strongest driver of climate engagement is proximity to decarbonization activities such as fleet electrification and fossil-free heating, which naturally raises awareness.

Partnerships along the value chain complement internal efforts. Decarbonization requires ecosystem collaboration. A precondition is climate literacy and a mutual recognition of the need to reduce GHG emissions, making Climate & Energy a substantive topic in supplier relationships. This shift is evident in core relationships with subcontracted road transport in Switzerland (Sect. 9.4.3). Over the past 4 years, Swiss Post and key subcontractors have increased electric truck kilometers by phasing in zero-GHG-emission service expectations and co-developing shared heavy vehicle charging, lifting the share of electric kilometers on key routes from zero to about 20 percent.

Cultural change is the execution engine of Swiss Post's climate ambition. By aligning everyday decisions with science-based targets and integrated steering, it turns intent into reliable delivery along the decarbonization pathway toward net-zero. Ultimately, transformation succeeds not only through governance, materiality, targets, and steering, but through people and partnerships who make the transition work—day by day, site by site, and together with the value chain.

9.5 Conclusion

The case of Swiss Post shows that sustainability transformation requires a shift in how an organization understands its purpose, governs itself, and organizes its operations. Credible change begins with a recognition of the institutional foundations that shape present-day choices—and with the willingness to redirect established trajectories when they are no longer aligned with long-term relevance.

This perspective is reflected in the Three-Pillar Model, which highlights three essential capabilities for navigating transformation: a clearly articulated Sustainable Purpose, the adaptability of a Travelling Organization, and the cross-boundary collaboration embodied in Connectivity. Swiss Post's experience illustrates how these capabilities enable an organization to operate under uncertainty while contributing to broader sustainability priorities, including the UN SDGs and the Swiss Sustainable Development Strategy.

A further insight from the case is the central role of integrated thinking as the connective tissue of sustainability transformation. By aligning financial and nonfinancial steering, creating shared information systems, and linking resource allocation to long-term value creation, integrated thinking ensures that sustainability becomes embedded in governance and managerial routines rather than remaining a parallel universe.

The Climate & Energy case study demonstrates how ambition can be translated into tangible outcomes. Science-based targets, a net-zero 2040 trajectory and operational levers—from fleet electrification to fossil-free heat and Scope 3 engagement—show what it takes to move from purpose to execution at scale. A recurring lesson is that progress becomes real only when top management commitment is explicit, continuous, and reflected in governance, decision-making, and resource allocation. This leadership dimension underpins a structured five-step model of sustainability transformation: from executive commitment to materiality integration, to science-based goal setting, to integrated steering, and finally to cultural change and partnerships.

Despite the transformation success illustrated by Swiss Post, the case also implicitly reveals inherent constraints that shape sustainability transformation. Climate & Energy benefit from comparatively mature metrics and established standards, whereas social and other environmental dimensions lack similar methodological clarity. This creates a bias: organizations tend to focus on what can easily be measured and audited while under-addressing areas that are societally important but harder to quantify. In addition, a universal service organization operates within dependencies it cannot fully control. Large-scale decarbonization hinges on the availability of technologies, the readiness of national infrastructure, regulatory developments, and political alignment—factors that determine what is feasible and at what pace. These external conditions make resilience and adaptive sequencing essential components of any credible transition pathway. Sustainability transformation also relies on stable financing and enduring legitimacy. Investments in electrification, buildings, or value chain measures require long-term commitment, yet their acceptance depends on leadership support, customer acceptance, and political confidence—all of which are sensitive to economic cycles and shifting public sentiment. This creates an inherent fragility even in well-designed transition programs. These critical reflections bring the central tensions in sustainability transformation to the forefront: organizations must pursue bold, long-term change while remaining constrained by what can currently be measured, controlled, and institutionally executed.

With electrification being the main driver for decarbonization in Scope 1 and 2, Swiss Post had the advantage of technological development supporting the change. During the first few years of replacement of delivery fleet, the lifecycle cost of electric vehicles declined in a way that reduced the decarbonization costs to zero. This cannot be expected in all measures or all technological changes. Hence, financial restrictions may very well limit sustainability initiatives in other circumstances.

Looking ahead, the Swiss Post's case opens two deeper questions that extend far beyond this single organization. First, how can institutions design transformation architectures that remain robust when essential sustainability impacts lie outside their direct control—in value chains, infrastructures, regulatory environments, and societal behavior? Second, what forms of leadership, collaboration, and incentive structures are needed to ensure that sustainability becomes a shared endeavor across ecosystems rather than an isolated corporate journey? Addressing these questions will determine whether organizations can move from transforming themselves to shaping the wider systems on which sustainability ultimately depends.

Vivien Büchler-Bezzola is Co-Head of Sustainability at Swiss Post. She leads corporate sustainability management and drives the transformation toward integrated performance management providing functional leadership in climate and energy, steering targets and initiatives to achieve the company's net-zero ambition. She has a background in corporate strategy, strategy advisory and execution, and large transformation projects in industrial organizations. She holds academic degrees in Economics as well as Accounting and Finance from the University of St. Gallen (HSG) and has expertise in business ethics and economic philosophy.

Christian Plüss has 20+ years' senior management experience in various organizations in the energy and mobility sectors as well as in public services. He has a track record in creating new organizations and implementing complex transformation projects in large organizations. He was Head of Mobility Services and a Member of Executive Management at Swiss Post and currently serves as Delegate of Executive Management for Sustainability at Swiss Post. He is also a member of the Board of Directors of several mid-sized companies. Christian holds a PhD in environmental sciences from the Swiss Federal Institute of Technology Zurich (ETH) and further developed his management skills at the University of Chicago, London School of Economics, and Harvard Business School.

10 Case Studies on Organizational Resilience and Sustainability Through Decision-Oriented Governance Systems

Christoph Schade

Abstract

Sustainability has increasingly been framed through Environmental, Social, and Governance (ESG) metrics, disclosures, and labels. While these instruments improve transparency, they often fail to guide real management decisions under uncertainty. This chapter argues that organizational resilience and with it, credible progress toward the United Nations (UN) Sustainable Development Goals (SDGs) emerges not from reporting, but from decision-oriented governance systems. To reach resilience the Three-Pillar (Mindset and Thinking) Model delivers a crucial support (see: https://www.peterwollmann.com/three-pillars). However, the Three-Pillar (Mindset and Thinking) Model needs on an operational level concrete additions to support purposeful decisions.

Building on the concept of the "PPP-Lens" for Decision-Making (People, Planet, Profit) as a decision enabler rather than a reporting framework, the chapter demonstrates how integrated governance, risk management, and compliance (GRC) and internal control systems (ICS) form the operational backbone of sustainable organizations. It shows how qualitative risk classifications are insufficient for strategic steering and how scenario-based risk quantification, Monte Carlo simulation, and distribution-based interpretation enable transparent trade-offs between safety, environmental impact, and financial resilience.

Disclaimer The views, interpretations, methodologies, models, and conclusions presented in this chapter are solely those of the author and do not necessarily reflect the official views, policies, positions, or strategic priorities of thyssenkrupp nucera AG & Co. KGaA, its affiliated entities, governing bodies, or employees.

C. Schade (✉)
Risk & Internal Control, Thyssenkrupp Nucera, Dortmund, Germany
e-mail: Christoph.schade@thyssenkrupp-nucera.com

© The Author(s), under exclusive license to Springer Nature Switzerland AG 2026
P. Wollmann et al. (eds.), *The Sustainable Organization – Volume 3*, Future of Business and Finance, https://doi.org/10.1007/978-3-032-28949-0_10

All case examples are purely illustrative and do not refer to specific ongoing projects, customers, suppliers, contractual arrangements, or internal company matters.

Any quantitative illustrations, simulations, probability assumptions, or financial impact estimations are provided exclusively for conceptual and academic demonstration purposes. They do not constitute actual corporate forecasts, formal risk assessments, internal evaluations, or forward-looking statements.

The chapter further illustrates how artificial intelligence (AI) and machine learning (ML) can augment—rather than replace—human judgment by automating preparation, interpretation, and assurance within risk-and-control processes. By connecting quantification, decision-making, and continuous assurance, the chapter reframes sustainability as a governance capability embedded in planning, investment decisions, and execution.

The contribution targets board members, executives, and governance professionals seeking practical guidance on how to operationalize sustainability objectives—across the 17 UN SDGs—through resilient, risk-informed decision-making rather than through disclosure alone.

10.1 Governance, Risk, and Sustainability: An Approach for Resilient Organizations

10.1.1 Why Resilience Needs More Than ESG Labels

From Labels to Leadership

Over the past decade, sustainability has been framed through a dense alphabet soup of standards, ratings, and ESG[1] labels. They look credible in annual reports but rarely guide day-to-day decisions. Resilient organizations, however, are not those with the best disclosure score—but those that can decide under uncertainty, balance short-term pressure against long-term viability, and allocate capital where it truly matters. That shift—from what we report to how we decide—marks the frontier of sustainable governance.

The Problem Behind the Labels

ESG frameworks were designed to make corporate behavior comparable. Yet in practice, ESG metrics often become compliance rituals rather than management tools. Boards still face the same unresolved dilemma: How much risk is acceptable to protect People? How much profit can we sacrifice for Planet? Without quantification and accountability, sustainability remains moral intent—not operational guidance.

A glossy ESG rating may signal virtue to investors, but it does not prevent downtime in a hydrogen plant, nor ensure the safety of those operating it. Executives increasingly recognize that resilience becomes measurable only through systems that connect risk, control, and decision-making. Resilience is therefore not a "pillar

[1] ESG—Environmental, Social, and Governance (ESG) refers to criteria used to evaluate a company's impact on society, the environment, and corporate governance practices. Source: Wikipedia, *Environmental, Social and Governance.* https://en.wikipedia.org/wiki/Environmental,_social,_and_governance

next to ESG"; it is the mechanism that turns purpose into performance—and it creates the conditions for measurable progress across sustainability objectives, including SDG 7 (Affordable and Clean Energy) and SDG 13 (Climate Action).[2]

The Three-Pillar Mindset and Thinking Model

The Three-Pillar Model[3] is based on three key building blocks (pillars) for organization design and leadership in the VUCA[4] world. The starting point of the developing global author community in 2019 was to design principles for a world with almost no principles left. The model was since then frequently applied—and its applicability and usefulness proven—in various global use cases in the private and public sector (so far >100 described in 5 books[5]).

1. The first Pillar: **Sustainable Purpose**
   Sustainable Purpose is the raison d'être of an organization; having a guiding objective means bringing orientation and certainty to the people for their joint endeavor and success. This is especially important in times of crisis and transformation. It is also the key source for the narrative of transformation. It is the north star for an organization and its people and stakeholders. An organization with a strong Sustainable Purpose and convinced people is more resilient than organizations which lack this.
2. The second Pillar: **Travelling Organization**
   Travelling Organization means for an organization accepting to be in a permanent state of flux and transformation, while interacting with the external requirements of its journey with rapid learning and adaptivity. Ideally, this is the mindset of loving the unknown and being well able to manage it. The Travelling Organization stands for all stakeholders to be actively or passively affected by the organization's Sustainable Purpose.
   One has an intuitive idea of a Travelling Organization from the metaphor of an expedition to Africa in former times. The participants of the Travelling Organization must be curious, open to learn, courageous, keen to experiment, and deal well with uncertainty, stress, unforeseen incidents in a totally new world. This mindset of a Travelling Organization is the core of a Learning Organization. It is impossible for a Travelling Organization not to learn. Learning is a question of survival. Being able to learn and adapt is a precondition for resilience.

[2] UN Sustainable Development Goals (SDGs)—United Nations, Department of Economic and Social Affairs. *The 17 Sustainable Development Goals (SDGs).* https://sdgs.un.org/goals

[3] Developed in Wollmann, P.; Kühn, F.; Kempf, M. (Eds.): Three Pillars of Organization and Leadership in Disruptive Times—Navigating Your Company Successfully through the twenty-first Century Business World. Cham: Springer Nature, © 2020

[4] Volatile, Uncertain, Complex, Ambiguous World

[5] See https://link.springer.com/search?query=&content-type=Book&dateFrom=&dateTo=&contributor=%22Peter+Wollmann%22&sortBy=relevance

3. The third Pillar: **Connectivity**
   Connectivity is all about the ability to interconnect on the journey with all needed parties and resources inside and outside the silos, therefore creating high efficacy and consistency and during this process building also shared learning networks. If the whole world is affected by a transformation, it has to connect with everybody. The ability to connect is also an important precondition for resilience.

As the Three-Pillar Model primarily operates on a meta-level by establishing a shared mindset and fundamental thinking logic for organizations, teams, and individuals, all actors building on the Three-Pillar Model additionally need a complementary concept to translate this orientation into concrete decision-making and action in specific situations.

Clarifying the Conceptual Layers: From Mindset to Decision-Making
Within this chapter, two conceptual layers are deliberately distinguished:

- The Three-Pillar Model serves as a mindset and thinking framework. It provides orientation, purpose, and leadership logic for organizations operating under uncertainty and transformation.
- The PPP Lens (People–Planet–Profit) serves as an operational decision lens. It translates purpose and mindset into structured, risk-informed management decisions by making trade-offs explicit and comparable.

The Three-Pillar Model answers why organizations act.
The PPP Lens supports how concrete decisions are evaluated and steered.
Both layers are complementary: mindset without operationalization remains abstract; operational tools without purpose lose direction.

The "PPP-Lens" for Decision-Making as Driver of Resilience and Sustainability When Connected with a Three-Pillar Model Application
In the case of applying the Three-Pillar Model in the context of Sustainability and UN SDG Coverage,[6] the perspectives "People, Planet, Profit" (abbreviated by "PPP-Lens" from now on) might offer a simple but powerful lens for steering decisions rather than labeling them. Instead of static categories, this concept introduces a practical evaluation logic for decision-making (with a strong connection to the organization's Sustainable Purpose):

- **People:** How does this decision protect competence, safety, and trust—especially in high-risk industrial environments?
- **Planet:** How does it influence material and energy efficiency across the value chain?

[6] Like in Wollmann, P., Pemler, D., Ndrevataj, M.: The Sustainable Organization—How Organizations Address the 17 UN SDGs Using the 3-P-Model; Cham 2025, SpringerNature (https://link.springer.com/book/10.1007/978-3-031-89549-4)

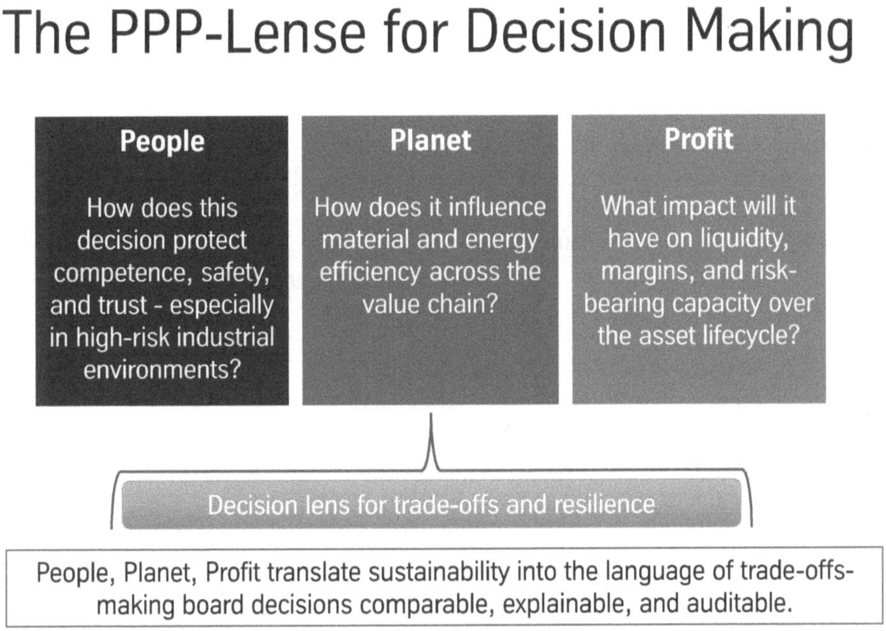

Fig. 10.1 The PPP-lens for decision-making as a decision enabler. Created by the author. Used with permission. All rights reserved

- **Profit:** What impact will it have on liquidity, margins, and risk-bearing capacity over the asset lifecycle?

Applied as a management concept, the PPP-Lens for Decision-Making forces transparency in trade-offs. It aligns risk data, control results, and strategic KPIs into one narrative: how each choice strengthens—or weakens—organizational resilience. In that sense, sustainability moves from a communication task to a governance capability (see Fig. 10.1).

10.1.2 Mini-Case Study: The Safety Investment Dilemma in Electrolysis (Illustrative)

The following case illustrates how quantitative risk assessment can transform an apparently ethical decision into a strategic investment in resilience. It demonstrates how applying the PPP-Lens for Decision-Making shifts management discussions from short-term cost considerations to long-term risk-adjusted value creation—one of the core principles of sustainability-oriented governance.

An aging pressure relief valve in a core module is due for replacement, requiring an investment of **€600,000**. Finance argues that margins are tight and replacement could be postponed for another year. Production warns that the component has reached the statistical end of its reliable service life and failure probability is rising steeply.

Applying the PPP-Lens for Decision-Making reframes the decision:

- **People:** After 10 years in operation, the likelihood of failure with safety consequences increases significantly. A malfunction could lead to overpressure and physical harm to personnel.
- **Planet:** A valve rupture could release hydrogen and coolant, potentially causing environmental damage and reputational exposure (e.g., exceeding **€5 million**).
- **Profit:** Postponement would save **€0.6 million** (≈ **0.2% of EBIT**). Yet a probabilistic loss model (e.g., Monte Carlo simulation[7]) estimates a **high-tail loss** of roughly **€7 million at the 95th percentile (P95)**[8] of the loss distribution, driven by downtime, repairs, penalties, and associated costs.[9]

The executive board now faces a quantified trade-off rather than competing opinions. The investment is no longer perceived as ethical overhead, but as resilience optimization, driven by economic efficiency reasons. Over a five-year horizon, avoided incidents, reduced downtime, and lower insurance premiums can translate into an internal rate of return of roughly **18%** (illustrative).

Resilience becomes measurable: the organization transforms uncertainty into a controllable variable—aligning operational integrity with strategic value creation and supporting SDG 9 (Industry, Innovation, and Infrastructure) through safe, reliable clean-tech operations (see Fig. 10.2).

10.1.3 From Reporting to Steering: The Resilience Loop

When sustainability indicators evolve into integrated risk-and-control metrics, sustainability becomes part of the management cycle:

Sense -> Quantify -> Act -> Assure

1. **Sense** emerging signals—economic, social, environmental.
2. **Quantify** likelihood and impact through scenarios and probabilistic modeling.
3. **Act** by allocating controls and budgets where marginal resilience gain is highest.
4. **Assure** effectiveness through monitoring, testing, and independent challenge.

[7] Monte Carlo Simulation—Bara, M. (2025). *Enhanced Monte Carlo simulation for project risk analysis: integrating cost and schedule impacts with time-shifted risks and dependency modeling.* Applied Operations and Analytics. https://www.tandfonline.com/doi/full/10.1080/29966892.2025.2552675

[8] More background see: https://en.wikipedia.org/wiki/Percentile

[9] Jorion, P. (2007). Value at Risk: The New Benchmark for Managing Financial Risk. McGraw-Hill.

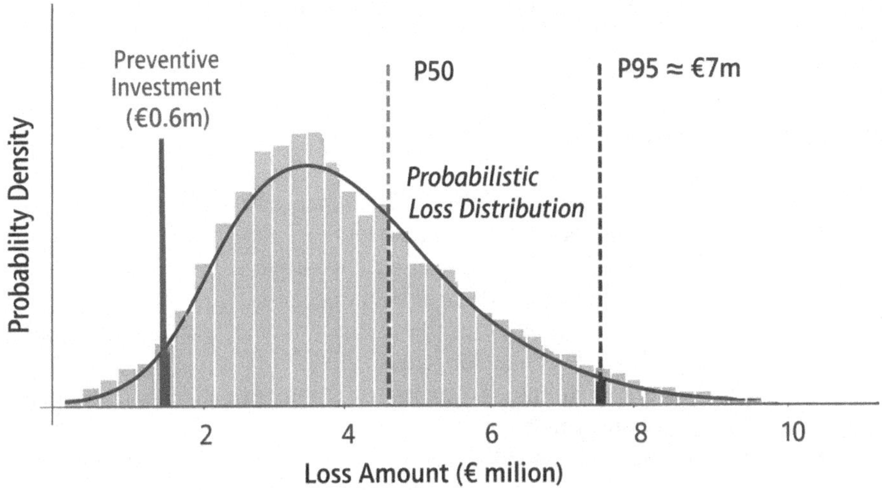

Fig. 10.2 Loss distribution and decision threshold. Created by the author. Used with permission. All rights reserved

Such systems transform board discussions from "what do we disclose?" into "how robust are we?" Executives can compare sustainability trade-offs in the same quantitative language they use for liquidity, safety, and quality.

10.1.4 Decision Value for Executives

At C-level, this shift creates three tangible benefits:

1. **Capital discipline:** Investments in safety, cybersecurity, or supplier diversification are prioritized by expected resilience return—not narrative urgency.
2. **Strategic clarity:** PPP-Lens for Decision-Making trade-offs reveal which initiatives strengthen the enterprise system rather than optimize one dimension at the expense of another.
3. **Audit readiness:** Quantified decisions provide evidence for regulators, investors, and assurance functions—without adding reporting overhead.

Resilience thus becomes a board capability—embedded in planning, budgeting, and performance management.

10.1.5 Mini-Conclusion

True sustainability is not about how green the report reads, but how resilient the organization behaves under stress. The PPP-Lens for Decision-Making transforms sustainability intent into governance logic: decisions are evaluated for their impact on People, Planet, and Profit—and optimized over the long run. Resilience is not the absence of risk; it is the ability to steer through it with data, structure, and accountability.

10.1.6 Key Takeaways (I)

1. ESG labels describe; governance decides. Resilience grows from integrated risk-and-control systems—not disclosure checklists.
2. PPP-Lens for Decision-Making is a steering lens, not a label. It clarifies trade-offs between People, Planet, and Profit in every decision.
3. Quantification enables credibility. Scenario analysis and probabilistic metrics translate sustainability into capital language.
4. Resilience is measurable. It links risk exposure, control effectiveness, and strategic outcomes in one steering cycle.
5. Boards own the lens. Sustainability becomes governance in action only when executives use PPP-Lens for Decision-Making to guide decisions—not to justify them.

10.2 GRC (Governance, Risk Management, Compliance) Foundations that Scale

10.2.1 Why Resilience Needs Architecture

Every sustainable transformation eventually collides with its own complexity. Hydrogen projects, renewable-energy supply chains, or circular-manufacturing programs operate across entities, regulations, and time zones. Without a unifying governance backbone, even the best sustainability strategy becomes a patchwork of initiatives. Scalable resilience therefore depends on governance architecture: a system that defines decision rights, control responsibilities, and assurance mechanisms.

10.2.2 Core Concepts: Governance as System, Not Hierarchy

Governance, Risk Management, and Internal Control are often treated as compliance silos. Scalability emerges only when they operate as an integrated system:

- **Governance** sets principles and accountability (the "why" and "who").
- **Risk Management** creates transparency on uncertainty (the "what").
- **Internal Control** ensures mitigation is designed and operating (the "how").

This triad aligns with established frameworks such as the Committee of Sponsoring Organizations of the Treadway Commission (COSO)[10] and the International Organization for Standardization (ISO) 31,000,[11] but in resilient organizations these are living interfaces—not manuals on a shelf. They define a shared decision language and an assurance loop that connects board intent to operational reality. Such integration strengthens corporate accountability and transparent decision-making, aligning—especially when used by organizations across the width—with SDG 16 (Peace, Justice, and Strong Institutions).

10.2.3 Mechanics: The Integration Loop

At scale, GRC is not a linear process chain but a feedback loop that balances design, execution, and evidence:

1. **Set Appetite:** The board defines which level of risk is acceptable to achieve strategic and sustainability objectives.
2. **Identify and Assess Exposure:** Risk owners evaluate exposure using consistent criteria (probability, impact, scenarios).
3. **Control and Monitor:** The ICS implements safeguards and monitoring, increasingly supported by automation and data-driven evidence.
4. **Assure and Improve:** Independent testing validates effectiveness and feeds lessons learned back into governance and risk criteria.

When these steps run in isolation, resilience fragments. When they operate as one loop, the organization gains a dynamic map of operational health and strategic exposure. This capability is a prerequisite for safe industrial innovation (SDG 9) and responsible production systems (SDG 12) (Fig. 10.3).

10.2.4 Practice: From Silos to Synergy

In industrial transformation, Risk, ICS, and Audit often use different data models, timelines, and vocabularies. Scaling GRC requires three pragmatic design principles:

1. **One taxonomy:** Risks, controls, and findings share a common identifier structure. A supply-chain risk in the risk register equals the same entity in the control matrix and the audit plan.

---

[10] COSO—Enterprise Risk Management Committee of Sponsoring Organizations of the Treadway Commission (COSO). *Enterprise Risk Management—Integrating with Strategy and Performance.* 2017. https://www.coso.org

[11] ISO 31000—Risk Management. International Organization for Standardization (ISO). ISO 31000:2018 - Risk management—Guidelines

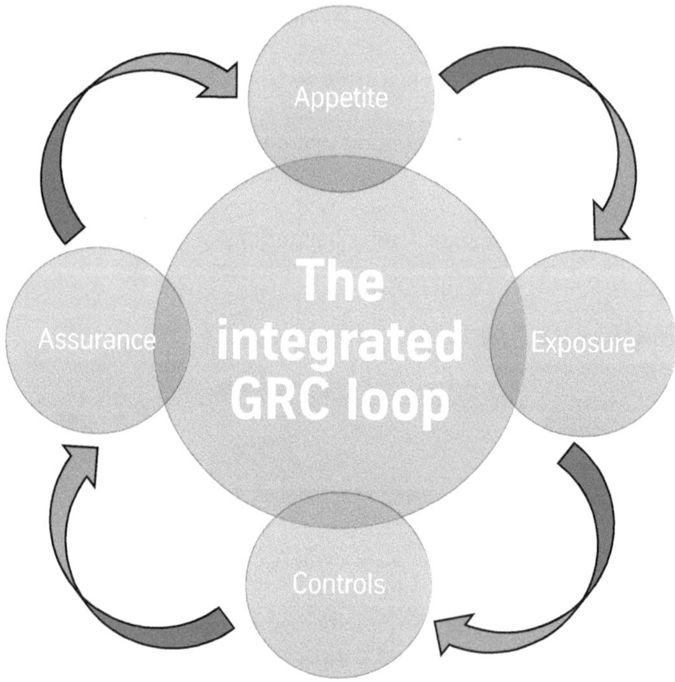

Fig. 10.3 The integrated GRC loop. Created by the author. Used with permission. All rights reserved

2. **Role clarity:** The risk owner manages exposure, the control owner manages prevention and detection, and audit validates assurance—no duplication, no gaps.
3. **Automated evidence:** Wherever possible, control results are drawn directly from operational data (logs, sensor outputs, ERP transactions) rather than manual checklists.

10.2.5 Mini-Case Study: Integrated Risk Cycle in a Hydrogen Plant

During commissioning of a large-scale electrolysis facility, a delay in membrane deliveries could trigger both financial and technical risks. In an integrated GRC system, the loop demonstrates its value:

- The risk register flags the event as critical exposure, linking supply-chain data to project risk categories.
- The control system triggers predefined responses (e.g., supplier qualification review, secondary sourcing, contract milestone re-planning).

- Audit and second-line functions later verify that the response was executed, documented, and effective.

Through this closed loop, potential downtime and cascading schedule penalties can be reduced. The architecture converts reactive problem-solving into structured resilience—an essential capability for industries driving the energy transition.

10.2.6 Decision Value: Scaling Without Bureaucracy

Executives often fear stronger governance means slower execution. In practice, clarity accelerates when decision rights, risk criteria, and control thresholds are codified, and discussions move from *who acts* to *how fast we act*. Benefits for leadership include:

1. **Consistency:** Comparable data across risk, controls, and audit enables like-for-like prioritization across plants and projects.
2. **Focus:** Resources concentrate on high-impact areas—capital, safety, cybersecurity—reducing "control theater."
3. **Scalability:** New markets and technologies can adopt the same GRC skeleton without reinventing procedures.

Governance becomes an enabler of transformation, not an administrative burden.

10.2.7 Link to the (UN) SDGs

Strong GRC systems translate the abstract goals of Agenda 2030 into operational reality:

1. **SDG 7 (Clean Energy):** risk and control transparency safeguards hydrogen and renewable projects.
2. **SDG 9 (Industry and Infrastructure):** integrated GRC supports safe, reliable industrialization and innovation.
3. **SDG 12 (Responsible Production):** risk-based control allocation minimizes waste and resource loss.
4. **SDG 16 (Strong Institutions):** governance integrity and assurance underpin stakeholder trust.

10.2.8 Mini-Conclusion

Sustainability scales only when governance scales with it. Frameworks like COSO and ISO 31000 matter not for documentation, but for the discipline they embed.

Integrated GRC connects purpose to process—turning resilience into a repeatable outcome rather than a coincidence.

10.2.9 Key Takeaways (II)

1. Integration beats addition. Risk, Control, and Audit must operate in one feedback loop—not as parallel functions.
2. Architecture drives resilience. Governance clarity enables faster, evidence-based decisions.
3. Automation supports assurance. Data-driven control evidence reduces manual compliance overhead.
4. SDGs need systems. Without strong GRC infrastructure, sustainable industry goals remain aspiration.
5. Scalable resilience is structured governance: what can be repeated can be sustained.

10.3 Quantifying Uncertainty: From Heatmaps to Decisions

10.3.1 Why Heatmaps Are Insufficient for Strategic Decisions

Qualitative risk heatmaps are widely used because they are intuitive and easy to communicate. Yet from a board-level perspective, they remain fundamentally descriptive. Heatmaps classify risks relative to one another but do not quantify exposure in a way that supports capital allocation, risk appetite decisions, or trade-offs between competing objectives.

Two risks marked as "high" may differ by orders of magnitude in potential financial loss. A visually "medium" risk may dominate downside exposure due to tail effects, while a "red" operational issue remains bounded. Heatmaps also remain static snapshots, compressing uncertainty into a single probability-impact combination and ignoring distributions, dependencies, and escalation dynamics.

As a result, heatmaps fail to answer the questions boards actually face:

- How much value is realistically at risk?
- How severe are plausible downside scenarios?
- What changes if mitigation is delayed or partially implemented?

Without quantitative answers, governance discussions shift from steering to persuasion.

10.3.2 From Classification to Quantification: A Conceptual Shift

Quantifying risk does not replace expert judgment; it makes it actionable. The decisive shift lies in moving from categorical labels to probabilistic representations of uncertainty.

Rather than asking whether a risk is "high" or "medium," quantification reframes the decision problem:

- What range of outcomes is plausible?
- How concentrated are losses in the tail?
- How does uncertainty aggregate across risks and time?

Applied through the PPP-Lens for Decision-Making (People, Planet, Profit), quantification enables transparent trade-offs between safety, sustainability-relevant externalities, and financial resilience. It transforms risk management from a reporting exercise into a decision infrastructure.

10.3.3 Beyond Monte Carlo Simulation: Unlocking the Full Potential of Probabilistic Analysis

Monte Carlo Simulation (MCS) is a well-established technique for modeling uncertainty by generating large numbers of potential futures based on probability distributions. Instead of producing a single expected value, it yields a distribution of possible outcomes, including tail events.

In practice, however, MCS is often applied in a conceptually narrow way. A single impact value is estimated, a "reasonable" distribution is selected, and one percentile—typically Value at Risk (VaR) at the 95% confidence level (P95)—is extracted as the decision anchor.

While mathematically correct, this approach significantly underutilizes the strategic potential of probabilistic analysis. The limitation lies not in the simulation technique itself, but in the logic of its inputs and interpretation.

10.3.4 Rethinking Monte Carlo Simulation: From Single Impacts to Expert-Driven Scenarios

The approach presented here deliberately reverses the traditional logic of Monte Carlo modeling.

Instead of starting with the question of which key performance indicators (KPIs) should be used to quantify impact, the simulation is preceded by a structured scenario analysis, developed collaboratively with subject-matter experts.

Each expert contributes a representative assessment of how the risk could realistically unfold, based on operational experience, technical understanding, and situational context. These assessments are captured as Best Case, Base Case, and Worst-Case scenarios, rather than abstract single-point estimates.

This design delivers two decisive advantages: ***Quality and breadth of judgment.*** Experts without statistical training can meaningfully contribute. The focus shifts from stochastic sophistication to domain expertise—where the most relevant insights reside.

10.3.5 Efficiency and Consistency

Supported by a purpose-built software solution, scenario input is intuitive, fast, and standardized. This enables broad participation without sacrificing methodological rigor.

The underlying principle is simple: *Monte Carlo simulation amplifies the quality of its inputs*. Scenario-first design ensures "best in, best out."

10.3.6 Modeling Risk Clusters and Differentiated Probabilities

The scenario-based approach further enables the modeling of risk clusters. Multiple individual risks can be aggregated into a single exposure profile while retaining differentiated probabilities of occurrence.

This is particularly relevant for concentration risks, systemic dependencies, or portfolio-level exposures. By assigning probabilities at scenario level rather than enforcing uniform assumptions, Monte Carlo Simulation becomes a tool for structural risk understanding, not merely numerical aggregation.

10.3.7 From Single Percentiles to Distribution Logic

The objective of this approach is explicitly not to identify one "correct" percentile. Instead, the analytical focus shifts to understanding relationships across the distribution:

- How steeply does impact escalate beyond the median?
- At what point do tail risks dominate outcomes?
- How sensitive are decisions to changes in scenario probabilities?

This perspective fundamentally changes management and board discussions. Rather than defaulting to "we manage at the Value at Risk at the 95th percentile (VaR95)," organizations can consciously assess risk appetite across People, Planet, and Profit.

A conservative posture may limit tail exposure to protect safety and environmental integrity. A more risk-tolerant stance may accept earnings volatility while maintaining non-negotiable thresholds for human and environmental harm.

10.3.8 PPP-Lens for Decision-Making as a Steering Logic for Probabilistic Decisions

Within this framework, the PPP-Lens for Decision-Making becomes the explicit interpretation and decision logic for Monte Carlo results:

- **People:** Which percentile range preserves safety, workforce trust, and operational stability?
- **Planet:** At what downside levels do environmental externalities become unacceptable?
- **Profit:** How much volatility can be absorbed without impairing liquidity or strategic flexibility?

Monte Carlo Simulation thus evolves from a reporting instrument into an active steering mechanism, enabling transparent and deliberate trade-offs.

10.3.9 Board-Level Decision Example: Distribution-Aware Steering in a Large-Scale Green Hydrogen Facility (Illustrative)

In a large-scale green hydrogen facility, management considers implementing an additional system-stability upgrade to mitigate operational volatility caused by grid fluctuations and project-related disturbances.

The following example is illustrative in nature. All figures, assumptions, and outcomes are simplified and anonymized and do not reflect actual projects, internal models, or confidential information.

10.3.10 Scenario Analysis First

Before applying quantitative simulation, a structured expert-driven scenario analysis is conducted. Specialists from operations, grid integration, maintenance, project controls, insurance, and finance jointly define Best Case, Base Case, and Worst-Case scenarios reflecting plausible operational developments rather than isolated point estimates.

This step is intentional. Monte Carlo Simulation does not generate insight by itself; it amplifies the quality of its inputs.

10.3.11 Monte Carlo Output: A Headline Number Without Context

Based on these scenarios, a Monte Carlo Simulation quantifies potential financial impacts from downtime, restart procedures, contractual penalties, and secondary effects.

The simulation produces a **VaR95 of €10 million** (explanation see below). Viewed in isolation, this number would typically trigger a conservative response, such as deferring or rejecting the upgrade.

10.3.12 Looking Beyond VaR (Value at Risk): Understanding the Distribution

Management deliberately avoids steering by a single percentile and analyzes the full loss distribution:

- The majority of simulated outcomes cluster between **€2 million and €3 million**, representing typical disruption patterns.
- The **90th percentile (P90)** lies around **€4 million**, indicating that most adverse scenarios remain bounded.
- The distribution tail is steep: losses escalate rapidly between P90 and P95, culminating in **VaR95 = €10 million**.
- Extreme outcomes beyond VaR95 exist but are rare by definition and driven by a narrow set of compounding conditions (Fig. 10.4).

The governance question shifts from "Is VaR95 high?" to "What drives VaR95—and how plausible and controllable are those tail conditions?"

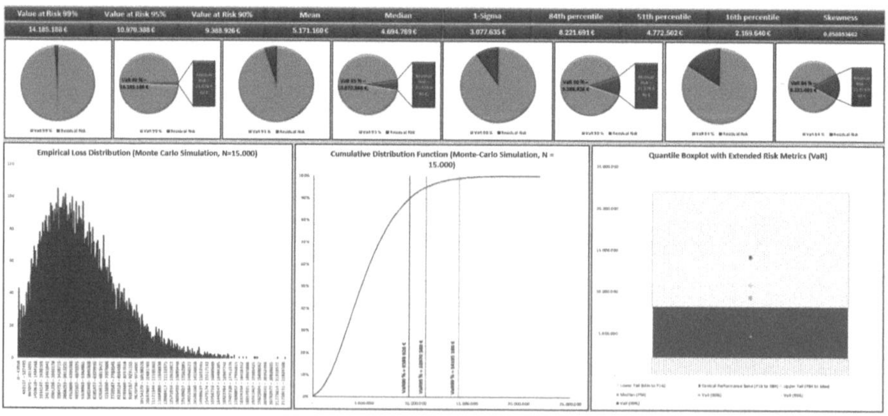

Fig. 10.4 Excerpt from a Monte Carlo simulation dashboard. Created by the author. Used with permission. All rights reserved

10.3.13 Decision Along the PPP-Lens for Decision-Making

The final decision is framed explicitly through People, Planet, and Profit:

- **People:** The upgrade improves safety margins and reduces stress situations during grid instability.
- **Planet:** Fewer unplanned shutdown–restart cycles reduce energy losses and environmental inefficiencies.
- **Profit:** Despite a high VaR95, the dominant distribution supports controlled financial volatility. Tail risk is consciously accepted where People and Planet thresholds are strengthened.

10.3.14 Outcome

The upgrade is approved. The decision is explicitly documented as distribution-based and scenario-grounded. Tail risks are understood and steered—not overridden by a single metric.

Key insight: Monte Carlo Simulation creates governance value only when expert-based scenario design and distribution logic replace rigid percentile fixation.

10.3.15 Mini-Conclusion and Outlook to the Next Chapter

Quantifying uncertainty is not about producing a single number; it is about making uncertainty governable. By combining expert-driven scenario analysis with Monte Carlo Simulation and distribution-based interpretation, organizations transform risk management into a core governance capability.

Interpreting probabilistic distributions, however, requires experience and statistical literacy—capabilities not universally available at board or management level. The following chapter therefore explores how AI-supported systems can automate interpretation, explanation, and prioritization of such analyses, translating advanced quantification into decision-ready insights without creating new governance barriers.

10.4 Artificial Intelligence (AI) and Machine Learning (ML) in Risk and Controls: From Automation to Augmented Governance

10.4.1 AI in GRC: From Technology Debate to Governance Capability

AI and ML are often discussed as disruptive technologies. In the context of governance, risk management, and internal control systems, however, this framing is misleading. The relevant question is not whether AI represents "new technology," but whether it enhances the organization's **ability to understand uncertainty, interpret risk, and steer decisions under complexity**.

Governance responsibility itself is not delegable. Boards, executives, and risk owners remain accountable for risk appetite, mitigation choices, and trade-offs. AI does not replace this responsibility. It augments it.

In this sense, AI-enabled risk-and-control systems do not "implement" individual SDGs. They provide the decision infrastructure that allows organizations to operationalize sustainability-related objectives across People, Planet, and Profit in a measurable and auditable way.

The guiding principle is therefore clear: ***AI prepares insight; humans make decisions.***

10.4.2 Reducing GRC Friction Without Losing Control

In many organizations, GRC functions consume significant capacity through repetitive administrative work:

- Drafting and updating risk descriptions
- Defining mitigation measures
- Specifying controls and test procedures
- Aligning risk registers, control matrices, and audit findings
- Preparing management and board documentation

These activities are necessary, but they rarely generate new insight. Their primary purpose is structure, not interpretation. This makes them well suited for AI-supported preparation.

Properly trained AI systems can:

- Generate structured risk-and-control drafts aligned with internal standards
- Identify inconsistencies, gaps, or redundancies across GRC artifacts
- Link risks, controls, and findings across systems and time horizons

Decision authority remains entirely with humans. AI does not decide—it reduces friction and frees resources for issues that genuinely require judgment.

10.4.3 From Signal Overload to Structured Early Warning

Beyond administrative relief, AI acts as a continuous sensing layer.

External signals from:

- Regulation and enforcement
- Geopolitics and sanctions

- Market and competitor developments
- Supply chain and technology disruptions

can be automatically captured, clustered, and mapped to existing risk structures.

The governance value does not lie in data collection itself, but in **contextualization**: where signals alter existing risk profiles, trigger reassessments, or indicate emerging risk clusters.

AI does not declare risks. It **directs attention**, enabling earlier and better-informed intervention.

10.4.4 AI-Supported Interpretation of Monte Carlo Simulations

Section 10.3.3 demonstrated that Monte Carlo simulations only become decision-relevant when distributions are understood—not when single percentiles are isolated.

In practice, this interpretation gap is one of the largest barriers to effective quantitative risk management.

Monte Carlo models are mathematically sound but cognitively demanding. As a result, decision-makers often default to seemingly precise indicators such as VaR95. This compresses rich distributional information into a single number, encouraging either excessive conservatism or poorly calibrated risk-taking.

AI fundamentally changes this dynamic.

Rather than generating new simulations, AI focuses on **interpreting existing results**:

- Explaining skewness, tail behavior, and clustering effects
- Separating operational "normal cases" from extreme scenarios
- Identifying the drivers behind high percentiles
- Highlighting sensitivities to scenario assumptions

This shifts discussions from *"What is the VaR?"* to *"What does this distribution actually tell us about the behavior of the risk?"*

10.4.5 Interactive Risk Steering Instead of Static Metrics

The real value emerges when interpretation becomes interactive.

AI enables structured dialogue with simulation results:

- *What happens if we deliberately accept more volatility within defined boundaries?*
- *Which assumptions drive tail escalation—and are they controllable?*
- *How does the decision profile change under alternative scenario weightings?*

These questions do not require recalculation, but reinterpretation.

Explanation depth can be adapted:

- Analytically for risk specialists
- Condensed for executive briefings
- Decision-oriented for boards and committees

This creates a shared understanding of risk across governance layers.

Such capabilities are particularly relevant for investment decisions in areas such as clean energy systems, infrastructure resilience, workforce safety, and resource efficiency—domains explicitly addressed by several UN Sustainable Development Goals, but only governable once uncertainty is made transparent and interpretable.

10.4.6 From Probabilistic Analysis to Auditable Decision Records

Interpretation must not become a black box.

Well-designed AI systems:

- Document assumptions and scenario logic
- Record questions, explanations, and conclusions
- Preserve approved analyses as immutable decision artifacts
- Clearly distinguish machine-generated insight from human approval

This closes a long-standing governance gap: probabilistic analysis becomes **decision-ready and auditable** without sacrificing analytical depth.

10.4.7 The Role of the PPP-Lens for Decision-Making: Decision Anchor, Not Permanent Lens

Within this chapter, the PPP-Lens for Decision-Making is not used as a constant framing device, but as a **targeted decision anchor**:

- Where do acceptable risk corridors lie in terms of operational stability and safety?
- At which loss levels do environmental or systemic impacts become unacceptable?
- How much financial volatility can be absorbed without impairing strategic resilience?

10.4.8 Conclusion

AI improves governance not by accelerating decisions, but by **improving understanding of uncertainty**.

When applied correctly, AI becomes an unrelenting preparation and interpretation layer: structuring, explaining, and documenting. Judgment, accountability, and steering remain firmly human.

In this way, the PPP-Lens for Decision-Making acts as a translation layer between abstract sustainability objectives—such as the UN SDGs—and concrete, risk-informed management decisions, without turning governance into a reporting exercise.

10.5 The Integrated ICS: From Documentation to Assurance

10.5.1 Why Documentation Is Not Assurance

Internal Control Systems (ICS) are often described as the backbone of corporate governance. In practice, however, many ICS implementations remain largely documentary in nature. Policies are defined, control matrices are maintained, and testing cycles are executed—yet management confidence remains limited.

The core issue is not a lack of controls, but a confusion of purpose. Documentation demonstrates intent. Assurance demonstrates effectiveness.

An ICS that merely documents controls answers the question *"What should happen?"* An ICS designed for assurance answers a different, far more relevant question: *"Is the organization actually operating within the boundaries it has defined?"*

For organizations pursuing long-term resilience—particularly in capital-intensive, safety-critical, or sustainability-relevant environments—this distinction is decisive.

10.5.2 From Control Catalogues to Risk-Driven Control Design

Traditional ICS approaches often start with predefined control catalogues. Controls are selected, mapped to processes, and tested periodically. While this creates formal completeness, it rarely creates decision relevance.

An integrated ICS reverses this logic.

Control design starts with **quantified risk exposure**, not with templates. Risks that materially affect operational stability, system integrity, or strategic objectives determine:

- Whether a control is required
- How strong it must be
- How frequently its effectiveness must be evidenced.

This shift has two important consequences:

1. **Key controls replace control proliferation.**
   Fewer controls, but with clear linkage to material risk drivers.

2. **Control relevance becomes transparent.**
   Management understands *why* a control exists—not just *that* it exists.

In this way, the ICS becomes a direct extension of risk-informed decision-making rather than a parallel compliance structure.

10.5.3 Design Effectiveness Versus Operating Effectiveness

A recurring weakness in many ICS implementations is the disproportionate focus on design effectiveness. Controls are logically sound on paper but insufficiently embedded in daily operations.

Design effectiveness asks: *Would this control mitigate the risk if it were applied as intended?* Operating effectiveness asks a more demanding question: *Is the control actually applied, consistently and reliably, under real operating conditions?*

True assurance emerges only at the level of operating effectiveness. An integrated ICS therefore prioritizes evidence of execution over formal descriptions. This requires moving beyond static attestations toward observable behavior.

10.5.4 From Periodic Testing to Continuous Control Evidence

Periodic testing remains a legacy of audit-driven ICS models. Annual or semi-annual testing cycles provide point-in-time comfort but fail to capture dynamic risk environments.

Integrated ICS architectures increasingly rely on **continuous control evidence**:

- System logs
- Transaction data
- Approval workflows
- Process exceptions
- Sensor or operational signals

Instead of asking whether a control was tested, management can observe whether it is continuously functioning within defined parameters.

This approach transforms controls from administrative artifacts into live governance instruments. It also aligns naturally with AI-supported preparation and interpretation layers, as discussed in Chapter AI/ML in Risk and Controls.

10.5.5 ICS as an Assurance Layer for Management and Board

An assurance-oriented ICS is not designed to satisfy auditors—it is designed to support executive steering.

At board and executive level, the system should provide clear answers to questions such as:

- Where are we currently operating outside defined risk tolerances?
- Which risks are insufficiently mitigated despite existing controls?
- Where are early signals indicating a deterioration of control effectiveness?

When these questions can be answered continuously and transparently, the ICS becomes a confidence engine for management decisions—particularly in environments where safety, reliability, and long-term viability are critical.

10.5.6 Mini-Case: From Audit Finding to Structural Assurance

Consider a recurring audit finding related to vendor master data changes. In the traditional approach, remediation consists of updated policies, training sessions, and follow-up testing in the next audit cycle.

In an integrated ICS, the control is redesigned:

- All vendor changes are logged automatically.
- Approval workflows are enforced systemically.
- Deviations trigger immediate alerts.
- Evidence is continuously available to management and audit alike.

The finding does not disappear because it was "fixed," but because the underlying risk is structurally controlled. Assurance is embedded in the process itself.

10.5.7 The Role of the PPP-Lens for Decision-Making in Control Prioritization

Within an integrated ICS, the PPP-Lens for Decision-Making serves as a **prioritization logic**, not as a reporting overlay.

- **People:** Controls are strengthened where human safety, workload, or accountability are most exposed.
- **Planet:** Control intensity increases where operational failures would cause disproportionate environmental or systemic impact.
- **Profit:** Control effort is calibrated to avoid overcontrol while safeguarding liquidity, asset integrity, and strategic flexibility.

This ensures that control investments are aligned with what truly matters for organizational resilience.

10.5.8 Mini-Conclusion

An integrated ICS is not a documentation exercise. It is an assurance system.

By linking quantified risk exposure, continuous evidence, and executive interpretation, the ICS becomes a core governance capability—providing confidence that strategic decisions are executed as intended.

In doing so, it creates the operational foundation that allows organizations to pursue long-term objectives—including sustainability-related ambitions—without relying on labels or reporting alone.

10.6 Overall Conclusion: Reaching the 17 UN SDGs Through Governance That Actually Works

Sustainable development goals are not achieved by declarations. They are achieved by execution, repeatedly, under uncertainty, across functions, and under real constraints.

This chapter has argued a pragmatic thesis: **the path to reaching the 17 UN Sustainable Development Goals (SDGs) is not primarily a reporting challenge, but a governance and operating challenge.** The SDGs describe *what* societies and organizations should move toward. The missing piece in many organizations is the system that determines *how* such ambitions become consistent decisions, funded priorities, controlled execution, and auditable outcomes.

In other words: the SDGs require a management system that can **navigate trade-offs, quantify uncertainty, allocate resources, and prove effectiveness without** collapsing into buzzwords or bureaucracy. This is exactly the role of an integrated Governance–Risk–Control architecture.

10.6.1 From Labels to Decisions: Why "Sustainability" Fails Without Resilience

The chapter opened by challenging the common pattern of sustainability initiatives: strong language, weak steering. Labels and ratings may create comparability, but they rarely create operational clarity. When organizations face real dilemmas—safety investments, supply chain disruptions, grid volatility, cyber threats, quality escapes, regulatory shocks—boards do not decide in labels. They decide in **risk appetite, capital allocation, timing, and accountability**.

That is where resilience becomes the decisive bridge: **resilience is the mechanism that turns purpose into performance.** It does not compete with sustainability ambitions; it makes them governable. If an organization cannot steer through uncertainty, it cannot credibly pursue multi-year sustainability transformations that depend on stable execution and trust.

10.6.2 GRC That Scales: Turning Ambition into Repeatable Execution

Sustainable transformation scales only when governance scales with it. Most SDG-relevant initiatives—clean energy projects, industrial decarbonization, resource-efficient manufacturing, safe and fair operations—span entities, geographies, suppliers, and regulatory regimes. Without a scalable backbone, organizations end up with fragmented initiatives: pockets of excellence surrounded by inconsistency.

The chapter therefore positioned **integrated GRC** not as a compliance add-on, but as an operating architecture. The essential move is from parallel silos (risk, controls, audit) toward one integrated loop:

- Appetite defines boundaries and priorities.
- Risk transparency makes uncertainty visible.
- Controls translate mitigation into execution.
- Assurance closes the loop with evidence and improvement.

This loop is the "invisible infrastructure" that enables sustainability objectives to survive real-world complexity. It reduces governance friction while increasing governance precision—exactly what large-scale SDG-oriented transformations require.

10.6.3 Quantification: The Missing Language Between Sustainability Intent and Capital Decisions

A major reason why sustainability goals remain aspirational is that **they are not consistently translated into capital language**. Heatmaps may help to communicate, but they do not enable decisions. Real steering requires quantification: distributions, tails, dependencies, scenario logic, and the ability to compare choices under uncertainty.

This chapter emphasized that probabilistic methods—especially scenario-driven Monte Carlo simulation—are not academic sophistication. They are decision instruments. They allow boards to answer the questions that determine whether SDG-relevant investments happen or stall:

- What range of outcomes is plausible?
- Where is the real tail exposure?
- What is the resilience return on a control or investment?
- Which risks dominate the portfolio when aggregated?

Equally important, the chapter reframed how quantification should be done: **not as "single-value in, percentile out," but as "expert-driven scenarios in, distribution logic out."** That shift upgrades risk quantification from technical output to governance insight. It reduces the risk of "false precision" and increases the credibility of decisions.

10.6.4 AI/ML: Making Advanced Governance Usable at Scale

Even the best quantification fails if only specialists can interpret it. A central barrier to reaching long-term goals is therefore not data, but *translation*: converting complex signals and probabilistic outputs into decisions that executives can make quickly, confidently, and consistently.

This is where AI/ML becomes relevant—not as a technology debate, but as a governance capability. The chapter positioned AI as an augmentation layer that reduces GRC friction and strengthens decision quality:

- Automating repetitive documentation tasks
- Scanning external signals and mapping them to risk structures
- Interpreting Monte Carlo distributions in decision-ready language
- Generating auditable decision records and preserving governance traceability

The key governance principle remains unchanged: **AI prepares insight; humans remain accountable and decide.** Used this way, AI does not replace governance; it operationalizes governance at scale.

10.6.5 Integrated ICS: From Paper Controls to Assurance That Enables Trust

Finally, this chapter argued that long-term objectives—including SDG-related ambitions—require more than good intentions and smart analytics. They require **proof of execution**. That is the role of the integrated Internal Control System: not documentation, but assurance.

An assurance-oriented ICS transforms controls from administrative artifacts into operational evidence:

- Fewer, risk-driven key controls
- Continuous evidence rather than periodic snapshots
- Observable operating effectiveness rather than formal design descriptions
- Shared transparency for management and audit

This is where sustainability becomes measurable in practice: when the organization can show, with evidence, that decisions are executed within defined boundaries—and that deviations are detected, corrected, and learned from.

10.6.6 A Practical Answer to "How to Reach the 17 UN SDGs"

This chapter does not claim that organizations "implement SDGs" by mapping activities to 17 boxes. Instead, it provides a practical answer to the more relevant executive question:

How do we build a system that makes SDG-relevant objectives executable, steerable, and provable under real constraints?

The answer is: by operating sustainability ambition through an integrated governance backbone **Sense -> Quantify -> Act -> Assure** supported by scalable GRC architecture, credible quantification, AI-supported translation, and assurance-oriented controls.

In that sense, the SDGs are reached not by labeling, but by **repeatable management discipline**.

10.6.7 Overall Key Takeaways (III)

1. **SDGs require execution systems, not slogans.** Sustainable ambition becomes real only when governance can steer trade-offs and prove outcomes.
2. **Resilience is the mechanism that turns purpose into performance.** It is not "next to ESG"; it is what makes long-term goals governable under uncertainty.
3. **Integrated GRC is scalable infrastructure.** Appetite, risk transparency, controls, and assurance must work as one loop—not as parallel functions.
4. **Heatmaps communicate; quantification decides.** Boards need distributions, tail exposure, and scenario logic to allocate capital responsibly.
5. **Scenario-first Monte Carlo makes risk governance credible.** Expert-driven inputs and distribution logic outperform rigid percentile fixation.
6. **AI/ML makes advanced governance usable.** It reduces GRC friction, translates complex analysis into decision language, and preserves auditability—without delegating accountability.
7. **An integrated ICS is an assurance system, not a document set.** Continuous evidence and operating effectiveness create trust and control integrity.
8. **PPP-Lens for Decision-Making is a targeted decision anchor.** It supports explicit prioritization where People and Planet thresholds constrain Profit-oriented volatility—not as a constant rhetorical frame.
9. **Reaching SDGs is a governance capability.** Organizations progress when they can sense signals, quantify uncertainty, act decisively, and assure effectiveness repeatedly.

Further Reading

United Nations & Sustainability Frameworks

United Nations. *Transforming our World: The 2030 Agenda for Sustainable Development.* UN General Assembly Resolution A/RES/70/1, 2015. Available at: https://sdgs.un.org/2030agenda (Accessed: 21 December 2025)

United Nations Framework Convention on Climate Change (UNFCCC). *The Paris Agreement.* **Available at: https://unfccc.int/process-and-meetings/the-paris-agreement/the-paris-agreement (Accessed: 21 December 2025)**

Governance, Risk Management, & Internal Control

Committee of Sponsoring Organizations of the Treadway Commission (COSO). *Enterprise Risk Management—Integrating with Strategy and Performance.* COSO, 2017.

Committee of Sponsoring Organizations of the Treadway Commission (COSO). *Internal Control—Integrated Framework.* COSO, 2013.

International Organization for Standardization (ISO). *ISO 31000:2018—Risk management—Guidelines.* **Geneva: ISO.**

International Organization for Standardization (ISO). *ISO 37301:2021—Compliance management systems—Requirements with guidance for use.* **Geneva: ISO.**

Quantitative Risk Analysis & Monte Carlo Simulation

Bara, M. (2025). *Enhanced Monte Carlo simulation for project risk analysis: integrating cost and schedule impacts with time-shifted risks and dependency modeling.* **Applied Operations and Analytics, 1(1), 1-15. https://doi.org/10.1080/29966892.2025.2552675**

AI / Machine Learning in Risk, Controls, & Governance

National Institute of Standards and Technology (NIST). *Artificial Intelligence Risk Management Framework (AI RMF 1.0).* **NIST, 2023. Available at: https://www.nist.gov/itl/ai-risk-management-framework (Accessed: 21 December 2025)**

International Organization for Standardization / International Electrotechnical Commission (ISO/IEC). *ISO/IEC 42001:2023—Artificial intelligence management system (AIMS).* **Geneva: ISO.**

- **Organisation for Economic Co-operation and Development (OECD).** *OECD Principles on Artificial Intelligence.* OECD, 2019. Available at: https://oecd.ai/en/ai-principles
  (Accessed: 21 December 2025)

Christoph Schade is a Senior Manager for Risk & Internal Control at **thyssenkrupp nucera**, where he is responsible for the global Enterprise Risk Management and Internal Control framework across the organization.

He brings more than 17 years of leadership experience, and over 9 years in Governance, Risk Management, and Internal Control. His work focuses on translating uncertainty into decision-relevant insights for executive management and supervisory bodies, particularly in capital-intensive and safety-critical environments. Christoph combines scenario-based risk analysis, Monte Carlo simulation, and assurance-oriented control design to support strategic decision-making beyond reporting and compliance.

A key element of his work is the digital transformation of GRC processes through automation, data analytics, and AI-supported early warning and interpretation systems. Prior to his current role, he led the internal audit department and held various leadership positions in operational and commercial functions.

Christoph holds professional certifications in internal auditing (IIA, DIIR) and completed advanced management studies at Steinbeis University, Berlin. His work focuses on building resilient governance capabilities that enable risk-informed, data-driven decision-making in complex organizational settings.

Remote Monitoring Advancing Sustainability in the MedTech Industry

11

Tom Günther

Abstract

This chapter explores how digital transformation in the MedTech sector can simultaneously improve patient care and enable measurable sustainability impact aligned with the United Nations Sustainable Development Goals. Remote monitoring in cardiac care is introduced as a digital service model that is reshaping care delivery processes within a highly regulated and resource-intensive industry. By replacing physical service interactions with connected, data-driven workflows, remote monitoring leads to measurable reductions in CO_2e emissions while enhancing clinical quality. It is highlighted that digitalization and sustainability can mutually reinforce one another by increasing process efficiency and making environmental impacts transparent. Applying the Three-Pillar Model, consisting of Sustainable Purpose, Traveling Organization and Connectivity, illustrates that sustainable value creation in digital health extends beyond technological efficiency alone. To embrace sustainability as an essential element of digital services, alignment around purpose-driven, adaptable, and cross-organizational models is needed. Overall, remote monitoring is positioned as a concrete example of digital health as a strategic enabler of sustainable healthcare systems.

11.1 Sustainability and Digitalization in MedTech Industry

The medical technology industry stands in a pivotal position between health innovations and environmental responsibility. Accounting for around 10 percent of most countries' gross domestic product (GDP), the healthcare and MedTech sector plays

T. Günther (✉)
BIOTRONIK SE & Co. KG, Berlin, Germany
e-mail: tom.guenther@biotronik.com

© The Author(s), under exclusive license to Springer Nature Switzerland AG 2026

P. Wollmann et al. (eds.), *The Sustainable Organization – Volume 3*, Future of Business and Finance, https://doi.org/10.1007/978-3-032-28949-0_11

an important social and economic role. But at the same time, the global healthcare system contributes approximately 4.4 percent of total greenhouse gas emissions.[1] Key drivers include energy-intensive operations, mobility and material consumption, and medical devices. With ongoing demographic change and increasing care demands, experts expect further expansion of healthcare systems which could triple their sectoral emissions by 2040.[2] This highlights the dual responsibility of the MedTech industry to balance healthcare innovation with environmental sustainability.

Meanwhile, regulatory frameworks such as the EU Green Deal, the Corporate Sustainability Reporting Directive (CSRD), and ESG (Environmental, Social, and Governance) standards are encouraging organizations to integrate sustainability into their core strategies and reporting practices. At the same time, society expects innovation in MedTech to be both efficient and environmentally friendly. For this purpose, the United Nations Sustainable Development Goals (SDG) provide targets for ethical and future-oriented development which connects business strategies with broader goals for peace, prosperity, and planetary protection.[3] These goals influence national policies and reshape strategic orientations across industries. For MedTech companies, those regulations are setting clear expectations for sustainable transformation within the sector.

In this context, digitalization constitutes a key enabling mechanism for sustainability by offering measurable reductions of greenhouse gas emissions while improving process efficiency. The integration of digital health solutions, such as telemedicine alternatives, allows a decrease of unnecessary patient transport and travel, leading to decarbonization.[4] However, those changes require a new approach to patient care models to emphasize more data-driven and resource-efficient models. Digitalization serves to support a transition toward a low-carbon healthcare infrastructure.

Aligning technical innovations with UN SDGs can demonstrate both corporate responsibility and social commitment of MedTech companies. A sustainable-oriented and balanced industry position is essential for a greener and more resilient healthcare system. In MedTech, sustainability can therefore represent a shared global responsibility and key component of future healthcare transformation.

[1] Or, Z., & Seppänen, A.-V. (2024). The role of the health sector in tackling climate change: A narrative review. Health Policy, 143, 1–9.

[2] Ibid.

[3] United Nations. (2024). The Sustainable Development Goals report 2024. United Nations.

[4] Purohit, A., Smith, J., & Hibble, A. (2021). Does telemedicine reduce the carbon footprint of healthcare? A systematic review. Future Healthcare Journal, 8(1), e85–e91.

11.2 Sector Dynamics and Environmental Pressures

11.2.1 Industry Outline

The medical technology (MedTech) industry is a specialized segment of the healthcare sector and creates technologies, devices, and software used for diagnosis, prevention, treatment, monitoring, and outpatient care. It combines engineering and medicine to create equipment such as pacemakers or MRI scanners to enhance medical procedures and patient care. Its primary goal is to improve healthcare delivery and patients' quality of life.[5]

But in recent years, the industry has experienced a structural change from hardware-driven product innovations to connected service models. On-site care is increasingly replaced by cloud-based diagnostic and digital platforms.[6] Therefore, MedTech is redefining its orientation. Value creation increasingly shifts from high-performance devices toward data-driven services and interconnected care ecosystems. These are expected to lead to higher clinical quality and lower environmental impacts. This change aims to address operational efficiency and decarbonization potential in the sector.

11.2.2 Measuring its Environmental Impact

Along with frameworks for ESG accounting and rising social pressure to embrace sustainability, other major barriers can be identified. One of the biggest challenges is the high energy and material consumption of the sector, as many products require specialized materials, sterilization, and resource-intensive manufacturing processes. Nearly 70–80% of the total industry footprint is caused along its value chain.[7] Particularly logistics, supplier activities, and service-related mobility are primary sources. These indirect emissions are difficult to assess due to limited data availability and dependence on external service providers.

Additionally, it is still challenging to apply consistent sustainability strategies due to insufficient integration of ESG structures and large amounts of waste from single-use products. Many MedTech companies still do not have standardized internal processes or a lifecycle-based product design. They still view sustainability more as an add-on than as a strategic driver.[8]

---

[5] MedTech Europe. (n.d.). What is medical technology? Retrieved November 21, 2025, from https://www.medtecheurope.org/about-the-industry/what-is-medical-technology/

[6] Ferrick, A. M. et al. (2023). 2023 HRS/EHRA/APHRS/LAHRS expert consensus statement on practical management of the remote device clinic. Europace, 25, 1–44.

[7] Or, Z., & Seppänen, A.-V. (2024). The role of the health sector in tackling climate change: A narrative re-view. Health Policy, 143, 1–9.

[8] Sepetis, A., Rizos, F., Pierrakos, G., Karanikas, H., & Schallmo, D. (2024). A sustainable model for healthcare systems: The innovative approach of ESG and digital transformation. Healthcare, 12(2), 156, 1–18.

Increasing stakeholder expectations present a challenge to the sector. Hospitals increasingly expect low-carbon service models and patients associate quality of care with responsible resource use. Investors, on the other hand, view ESG performance as essential for long-term competitiveness.[9] Thus, sustainable transformation depends on this standardized data and robust impact measurement. Without consistent metrics and transparent emission baselines, companies cannot manage decarbonization pathways or align innovation with sustainability objectives.

11.2.3 Key Sustainability Challenges

To assess the emissions generated by products or processes, organizations require comparable accounting frameworks. The Greenhouse Gas Protocol (GHG) of the World Resources Institute and World Business Council for Sustainable Development is internationally recognized as the leading guideline for emissions assessments. It divides emissions into three categories and offers standardized rules for capturing direct and indirect emissions to help organizations define, track, and evaluate climate targets.[10]

Scope 1 covers all direct greenhouse gas emissions generated within a company. The company owns or controls the corresponding emission sources, such as company facilities or service vehicles.[11]

Emissions from purchased energy or external service providers are distributed among the other influencing factors. Scope 2 includes indirect emissions from the production of energy that has been purchased or supplied (such as heat, cooling, and electricity). Since the company only uses the energy and does not produce it, it is not responsible for these emissions.[12]

Scope 3 includes all other indirect emissions along the value chain. This covers all CO_2e emissions from upstream or downstream areas of the company's business activities. This also includes emissions that the company has indirect influence on, but do not originate from the company's own property or control, such as supplier activities, patient transport, or a product's use phase.[13]

[9] Sepetis, A., Rizos, F., Pierrakos, G., Karanikas, H., & Schallmo, D. (2024). A sustainable model for healthcare systems: The innovative approach of ESG and digital transformation. Healthcare, 12(2), 156, 1–18.

[10] World Resources Institute, & World Business Council for Sustainable Development. (2004). The greenhouse gas protocol: A corporate accounting and reporting standard (Revised ed.). World Resources Institute.

[11] Ibid.

[12] Keil, M., & Grün, L. (2024). Greenhouse gas emissions of an outpatient care service: A cost-based approach. Zeitschrift für Evidenz, Fortbildung und Qualität im Gesundheitswesen, 175, 90–95.

[13] World Resources Institute, & World Business Council for Sustainable Development. (2004). The greenhouse gas protocol: A corporate accounting and reporting standard (Revised ed.). World Resources Institute.

In the MedTech industry, Scope 3 emissions are particularly relevant. They account for a majority of the total sector footprint, driven largely by logistics-intensive supply chains, service-related mobility, and energy-intensive material flows.[14] Even aside from difficulties in quantification, they remain essential for understanding the true environmental impact of MedTech operations.

Frameworks like the GHG protocol play a crucial role in meeting requirements of ESG guidelines as the CSRD. To evaluate the environmental impacts across the entire product lifetime, additional methods like the Life Cycle Assessment (LCA) need to be considered. Even though this requires more data and accounting models, it also offers a more holistic assessment of device and service sustainability.[15]

11.2.4 Remote Monitoring as a Digital Care Model

Measurement standards outlined above highlight that a significant amount of the environmental footprint in MedTech is caused within their value chain from transport- and service-related Scope 3 activities. Therefore, the sector needs to find solutions to lower emissions without compromising clinical quality and operational reliability.

Digitalization opens opportunities for processes to become more efficient in many ways. Unnecessary on-site activities are often replaced by digital alternatives, optimizing processes and reducing greenhouse gas emissions. Remote monitoring (RM) represents a relevant example of how digitalization can reduce environmental impacts within the MedTech industry. It represents an approach for lowering emissions associated with physical service interactions by enabling secure digital transmission of data from implantable cardiac devices. This allows clinicians to access diagnostic information without requiring on-site technician visits.[16] Replacing travel-intensive services with connected digital processes targets the Scope 3 emissions that dominate the MedTech sector's footprint.

Digital service approaches like remote monitoring are able to reshape established workflows, deliver environmental benefits, and enhance process efficiency. As MedTech companies increasingly aim to operate more climate-conscious and resource-efficient, remote monitoring approaches highlight how digitalization can become a catalyst for a more sustainable MedTech ecosystem.

[14] Bawa, D. et al. (2023). Impact of remote cardiac monitoring on greenhouse gas emissions: Global Cardiovascular Carbon Footprint Project. JACC: Advances, 2(3), 100286.

[15] Schmidt, L., & Bohnet-Joschko, S. (2022). Planetary health and hospitals' contribution: A scoping review. International Journal of Environmental Research and Public Health, 19(20), 13536.

[16] Slotwiner, D. J. et al. (2015). HRS expert consensus statement on remote interrogation and monitoring for cardiovascular implantable electronic devices. Heart Rhythm, 12(7), e69–e100.

11.3 Digital Transformation Toward Remote Monitoring in Cardiac Care

11.3.1 Clinical Context

A substantial share of environmental impact in the MedTech sector is produced by travel-intensive service processes and logistical dependencies. Cardiac care provides only one example of this challenge.

The field of Cardiac Rhythm Management (CRM) relies on cardiac implantable electronic devices (CIED). That includes devices like pacemakers, implantable cardioverter-defibrillators, or implantable cardiac monitors which have different roles in stabilizing heart rhythm, preventing arrhythmias, and monitoring cardiac activity.[17] In emergencies or even in regular checks, those devices need to be interrogated. Regular assessments of recorded arrhythmias, lead performance, and pacing thresholds are mandated not only for medical reasons but also to mitigate liability risks.

11.3.2 From On-Site Interrogation to Remote Monitoring

Traditionally, these interrogations follow a travel-based service workflow. When a patient presents in a hospital or emergency department, a trained technician must be physically present, bringing dedicated programming equipment in order to retrieve diagnostic data from implanted devices. Therefore, those CIED interrogations require time-consuming coordination, technician availability, and travel which leads to delayed clinical interventions, higher waiting times, and consumption of unnecessary resources. Moreover, every service trip contributes to Scope 3 mobility emissions, making it one of the largest drivers in the total footprint of the MedTech industry.

Nowadays, remote monitoring has become a central component of cardiac care which is supported by the digitalization of health systems and connected medical devices. As part of remote monitoring, remote interrogation enables secure transmission of data from CIEDs, reducing the need for in-clinic visits.

Modern CIED ecosystems consist of remote monitoring processes. In cases where clinical staff must review device parameters, remote interrogation allows hospital teams to perform it directly through a digital interface without the need for specialized training or on-site technicians. CIEDs are assessed via secure wireless telemetry with automated data transmission to a cloud infrastructure.[18] In parallel, remote monitoring workflows automate crucial parts of device follow-ups. CIEDs are able to periodically transmit device data from a patient's home environment via a home transceiver which forwards the encrypted information to a cloud-based

---

[17] Maleki, M., Alizadehasl, A., & Haghjoo, M. (2018). Practical cardiology (1st ed.). Elsevier.

[18] Slotwiner, D. J. et al. (2015). HRS expert consensus statement on remote interrogation and monitoring for cardiovascular implantable electronic devices. Heart Rhythm, 12(7), e69–e100.

platform. Healthcare professionals can access this data, allowing ongoing patient oversight without any on-site interrogations.[19]

Instead of coordinating technician availability, equipment transport, and manual data extraction, hospitals receive structured device diagnostics which include arrhythmia episodes, lead integrity parameters, pacing thresholds, and battery status of CIEDs. Remote monitoring improves patient care processes by providing automated data uploads and real-time alerting. Therefore, in-person routine check-ups are no longer needed, and a triage of abnormal cardiac events can happen immediately. Those changes mark a fundamental transformation of traditional and travel-dependent healthcare processes.

11.3.3 Sustainability Impacts and Emission Reduction Potential

A major share of emissions in cardiac care arises from travel-based Scope 3 emissions. Each on-site visit traditionally required a trained technician and therefore produced mobility-related greenhouse gas emissions. By replacing these trips with digital access to device diagnostics, remote interrogation removes the need for physical travel and therefore avoids unnecessary fuel consumption and CO_2e output.[20]

Each avoided in-person interrogation results in measurable emission reductions. Online consultations for device check-ups can save between 0.7 and 372 kg of CO_2e per session.[21] Considering that over a million CIEDs are implanted each year, the total potential to reduce the carbon footprint of cardiac follow-up care is substantial.

This shift toward remote processes also improves operational efficiency. Routine follow-ups can be automated, reducing administrative workload and freeing up staff capacities for other medical priorities. Remote monitoring can make clinical workflows faster, more predictable, and easier to scale in the context of growing patient volumes.[22]

Furthermore, remote monitoring can enhance the clinical quality of cardiac rhythm management. Continuous data uploads enable clinicians to detect abnormalities earlier than with traditional in-person follow-ups.[23] This earlier visibility supports proactive interventions and can therefore lower the risk of complications.

[19] Ferrick, A. M. et al. (2023). 2023 HRS/EHRA/APHRS/LAHRS expert consensus statement on practical ma-nagement of the remote device clinic. Europace, 25, 1–44.

[20] Nissinen, A. et al. (2025). The climate impacts of healthcare digitalization: A scoping review. Digital Health, 11, 1–11.

[21] Purohit, A., Smith, J., & Hibble, A. (2021). Does telemedicine reduce the carbon footprint of healthcare? A systematic review. Future Healthcare Journal, 8(1), e85–e91.

[22] Slotwiner, D. J. et al. (2015). HRS expert consensus statement on remote interrogation and monitoring for cardiovascular implantable electronic devices. Heart Rhythm, 12(7), e69–e100.

[23] Hindricks, G. et al. (2014). Implant-based multiparameter telemonitoring of patients with heart failure (IN-TIME): A randomised controlled trial. The Lancet, 384(9943), 583–590.

As a result, unnecessary hospital visits are avoided and the quality and safety of patient care are improved.

11.3.4 Organizational Implications and Workforce Transformation

As digital processes replace travel-dependent services, healthcare institutions and MedTech businesses must adapt their workforce, governance, and performance systems to provide continuous data-driven care. So, this shift requires new skills and redesigned collaboration models.

Firstly, new digital competencies across clinical teams must be developed. To work with dashboards, handle secure data flows, and interpret clinical alerts, healthcare professionals require strong digital literacy. Through the integration of digital workflows into daily routines, clinical roles are transformed into more connected and data-driven care delivery.[24]

Remote monitoring also strengthens collaboration between different professional groups. Aligning the needs of cardiologists and emergency clinicians with IT specialists and MedTech service staff can support creating dedicated remote care units. These interdisciplinary structures help embed digital device management into existing clinical processes.[25] Therefore, it fosters connected and resilient partnerships between manufacturers and care providers.

Nevertheless, digital service processes generate data that can be used to track sustainability performance. Indicators such as avoided travel distance, CO_2e savings, and the share of remote versus on-site services support ESG and CSRD reporting. By embedding these metrics in management and innovation decisions, sustainability is established as a part of digitally supported care.[26]

Remote monitoring therefore acts as a catalyst for organizational transformation. It influences how teams work, communicate, and measure the value of modern cardiac care. It also aligns with UN Sustainable Development Goals and supports the sector's transition toward more efficient and low-emission service models.

11.3.5 Outlook: Remote Programming in Future Patient Care

Remote programming (RP) represents the next stage in the evolution of digital CIED care, expanding remote monitoring (RM) from one-way data transmission to bidirectional therapy management. Programming involves actively adjusting device

---

[24] Varma, N. et al. (2023). Remote monitoring of cardiac implantable electronic devices and disease management. Europace, 25(9), 1–11.

[25] Slotwiner, D. J. et al. (2015). HRS expert consensus statement on remote interrogation and monitoring for cardiovascular implantable electronic devices. Heart Rhythm, 12(7), e69–e100.

[26] Nissinen, A. et al. (2025). The climate impacts of healthcare digitalization: A scoping review. Digital Health, 11, 1–11.

parameters such as pacing strategies, detection thresholds, or therapy delivery to ensure that implants operate optimally for a patient's condition.[27] While remote monitoring enables interrogation and data upload, remote programming extends the process toward direct therapeutic modification based on real-time clinical signals. The technical concept relies on authenticated, encrypted communication between the implanted device and the healthcare provider, allowing verified and traceable parameter changes.[28] Although still in clinical pilot programs, remote programming is already recognized as an emerging approach that broadens the scope of digital CIED care.[29]

The technological foundations for remote programming were initially established through remote monitoring infrastructures and are now evolving into fully bidirectional systems. Clinically, remote interventions enable real-time response to device events, for example the immediate correction of inappropriate shocks or unstable detection settings. Early clinical experience demonstrates that these interventions can be safely executed remotely.[30]

Sustainability benefits result from eliminating technician travel for routine programming, which reduces a dominant Scope 3 emission driver in current service models. The primary advantage over remote monitoring is the ability to intervene immediately based on device data rather than waiting for a scheduled visit or a physical programmer. Strategically, remote programming advances MedTech toward decentralized, data-driven therapy management, positioning virtual care as the future standard for CIED services.[31]

11.4 Embedding Sustainability in Digital Healthcare

Twin transformation refers to the combined development of digitalization and sustainability, where both reinforce one another rather than progress separately. By creating transparency, improving process efficiency, and making environmental impacts measurable, digital technologies are able to promote sustainability.[32] At the same time, sustainability targets digital innovation and investments toward

[27] Clarke, J.-R. D., & Kramer, D. B. (2023). Remote interrogation and reprogramming for cardiac implantable electronic devices. Heart Rhythm, 20(4), 553–556.

[28] Ferrick, A. M. et al. (2023). 2023 HRS/EHRA/APHRS/LAHRS expert consensus statement on practical ma-nagement of the remote device clinic. Europace, 25, 1–44.

[29] Seiler, A., Rosemas, S., Zhou, J., Franco, N., & Rogers, J. (2024). Utilization of remote reprogramming to manage insertable cardiac monitor arrhythmia alert burden. Journal of Cardiovascular Electrophysiology, 35, 1–5.

[30] Ibid.

[31] Ferrick, A. M. et al. (2023). 2023 HRS/EHRA/APHRS/LAHRS expert consensus statement on practical ma-nagement of the remote device clinic. Europace, 25, 1–44.

[32] Siedschlag, I., Mohan, G., & Yan, W. (2024). Twin transitions across enterprises: Do digital technologies and sustainability go together? Journal of Cleaner Production, 481, 144025, 1–17.

solutions with lower resource intensity and long-term climate alignment.[33] In this way, twin transformation provides a strategic framework for advancing the UN Sustainable Development Goals (SDGs) by linking digital progress with environmental and social impact.

The core mechanisms include[34]:

- **Data-driven decision-making** on energy use, emissions, and resource distribution across the value chain.
- **Process automation** by reducing travel, manual work, and waste in service delivery.
- **Platform technologies** that connect stakeholders within and across sectors and reduce system fragmentation.
- **Strategic alignment**, where ESG goals shape the direction of digital innovation.

For the MedTech sector, twin transformation is particularly relevant due to its highly regulated and resource-intensive nature. It shows how linking digital innovation with sustainability performance can become a central strategic focus for the future development of MedTech. The remote monitoring use case operationalizes the core mechanisms of twin transformation, linking clinical performance improvements with measurable climate impact through structural system redesign.[35] By removing physical presence for diagnostics, remote monitoring eliminates on-site requirements and travel-related emissions from routine follow-up interrogations. This directly supports the UN SDG 13 for Climate Action.

Cloud-based analytics further reduce environmental impact by eliminating repeated logistics loops, including the transport of diagnostic equipment and consumables.[36] This enables a transition from resource-intensive to information-intensive models, replacing material with data flow.

Additionally, improvements in patient care can be seen as a contribution to UN SDG 3 for Good Health and Well-Being. Real-time diagnostics allow preventive interventions that reduce emergency visits, inefficient staff mobilization, and resource waste while improving patient outcomes.

Connectivity also functions as a decarbonization lever. Implanted devices act as distributed diagnostic components, enabling continuous data transmission without travel. Secure digital infrastructures allow remote service models to scale and

[33] Vermeulen, B. (2024). The twin digital and green transition: Paradigm shift or tech fix?. artec Research Center for Sustainability, University of Bremen, 231.

[34] Ernst & Young GmbH, & Fraunhofer Institute for Industrial Engineering IAO. (2023). Building a digital and sustainable future: How companies can pioneer twin transformation. EY.

[35] Muench, S., et al. (2022). Towards a green and digital future: Key requirements for successful twin transitions in the European Union (EUR 31075 EN). Publications Office of the European Union, JRC129319.

[36] Siedschlag, I., Mohan, G., & Yan, W. (2024). Twin transitions across enterprises: Do digital technologies and sustainability go together? Journal of Cleaner Production, 481, 144025, 1–17.

reduce dependence on manual, travel-based workflows.[37] Moreover, it creates a stronger connection between the clinic, patients, and service providers as stakeholders.

The ecological benefits of remote monitoring are measurable. CO_2e reductions per interrogation can be quantified through avoided mileage and standardized emission factors. Remote monitoring has the potential to deliver CO_2e savings of up to 300 kg per session. Reduced fuel demand and material use further decrease the environmental footprint produced by processes of the MedTech sector.

Overall, remote monitoring exemplifies twin transformation by reducing environmental impact while improving clinical quality, creating a reinforcing cycle between Health & Well-Being (SDG 3) and Climate Action (SDG 13).

11.5 Applying the Three-Pillar Model to Remote Monitoring

The Three-Pillar Model describes a framework to assess how organizations can pursue innovation and adaptability while providing long-term sustainability outcomes. The model is divided into three different pillars: Sustainable Purpose, Traveling Organization, and Connectivity. In the highly regulated and resource-intensive MedTech sector, this framework is particularly suitable for analyzing how digital health solutions generate sustainable value beyond their technological improvements.

11.5.1 Sustainable Purpose

Sustainable Purpose defines the vision and mission of an organization. It provides orientation, aligns collective action, and creates identity, especially in times of crisis and transformation. It links economic success with long-term sustainability and societal contribution.[38]

In the MedTech context, this purpose emerges from the integration of patient benefit, organizational operability, and environmental responsibility due to the high emission rates. Remote monitoring aligns these dimensions by improving clinical outcomes while simultaneously reducing the environmental footprint of cardiac care. It enhances patient safety and reaction speed by enabling continuous oversight and earlier intervention. At the same time, it reduces travel-related emissions and resource consumption by eliminating the need for routine on-site visits. Consequently, remote monitoring contributes directly to UN SDG 3 (Good Health & Well-Being) and UN SDG 13 (Climate Action). Sustainability thus becomes an integral part of the service logic rather than an external objective. This

[37] Ernst & Young GmbH, & Fraunhofer Institute for Industrial Engineering IAO. (2023). Building a digital and sustainable future: How companies can pioneer twin transformation. EY.

[38] Wollmann, P., Kühn, F., Kempf, M., & Püringer, R. (Eds.). (2023). Organization and leadership in disruptive times: Design and implementation of the 3-P model. Springer Nature.

purpose-driven digital innovation reflects the organization's commitment to delivering state-of-the-art patient care while reducing environmental impact.

11.5.2 Traveling Organization

The concept of the Traveling Organization captures the ability of organizations to continuously adapt and learn while navigating in uncertain and evolving transformation paths. While developing innovations, this permanent state of change should embrace agility and decision-making.[39]

As in many industries, digitalization reshapes workflows and processes to enhance efficiency. In cardiac care, this transformation takes place as technological innovation and regulations while sustainability expectations evolve in parallel. Remote monitoring reflects an organization's ability to operate in this dynamic environment through agility and a strong learning orientation. Continuous improvement of digital workflows, supported by real-time clinical and operational data, enables organizations to improve care processes and respond to emerging needs. Moreover, the evolution from remote monitoring toward remote programming reflects growing acceptance of digital care and enables organizations to expand their service models. The integration of sustainability metrics such as CO_2e emissions into performance management further characterizes the Traveling Organization. Therefore, sustainability is embedded in decision-making and innovation, translating market signals and demands into scalable digital solutions.

11.5.3 Connectivity

Connectivity enables collaboration and exchange across organizational boundaries and stakeholders, supporting shared learning and coordination in sustainability-oriented change.[40]

In digital cardiac care, connectivity functions as a structural enabler of sustainable service delivery. Via secure cloud-based connections, hospitals, clinicians, and service experts are linked, enabling real-time data access and low-emission care delivery. At the ecosystem level, those collaborations of healthcare providers are essential for standard development and the safe scaling of digital services. Partnerships that align stakeholders around common sustainability goals contribute to UN SDG 17 (Partnerships for the Goals) and therefore boost the achievements of SDG 3 and SDG 13.

The application of the Three-Pillar Model demonstrates that remote monitoring and its further development toward remote programming represent more than a technological upgrade. The use case from MedTech is able to generate sustained clinical, organizational, and environmental value. This highlights the potential of

[39] Wollmann, P., Pemler, D., & Ndrevataj, M. (Eds.). (2023). The sustainable organization: How organizations address the 17 UN SDGs using the 3-P model (Future of Business and Finance). Springer Nature.

[40] Ibid.

digital health solutions to support long-term stability in the MedTech sector while contributing to improved patient outcomes and broader sustainability goals.

11.6 Conclusion and Takeaways

Remote monitoring in cardiac care highlights how digital health solutions address multiple challenges in healthcare delivery. By replacing travel-intensive, on-site service processes with digitally enabled care models, remote monitoring is able to improve clinical outcomes while reducing resource consumption and emissions. These effects illustrate the relevance of digital care models for sustainability-oriented change in the MedTech sector.

The transition from on-site service delivery to remote monitoring can directly reduce Scope 3 emissions across healthcare value chains while maintaining or enhancing clinical quality and operational coordination. Looking ahead, the evolution toward remote programming points to the next phase of digital cardiac care. This expands the scope of remote services from observation to therapeutic intervention, enabling an even greater contribution to the sustainable development of the sector.

Recognizing digitalization and sustainability as reinforcing elements helps explain how both become mutually enabling dimensions of organizational change. Applying the Three-Pillar Model further clarifies that sustainable value creation extends beyond technological efficiency alone. It demonstrates how an alignment between sustainable purpose, adaptive operating models, and cross-organizational connectivity shapes the sustainability impact of digital health solutions. When digital services are embedded in these dimensions, sustainability becomes an integral element of service models and management. Furthermore, those sustainable innovations contribute to several UN Sustainable Development Goals by aligning digital healthcare delivery with broader societal and environmental priorities.

Overall, integrating digital health solutions into core service delivery processes contributes to improved patient outcomes, more efficient healthcare operations, and reduced environmental impact. By embedding digital care models within organizational structures and healthcare ecosystems, remote monitoring lays the foundation for a resilient and sustainable health system, setting the direction for future-oriented and connected models of patient care.

Tom Günther is Product Manager at BIOTRONIK, a leading manufacturer of cardiac medical technology. With several years of experience in the MedTech industry, he specializes in data infrastructure and services within Cardiac Rhythm Management. He holds a Bachelor of Science in Business Information Systems from the Berlin School of Economics and Law (HWR Berlin). Working with implantable devices and clinical systems has deepened his understanding of how medical technology improves patient care and long-term health outcomes. His professional interests are driven by the transition toward a more sustainable health ecosystem.

Driving the Circular Economy: How Multi-Cycle Business Models Turn Circularity into Competitive Advantage

12

Martin Hewing and Phillip Brandts

Abstract

Climate change and resource volatility increasingly expose the fragility of linear production models. At the same time, sustainability policy—particularly in Europe—is shifting from compliance expansion toward competitiveness, raising the question of whether circular economy strategies can deliver economic advantage beyond regulation. This chapter argues that circular economy implementation becomes economically viable when circularity is embedded into core business models and treated as a transformation challenge. Building on lessons from digital transformation, it shows how firms can manage uncertainty, delayed returns on investment, and ecosystem interdependence through coherent product design, multi-cycle business models, and financing structures aligned with long-term value creation. Empirically, the chapter presents a qualitative case study of a German mid-sized water-meter manufacturer that has implemented a circular product service over more than two decades. The findings indicate that the main constraints on circular economy implementation are organizational and temporal, rather than technological or economic.

12.1 Introduction

The circular economy has moved from a niche sustainability discourse to a central element of industrial strategy. Climate change, biodiversity loss, and resource depletion have exposed linear "take–make–waste" models as increasingly fragile. At the same time, the European policy environment has entered a phase of political recalibration. Central initiatives such as the Ecodesign for Sustainable Products

M. Hewing (✉) · P. Brandts
Berlin, Germany
e-mail: ma.hewing@gmail.com; phillip.brandts@gmail.com

© The Author(s), under exclusive license to Springer Nature Switzerland AG 2026

P. Wollmann et al. (eds.), *The Sustainable Organization – Volume 3*, Future of Business and Finance, https://doi.org/10.1007/978-3-032-28949-0_12

Regulation (ESPR) are being operationalized through work plans and implementation structures; other elements of the sustainability regulatory architecture have been delayed, simplified, or narrowed in scope in response to competitiveness concerns and perceived economical disadvantages.

This dual movement—continuation of circular product policy but softened corporate obligations—creates a risk of misinterpreting circularity as primarily a compliance exercise. The more consequential question for organizations is therefore not whether regulation compels circularity, but whether circularity can be structured as a superior operating model—one that reduces long-run costs, stabilizes margins under input volatility, and thereby reduces risks in fragile global supply chains.

Despite growing attention, implementation remains limited. Many corporate initiatives remain confined to pilots or isolated product lines and fail to scale because they are decoupled from the core business logic, lack traceability and operational infrastructure, or remain organizationally siloed—more as a burden than a promising route for value creation.

Accordingly, this chapter reframes the circular economy transition as an organizational transformation problem that can be accelerated through lessons from digital transformation—specifically, business model innovation, product and lifecycle design, and ecosystem coordination. The chapter focuses on ways in which profitability and sustainability reinforce each other even without regulatory pressure or political mandates, while acknowledging that achieving them requires a shift in ownership logic, a longer managerial time horizon, and greater financing courage to invest in designs that pay off across multiple lifecycle periods.

Empirically, the chapter integrates a case study of a German mid-sized water-meter manufacturer that has implemented design-for-remanufacturing over more than two decades. The case demonstrates how circular design, take-back access, and upgradeable digital capabilities can jointly generate economic value (via reduced material costs and resilient supply) and ecological benefit—providing an actionable template for structuring the sustainable organization.

The findings suggest that the primary constraints of circular economy implementation are organizational and temporal rather than technological or economic. When these constraints are addressed, circular business models can function as competitive operating systems that combine ecological benefits with long-term economic resilience.

12.2 Circular Economy and the Three-Pillar Model

12.2.1 Definition of Circular Economy

The circular economy can be defined as an economic system that aims to decouple value creation from resource extraction by designing out waste, keeping products and materials in use, and regenerating natural systems.[1] Conceptually, circularity moves beyond end-of-life recycling. It is a system that intervenes in product design, business models and infrastructures, and governance, to keep products at high utility at all times.

In contrast to the linear model, the circular economy emphasizes the closing, slowing, and narrowing of material and energy flows throughout the entire lifecycle of products. The organization of these principles can be implemented at multiple levels, ranging from individual products and companies to industrial networks and national economies. It requires a quite severe change in common technological, organizational, and institutional changes.[2]

12.2.2 Circular Economy and the Three-Pillar Model

Viewing the circular economy through the lens of the Three-Pillar Model[3] is analytically fruitful, as circular transformation entails:

1. The development of sustainable and economically competitive business models, an objective increasingly emphasized in European policy frameworks. This relates to the *Sustainable Purpose* of an organization.
2. The development of a capability and willingness to fundamentally transform established organizational routines, especially the core business logic. This relates to the pillar aptly coined as the *Travelling Organization.*
3. The development of new forms of inter-organizational collaboration, including network-based cooperation and selective collaboration with competitors—meaning a new way of *Connectivity.*

The Sustainable Purpose

As a systemic alternative to linear economic models,[4] circularity enables organizations to articulate a credible and action-guiding purpose. It contributes to the ecological interpretation of sustainability, by maximizing material efficiency through closed loops. With regard to the economic interpretation of sustainability, many circular pilot initiatives continue to struggle to establish stable and coherent systems of value creation. This is due to a set of interdependent barriers, including higher input costs, technological gaps that prevent effective loop closure, and misaligned

---

[1] This definition is widely cited in academic and practitioner literature and originates from the Ellen MacArthur Foundation (2015), *Towards a Circular Economy: Business Rationale for an Accelerated Transition.*

[2] Kirchherr, J., Reike, D., and Hekkert, M., "Conceptualizing the Circular Economy: An Analysis of 114 Definitions," *Resources, Conservation & Recycling* 127 (2017): 221–232.

[3] Wollmann, P., Kühn, F., and Kempf, M., eds., *Three Pillars of Organization and Leadership in Disruptive Times: Navigating Your Company Successfully through the 21st Century Business World* (Cham: Springer Nature, 2020).

[4] See Reference 2.

take-back systems. Where these challenges are successfully addressed, however, circular solutions can result in highly resilient economic systems, which leads over to the second Pillar.

The Travelling Organization

Circular economy implementation requires the mindset of a Travelling Organization, as it unfolds through continuous experimentation. Established products, processes, and business models must be questioned, making organizational learning an ongoing necessity. Exploration under uncertainty becomes a core capability.

Connectivity

The circular economy is inherently relational, relying on networks of value creators across value chains, industries, and even institutional contexts. Circular value creation depends on collaboration with a broad set of stakeholders to close and slow material loops. Recent research suggests that making circularity a reality requires firms to engage in cooperative relationships with competitors, where collaboration on pre-competitive activities such as shared standards or joint reverse-logistics systems enables systemic circular outcomes that can't be reached by competition alone.[5] However, in this domain, the economy still has considerable ground to cover, as persistent barriers continue to limit participation in circular economy practices.

12.3 Introduction to the Coverage of UN Sustainable Development Goals (SDG) by the Circular Economy

The 2025 UN SDG Progress Report confirmed that only 18% of goals are on track.[6] As doubts grow regarding the feasibility of achieving the 2030 Agenda for Sustainable Development, the circular economy is emerging as a promising pathway for future solutions. There's substantial academic literature explicitly discussing how circular economy[7] and the UN Sustainable Development Goals relate.[8] The SDGs are typically treated as a normative goal framework, describing what outcomes society wants, while the circular economy is treated as a transition concept, focusing on how a system needs a profound redesign to reduce resource throughput.

[5] Harala, L., Aarikka-Stenroos, L., and Ritala, P., "Coopetition for a Circular Economy: Horizontal Initiatives in Resolving Collective Environmental Challenges," in *Stakeholder Engagement in a Sustainable Circular Economy: Theoretical, Methodological and Practical Perspectives*, ed. Johanna Kujala, Anna Heikkinen & Annika Blomberg (Cham: Springer Nature, 2023), 311–362.

[6] United Nations Department of Economic and Social Affairs (UN DESA), *The Sustainable Development Goals Report 2025* (New York: United Nations, 2025), https://unstats.un.org/sdgs/report/2025

[7] Ortiz-de-Montellano, C. G.-S., Samani, P., and van der Meer, Y., "How Can the Circular Economy Support the Advancement of the Sustainable Development Goals (SDGs)? A Comprehensive Analysis," Sustainable Production and Consumption 40 (2023): 352–362.

[8] Schröder, P., Anggraeni, K., and Weber, U., "The Relevance of Circular Economy Practices to the Sustainable Development Goals," *Journal of Industrial Ecology* 23, no. 1 (2019): 77–95.

Mapping studies generally find circular economy is most directly linked to SDG 12—Responsible Consumption and Production. In general, the circular economy is contributing directly or indirectly to many of the goals.[9] As the case study presented later in this chapter focuses on a physical product innovation, the following three SDGs were selected as anchor goals:

12.3.1 SDG 12: Responsible Consumption and Production

Circular economy strategies directly reduce virgin material demand, increase product lifespan, and enhance repair, reuse, and remanufacturing systems. With the right to repair initiatives, for example, there are initiatives in place which allow easy access to these practices for consumers on the one hand, but also new business practices and income streams for the economy on the other.

12.3.2 SDG 9: Industry, Innovation, and Infrastructure

The circular economy can enhance and drive the development of the crucial infrastructure necessary for the management and recovery of resources. The increasing demand for reverse-logistics infrastructure—which allows used materials to be returned to sellers or manufacturers—aligns with SDG 9's goal of creating resilient, sustainable infrastructure and promoting innovation. Advancements in circular technologies and industrial processes hold significant potential to contribute to SDG 9 by fostering innovation.

12.3.3 SDG 13: Climate Action

The circular economy advances SDG 13 by reducing emissions embedded in material extraction, processing, and production. By reusing and remanufacturing products, circular systems preserve embodied carbon and lower demand for energy-intensive virgin materials. This material-focused approach addresses emissions that cannot be eliminated through energy transition alone. The climate effect is particularly significant in material-intensive industries such as construction and electronics.[10]

---

[9] Rodríguez-Antón, J. M., L. Rubio-Andrada, M. S. Celemín-Pedroche, and S. M. Ruíz-Peñalver, "From the Circular Economy to the Sustainable Development Goals in the European Union: An Empirical Comparison," *International Environmental Agreements: Politics, Law and Economics* 22, no. 1 (2022): 67–95.

[10] De Pascale, A., Di Vita, G., Giannetto, C., Ioppolo, G., Lanfranchi, M., Limosani, M., and Szopik-Depczyńska, K., "The Circular Economy Implementation at the European Union Level: Past, Present and Future," *Journal of Cleaner Production* 423 (2023): 138658.

The latter case study shows how by retaining control over materials across multiple lifecycles, circular business models also reduce exposure to supply disruptions, embedding climate action directly into operational strategy. Circularity is thus not an alternative to the SDGs but a strategic mechanism to operationalize them within business and industrial systems. Therefore, this chapter examines what the circular economy can learn from a transformation process that began earlier and is often regarded as its closest counterpart: digital transformation.

12.4 Learning from Digital Transformation for Accelerating Circular Economy Implementation

The transition toward a circular economy confronts organizations with challenges that closely resemble those encountered during earlier waves of digital transformation: high uncertainty regarding viable business models,[11] uncertain investment horizons,[12] unclear performance metrics,[13] and the need to coordinate change across organizational and ecosystem boundaries.[14] In digital transformation, these challenges were progressively addressed not through technology alone, but through the redesign of business models, the adoption of experimentation-based approaches to reduce uncertainty, and the alignment of governance and financing structures with long-term value creation. This section builds on these parallels to examine how lessons from digital transformation can inform circular economy implementation, with particular attention to business model design, mechanisms for managing uncertainty, financing logic, and organizational governance. The focus is not on abstract principles but on structural decisions that embed circularity into the core of value creation, decision-making, and core processes. By reframing circular economy initiatives through the lens of transformation management rather than sustainability compliance, this section outlines practical pathways by which organizations can align profitability, resilience, and material efficiency.

---

[11] Stingl, V., T. Kesting, J. Comes, and C. Fröhling, "Uncertainty Management in Circular Business Model Innovation: The Case of Circular Plastics," *Proceedings of the Design Society* 3 (2023): 83–92.

[12] Stewart, R. A., and M. Niero, "Circular Economy in Corporate Sustainability Strategies: A Review of Corporate Sustainability Reports in the Fast-Moving Consumer Goods Sector," *Business Strategy and the Environment* 28, no. 3 (2019): 504–518.

[13] Kirchherr, J., L. Piscicelli, R. Bour, E. Kostense-Smit, J. Muller, A. Huibrechtse-Truijens, and M. Hekkert, "Barriers to the Circular Economy: Evidence From the European Union (EU)," *Resources, Conservation & Recycling* 153 (2019): 220–232.

[14] Tóth-Peter, A., R. Torres de Oliveira, S. Mathews, L. Barner, and S. Figueira, "Industry 4.0 as an Enabler in Transitioning to Circular Business Models: A Systematic Literature Review," *Journal of Cleaner Production* 393 (2023): 136284.

12.4.1 Focus on a Holistic and Coherent Business Model

Digital transformation refers to the pervasive integration of digital technologies into products, processes, and business models, leading to fundamental changes in how organizations create and capture value. While often associated with new technologies, the decisive impact of digital transformation has been its effect on business model logic. Organizations that succeeded did not merely digitize existing processes, but redesigned how value is created, delivered, and monetized.

A central lesson from digital transformation is that value no longer resides primarily in the single transaction, but in the continuous relationship with the customer and the ability to orchestrate systems over time. Many digitally transformed firms shifted from one-off product sales toward service-oriented, subscription-based, or platform-driven models. This shift required integrating products, services, data, and governance into a coherent system rather than optimizing individual components.

A circular business model now is one that besides economical sustainability also fosters ecological sustainability. In particular, circular business models aim to generate profits through a continuous flow of reused materials and products over time.[15] They move away from the traditional concept of selling a product once and ending the transaction. Instead, they repeatedly capitalize on the value embedded in existing products, leveraging them multiple times to extend their utility and profitability.[16]

Important for a circular business model is to shift the focus away from what is offered—for instance the product—to the system into which the product is embedded. The system makes sure that a product or parts of a product actually can circulate, so that the environmental impact of each product instance is reduced by using better and less materials (Circular Materials), by leveraging efficiencies in production (Circular Production), by the way it is delivered to the customer (Circular Logistics), by the use and reuse (Circular Usage) and how it is brought back to an instance that knows how to decompose and remanufacture or recycle it.[17] That means circular models seek to preserve the inherent value of products and materials as long as possible by either slowing down or closing resource loops.[18] To effectively reduce resource consumption and create closed-loop systems, businesses primarily employ two strategies: designing products for longevity and integrating services that support circularity.

---

[15] Bakker, C., Wang, F., Huisman, J., and den Hollander, M., "Products That Go Round: Exploring Product Life Extension Through Design," *Journal of Cleaner Production* 69 (2014): 10–16.

[16] Linder, M., and M. Williander, "Circular Business Model Innovation: Inherent Uncertainties," *Business Strategy and the Environment* 26, no. 2 (2017): 182–196.

[17] Tukker, Arnold, "Product Services for a Resource-Efficient and Circular Economy—A Review," *Journal of Cleaner Production* 97 (2015): 76–91.

[18] Bocken, N. M. P., I. de Pauw, C. Bakker, and B. van der Grinten, "Product Design and Business Model Strategies for a Circular Economy," *Journal of Industrial and Production Engineering* 33 (2016): 308–320.

First, products are crafted to be both emotionally appealing and technically durable, ensuring they remain desirable and functional over extended periods. These products are engineered to be easily upgraded, maintained, repaired, refurbished, remanufactured, and recycled,[19] which extends their usable life and reduces waste. Second, companies incorporate circular services into their business models to support sustainable products. They offer customers options like product upgrades and refurbishments, making it convenient to keep products in use longer.

The learning from digital transformation becomes more concrete: successful transformation depends on system-level coherence and clarity. Just as early digital business models integrated customer interaction, technology, and new monetization logics, circular business models must combine product and material selection, reverse logistics, and a service design with the right incentives for customers to actually return or use the embedded circular services. So, as a design principle, that might require that a product is built in a way that it keeps its functionality and desirability (also emotional desirability) across cycles—analogous to digital products that evolved through updates and feature extensions. Also, services need to be designed around the product itself to help keep it working and emotionally desirable or being able to hand it over to a next cycle. Just like in digital transformation, this seems to become successful when companies redesign the entire value-creation logic.

12.4.2 Managing Execution Uncertainty Through Scalable Experimentation

A central parallel between digital transformation and the circular economy lies in the challenge of developing viable business designs under conditions of high uncertainty, particularly when digital transformation entails uncovering innovative business models (rather than shifting manual processes to a digital one). In both domains, uncertainty does not primarily stem from technological feasibility but from the question of whether a new business model can become economically viable at scale. In the case of circular economy initiatives, this uncertainty is amplified by longer time horizons, questioning if higher upfront investments in product design, materials, and infrastructure will pay off once materials and products circulate across multiple usage cycles.[20]

As in digital transformation, this uncertainty calls for an experimental, hypothesis-driven approach. Rather than seeking immediate profitability, early stages focus on generating evidence regarding business model viability under realistic

[19] Ellen MacArthur Foundation, *Towards the Circular Economy: Accelerating the Scale-Up Across Global Supply Chains*, Vol. 2 (Cowes, UK: Ellen MacArthur Foundation, 2015), https://content.ellenmacarthurfoundation.org/m/50c85a620a58955/original/Towards-the-circular-economy-Vol-2.pdf

[20] Hopkinson, P., M. Zils, P. Hawkins, and S. Roper, "Managing a Complex Global Circular Economy Business Model: Opportunities and Challenges," *California Management Review* 60, no. 3 (2018): 71–94.

conditions. Drawing on established transformation practice, experimentation is structured along three interdependent dimensions:

Desirability

Desirability concerns whether customers are willing not only to adopt a new offering (digital or circular), but also to change their behavior. In earlier waves of digital transformation, customer acceptance was frequently driven by immediate benefits, such as increased speed and convenience or cost reductions. In circular economy contexts, by contrast, adoption may entail the introduction of additional customer responsibilities, requiring users to actively perform behaviors—especially at the end of the usage horizon—that enable the closing of loops (e.g., returning products or engaging in repair and reuse services). But still, achieving this behavioral shift requires the careful design of incentives embedded within core business processes, alongside systematic testing of their effectiveness.

Feasibility

Feasibility concerns the technical robustness of the solution, the viability of associated operational processes, and the ability to establish crucial partnerships to deliver the desired value. Often circular economy initiatives involve physical products and material flows—rather than purely digital processes, as in many digital transformation initiatives—operational processes must be developed through iterative prototyping, testing, and refinement, before investing. Critically, considerations of scalability need to be incorporated into process design from the outset, ensuring that solutions can be expanded beyond pilot phases.

Viability

Viability concerns whether a business model can generate positive economic outcomes over time. In many digital transformation initiatives, economic benefits materialize relatively quickly, as customer desirability often directly drives value creation and revenue generation. Circular economy initiatives, by contrast, require a longer-term perspective, accounting for multi-cycle use, remanufacturing and recovery costs, and thereby delayed returns on investment.

The viability of circular business models is rarely demonstrable within a single product lifecycle. Initiatives may therefore fail not because the underlying concept is economically unsound, but because experimentation is discontinued before sufficient circulation cycles have occurred to validate the economic logic. Consequently, experimentation in the circular economy must be intentionally designed for both scale and duration, enabling organizations to assess whether repeated reuse, refurbishment, or remanufacturing can ultimately offset initial investments and generate sustainable returns.

12.4.3 Ecosystem Collaboration

Digital transformation has shown that platform-based value creation depends on standards, interfaces, and coordination across organizational boundaries—sometimes including collaboration between competitors. Such ecosystem logics are maybe even more essential for the circular economy. Circular business models rely on interconnected actors across design, production, logistics, use, return, and remanufacturing, making isolated firm-level solutions structurally insufficient.[21]

This interdependence introduces both opportunity and risk. On the one hand, shared infrastructures and standards—such as take-back systems or material identification schemes—lower transaction costs and enable scale. On the other hand, uncertainty regarding participation, cost allocation, and value appropriation can inhibit investment, particularly when returns depend on multi-cycle circulation beyond a single organization's control.

Building on lessons from digital transformation, circular economy initiatives become viable at scale only when organizations deliberately design ecosystem-level systems, align incentives among partners, and establish trusted data-sharing mechanisms. In this sense, ecosystem collaboration is a core design element that determines whether circular business models can move beyond pilots and achieve durable economic viability.

12.4.4 Short-Term vs. Long-Term Financing and Incentivization

Digital transformation initiatives typically allow for rapid experimentation and short learning cycles. Digital services can often be tested within short timeframes, enabling organizations to quickly assess customer demand, pricing structures, and sometimes product-market fit. Financial viability can thus be evaluated within months rather than years. Still, several prominent digital companies have demonstrated that long-term strategic orientation can constitute a decisive competitive advantage, therefore accepting prolonged periods of limited or negative profitability in order to build scalable infrastructures and dominant market positions. This temporal logic bears some resemblance to circular economy transformation.

Many circular economy business models are inherently tied to the physical lifecycle of products. While circular initiatives may adopt experimental approaches conceptually, their learning phases are structurally elongated. Learning would only occur once products complete one or more usage cycles, which may take several years or, in some industries, even decades. As with those mentioned digital platforms, many circular business models require upfront investment and patience before economic benefits materialize.[22] The core challenge therefore lies not in the

---

[21] Mosgaard, M. A., H. S. Kristensen, and N. Bocken, "Competition or Coopetition? Collaboration Strategies for the Circular Economy," *Journal of Cleaner Production* 429 (2025): 145686.

[22] Fallahi, S., A. C. Mellquist, O. Mogren, E. Listo Zec, P. Algurén, and L. Hallquist, "Financing Solutions for Circular Business Models: Exploring the Role of Business Ecosystems and Artificial Intelligence," *Business Strategy and the Environment* 32, no. 6 (2023): 3233–3248.

economic viability of circular models per se, but in aligning financing structures, incentive systems, and performance metrics with the longer learning cycles inherent to circular value creation.

In the following, we present a case study of an organization that applies circular practices as its primary logic for generating both ecological and economic value. The case study is used to qualitatively explore the propositions developed above, examining business model design, experimentation under uncertainty, ecosystem collaboration, and financing logics in practice.

12.5 Research Design

12.5.1 Methods and Case Selection

This chapter adopts a qualitative, case-based research design using an explorative interview with the managing director of a manufacturing company. Given the exploratory nature of the research question, we look at a case study of an organization in which circular practices already work in practice (e.g., durable products, stable customer relationships, clear take-back routes).

The empirical basis consists of an in-depth, semi-structured interview with a senior executive of a German mid-sized manufacturer of water meters. The interview covered the history of the circular initiative, design and material decisions, business model logic, financial considerations, digital enablement, and organizational implications.

The interview is treated as a retrospective, practice-based narrative that allows reconstruction of key mechanisms and decision logics. The analysis focuses on how circular practices were conceived, implemented, and refined over time and how digital capabilities were integrated into this trajectory.

12.5.2 Interviewee Profile

The interview was conducted with Wilhelm Mauß. He is the founder and managing director of Lorenz GmbH & Co. KG, a German manufacturer of water meters. He initiated the development of the company's circular business model and has been actively involved in remanufacturing-based product and business model innovation in industrial manufacturing for more than two decades.

12.6 Case Study: Circular Business Models in Industrial Manufacturing—The Case of Lorenz Meters

Lorenz GmbH & Co. KG is a German medium-sized manufacturer of water meters operating in a highly regulated market. Water meters are subject to legally defined calibration periods, after which they must be replaced. This regulatory setting, combined with high material intensity and long product lifespans, makes the company a particularly relevant case for examining circular economy implementation beyond pilot projects.

Unlike many circular initiatives that remain peripheral to core operations, Lorenz embedded circular principles directly into its business model. Circularity is not framed as a sustainability add-on, but as a strategic response to competitive pressure, regulatory requirements, and supply chain risk—closely aligning with the organizational transformation logic discussed in this chapter.

12.6.1 Initial Situation and Strategic Turning Point

A central challenge in adopting a circular business model was the acceptance of delayed economic returns. The circular model required investments in high-quality materials and modular product designs, despite the fact that no contribution margin would be generated in the initial production cycle. This stood in contrast to prevailing managerial logics focused on short-term profitability and annual performance targets.

As Mauß emphasizes, the decision to proceed required a deliberate shift in perspective. Mauß explains that the first cycle should merely break even, with margins realized in later cycles: "If you focus only on immediate margins, you will never start."

This long-term orientation was facilitated by the company's ownership structure as a family-owned business, which made it easier to pursue strategic decisions with extended time horizons. In organizations with more dispersed ownership structures, such long-term commitments might have been more difficult to implement due to stronger short-term performance pressures.

Traditionally, calibration requirements led to the full replacement of water meters, even though many components remained technically functional. High-quality materials were thus removed from use after a single lifecycle, despite retaining substantial value. At the same time, Lorenz faced growing competition from low-cost manufacturers relying on inexpensive materials and offshored production.

Rather than engaging in price-based competition, the company adopted a fundamentally different strategic logic. As Mauß describes the impulse for this shift emerged more than two decades ago from a simple observation:

> The idea came to me when I privately dropped off my rubbish at a recycling center. I thought it cannot be right to throw these things away. We should reuse them and recondition them.[23]
>
> If cheap materials and low production costs become the dominant logic, we lose production and development capabilities in Europe.

[23] Wilhelm Mauß, interview by authors, 27th November 2025—the same applies to all other direct quotes. Used with permission.

Circular economy thus became a means of safeguarding long-term value creation and industrial competence.

12.6.2 Circular Business Model and Design Principles

The circular business model implemented by Lorenz is based on the systematic take-back of water meters after the calibration period. Returned devices are disassembled, inspected, cleaned, and remanufactured (where feasible). Central to this approach is a product development philosophy explicitly geared toward circularity.

Mauß refers to this as Design-for-Remanufacturing:

> Products must be designed so that they are easy to repair, easy to disassemble, and easy to reuse.

This principle remained central even as the industry transitioned from mechanical to digital water meters. While product architectures became increasingly digital, hardware components were deliberately designed for long-term use, with functionality extended through software updates:

> We use communication chips and memory capacities that are much larger than we currently need, so we can take back the same hardware and deploy new software.

The product strategy is complemented by service- and use-oriented business models that ensure reliable product return and lifecycle control.

12.6.3 Economic Logic and Organizational Effects

A defining feature of the model is its multi-cycle economic logic. Higher-quality materials and modular designs increase costs in the first lifecycle, but generate economic benefits in subsequent cycles:

> The contribution margins are lower in the first usage phase. You realize them in the second and third cycle.

This requires a shift from short-term unit-cost optimization toward a total cost of ownership perspective. This economic leverage is rooted in the high base cost share of the product, where high-quality components and their inherent value account for approximately 60–70% of the total product value. Recovering these assets in the second lifecycle nearly eliminates the need for new material investments. From the second lifecycle onward, both manufacturer and customer benefit economically:

> Both the customer and the supplier gain economic advantages.

In addition, the circular model increased organizational resilience. During global supply chain disruptions, Lorenz was able to rely on remanufactured components:

> During the Suez Canal crisis and the pandemic, we had zero restrictions. That's when you become resilient.

12.6.4 Transferability and Case Implications

The transition toward a circular business model was not linear. Early iterations revealed technical, logistical, and organizational challenges, particularly with regard to return logistics and coordination between development, production, and service units. Several assumptions concerning reuse rates and component durability had to be revised over time.

Mauß describes the process as one of continuous learning.

Many things only become clear once you actually do them.

This learning-by-doing dynamic required new routines and closer collaboration across organizational boundaries, reinforcing the view that circular transformation is fundamentally an organizational learning process.

According to Mauß, the primary challenges of circular transformation are organizational rather than technological. Long-term thinking, learning by doing, and tolerance for initial uncertainty are critical success factors:

> If I hadn't started, I wouldn't have learned.

The approach is considered transferable across industries where products have long lifespans, high material value, and controlled usage cycles (The following Figs. 12.1, 12.2, 12.3 show the difference between linear and circular models):

> From my perspective, this approach is unavoidable and easily transferable. The core principle must be reuse and maximum remanufacturing.

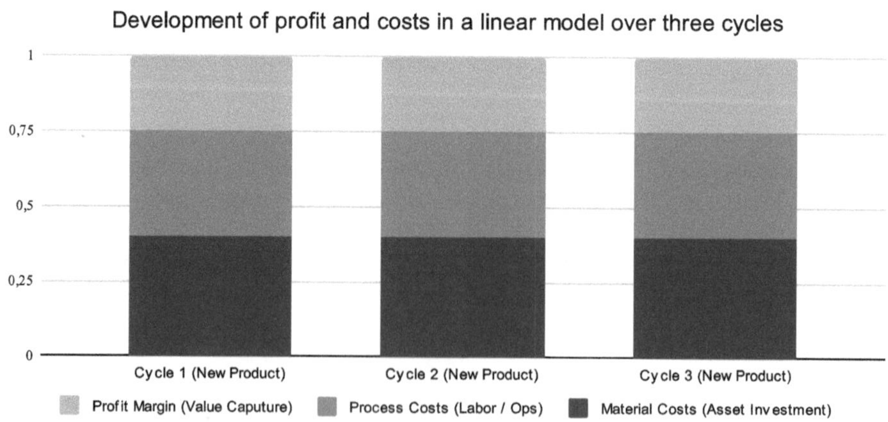

Fig. 12.1 The figure illustrates a conventional linear production model with stable production costs and a constant contribution margin across product cycles. Developed by the authors. Used with permission. All rights reserved

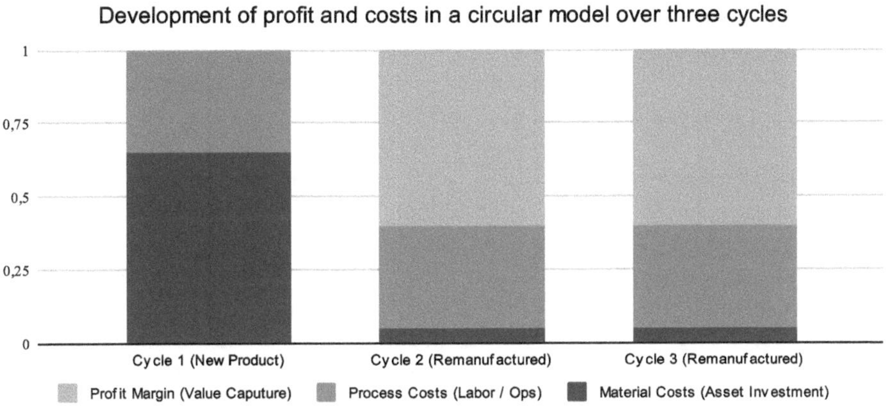

Fig. 12.2 The figure illustrates the circular business model logic described in the case. While no contribution margin is generated in the first production cycle, margins increase in the following phases. Developed by the authors. Used with permission. All rights reserved

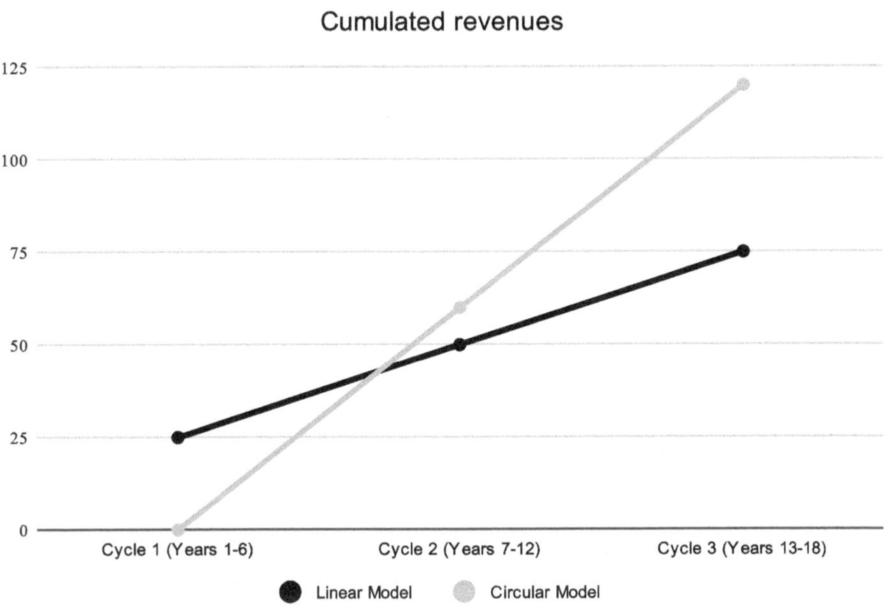

Fig. 12.3 This figure shows the cumulated revenues. While the circular business model initially underperforms, it slightly overperforms in the medium term and clearly outperforms in later stages due to cumulative value creation. Developed by the authors. Used with permission. All rights reserved

12.6.5 Learnings from the Lorenz Meters Case

The Lorenz Meters case provides three central learnings for advancing circular economy implementation beyond pilot initiatives.

Multiple-Cycle Endurance
First, circular business models can be economically superior to linear production logic when assessed across multiple product lifecycles. While no contribution margin is realized in the initial production cycle, subsequent remanufacturing phases enable substantially higher margins. In the Lorenz case, from the second usage cycle onward, circularity outperforms linear production across key economic dimensions, creating a clear win-win situation for both manufacturers and customers. This finding challenges the widespread assumption that circular economy models primarily entail economic trade-offs and demonstrates that their profitability is fundamentally linked to multi-cycle value creation.

Resilience Spillovers
Second, circular economy strategies address market-driven challenges that extend well beyond regulatory compliance. The case shows that circularity directly responds to acute operational problems such as material availability, supply chain volatility, and competitive pressure from low-wage economies. By reducing dependence on virgin raw materials and global sourcing, circular business models enhance organizational resilience and mitigate strategic risk. Circularity thus emerges not only as a sustainability response, but as a competitive strategy in an increasingly resource-constrained and geopolitically fragmented global economy.

Financing Long-Term Risk and Uncertainty
Third, the primary challenge of circular economy implementation lies in its temporal logic rather than its technical or economic feasibility. Circular business models unfold over longer investment and learning cycles than most digital or linear initiatives. Their benefits materialize only after one or more product lifecycles have been completed. This temporal mismatch with prevailing short-term performance metrics and investment horizons suggests the need for complementary actors and financial mechanisms capable of absorbing long-term risk. Analogous to reinsurance markets that manage low-probability, high-impact risks, specialized investment vehicles or intermediary business models could emerge to bridge the gap between upfront investment and delayed economic returns. This represents a promising hypothesis for future research and experimentation.

Taken together, these learnings reinforce the central argument of this chapter: circular economy transformation is not primarily constrained by technological readiness or economic viability, but by the alignment of time horizons, incentive structures, and governance models. When these elements are aligned as a holistic business model—as demonstrated in the Lorenz Meters case—circular economy principles can move beyond pilot projects toward resilient, scalable, and economically compelling business systems.

12.7 Overall Conclusion

This chapter examined how circular economy implementation can move beyond pilots and compliance toward durable competitive advantage. Across the conceptual framing on the Three-Pillar Model, the comparison with digital transformation, and the case evidence, one conclusion stands out: circularity becomes economically compelling when it is designed into the core logic of value creation and managed as a transformation with long horizon perspectives—not added as a sustainability layer.

The comparison with digital transformation highlights four recurring challenges that also shape circular economy implementation: business model uncertainty, delayed investment returns, unclear performance metrics, and cross-boundary coordination needs. Digital transformation addressed these constraints not through technology alone, but through business model redesign, structured experimentation, and governance and finance aligned with long-term value creation. A similar approach might be required for circular economy strategies.

The case of the German water-meter manufacturer provides empirical support for these parallels. Over two decades, the firm embedded circularity into a digitally enabled product-service system by combining design-for-remanufacturing, lifecycle control via take-back, and upgradeable digital architectures. The result was not only ecological benefit but also clear economic outcomes: lower long-term material costs, more stable margins, and greater resilience under supply disruptions. The central mechanism was multi-cycle value capture—enabled by product architecture and reinforced by organizational routines and financing patience.[24]

Overall, the main barriers to circular economy implementation are less technological than organizational and temporal: misaligned incentives, short-term evaluation logics, and insufficient coordination across ecosystems. When these constraints are addressed—through coherent business model design, experimentation that matches multi-cycle realities, and financing structures that absorb long-term risk—circularity can function as a competitive operating model.

[24] **Disclaimer:** During the preparation of this work, the author(s) used *ChatGPT* and *Claude* to support language refinement and stylistic fine-tuning. All content generated with the assistance of this tool was subsequently reviewed and edited by the author(s).

Dr. Martin Hewing has over 15 years of experience spanning academic research, start-up founding, and technology development. He has co-founded and built technology-driven ventures and worked in R&D-oriented organizations, translating emerging technologies into innovative products and sustainable business models. He holds a PhD in Innovation Management & Entrepreneurship and has published extensively in academic journals and books. He has lectured at leading universities and remains actively engaged in research-informed innovation practice.

Phillip Brandts has spent more than 20 years supporting corporations in digital transformation and innovation. He began his career in the media industry and later worked as a Partner at a digital and innovation consultancy, advising companies on adapting their business models to the digital age. He subsequently founded a start-up in the field of circular economy and today leads the consulting practice at PHINEO gAG, where he advises corporations on impact-oriented strategies at the intersection of digital transformation, circular economy, and long-term value creation.

Collaborative Sourcing: A Three-Pillar-Based Approach to Meeting the 17 SDGs (UN Sustainable Development Goals)

13

Jürgen Scherer

Abstract

In pursuit of contributing to their respective Sustainable Purpose, Traveling Organizations have recognized a need for intensified collaboration, for better Connecting Resources within and across their value chain partners. Removing organizational boundaries, internally between divisions, regions, and functions, and externally between a focal firm and its up- and downstream business partners has led to new and emerging concepts: open organization, open strategy, open innovation, and open sourcing.

The sourcing of materials and services, in most for-profit business organizations, representing between 50% and 70% of the cost of goods sold (COGS), is a key strategic task. Procurement functions define the best mix of supply, optimizing total costs, innovation, quality, risks, and sustainability of outsourced goods, and services. Especially the sourcing of sustainable and ethically responsible products has gained momentum and is increasingly recognized as a major contributor to an organization's environmental stewardship, social responsibility, and corporate governance.

In this chapter, we will showcase, by different use cases, how business organizations across a multitude of industries have not only developed responsible sourcing in their own rights, but moreover have joined forces to collaboratively help sustainable and ethical supply solutions to leverage "ecologies of scale." The various sourcing initiatives, meeting many of the 17 UN Sustainable Development Goals, are examples of successful collaborations between for-profit, non-profit, and governmental organizations.

J. Scherer (✉)
BxB-Exchange, Dieblich, Germany
e-mail: js@bxb-exchange.com

© The Author(s), under exclusive license to Springer Nature Switzerland AG 2026

P. Wollmann et al. (eds.), *The Sustainable Organization – Volume 3*, Future of Business and Finance, https://doi.org/10.1007/978-3-032-28949-0_13

13.1 Introduction

With the burgeoning of the Triple Bottom Line (TBL or 3BL) framework since 1994,[1] the focus of business organizations has started to shift from a predominantly economic, financial performance (profit) toward a more balanced approach, targeting equally environmental stewardship (planet) and social responsibility (people). Consequently, the sourcing of goods and services (typically 50%–70% of cost of goods sold), aiming for the best economic mix of total costs, supply continuity, and business growth, is striving as well for more sustainable and ethically responsible (out-)sourced products and services.

The Greenhouse Gas (GHG) protocol,[2] issued in 2001 by the World Resource Institute (WRI) and the World Business Council for Sustainable Development (WBCSD), provides a reporting guideline on Carbon Footprint (CFP). Beyond direct (scope 1) and indirect (scope 2) emissions, it requires business organizations to incorporate their value chain partners, i.e., activities with upstream suppliers and downstream customers (scope 3), into their total CFP exposure. Thus, procurement is responsible for creating transparency, collecting data, vetting, and developing the supplier base to collectively achieve the decarbonization targets of the respective organization.[3]

The 17 United Nation's Sustainable Development Goals (SDGs, Fig. 13.1), published in 2015, detailed in 169 specific targets and based on the TBL framework, are a global initiative to end poverty, protect the planet, and ensure prosperity by 2030. Besides the explicit call for "Partnerships for the Goals" (SDG # 17) with an extended set of stakeholders, the purchasing activities of an organization largely contribute to the Environmental, Social, and Governance (ESG) performance, in pursuit of simultaneously improving and meeting various SDGs.

These three fundamental milestones have changed the corporate agenda of many business organizations over the course of the recent 30 years, and as a result also the operating model[4] "Implementing sustainable sourcing—Does purchasing need to

[1] The triple bottom line (or otherwise noted as TBL or 3BL) is an accounting framework with three parts: social, environmental (or ecological), and economic. See: Elkington, John: 25 years ago, I coined the phrase "Triple Bottom Line." Here's why it's time to rethink it, www.hbr.org, 25.06.2018.

[2] https://ghgprotocol.org, see also Sarmad, O.: "Greenhouse Gas Protocol"; Chap. 12 in: Wollmann, Peter, Pemler, Doris., Ndrevataj, ersida.: "The Sustainable Organization—How Organizations Address the 17 UN SDGs Using the 3-P-Model"; Cham 2025, SpringerNature (https://link.springer.com/book/10.1007/978-3-031-89549-4)

[3] Bricheux, Charlotte; Lehr, Jonas; Ponbauer, Lucas; Gatzer, Sebastian: Tackling Scope 3 emissions through supplier collaboration, www.mckinsey.com, 17.01.2024.

[4] Knight, Louise; Frederik Vos, Frederik; Meehan, Joanne: New procurement directions in supply chain manage-ment, in: Sweeney, Edward; Waters, Donald (Eds.): Global Logistics, 8th ed., London et al. 2021: 95–118. Dassu, Imran; Kumari, Priya; Raju, Adithi; Powell, Richard: Sustainable procurement demands a new operating model—and more collaborative supplier partnerships, www.kearney.com, 18.04.2023.

Fig. 13.1 UN SDG (source: https://www.un.org/sustainabledevelopment/), used with friendly permission of UN. The content of this publication has not been approved by the United Nations and does not reflect the views

change?,[5] not too long ago still a valid, meanwhile more of a rhetorical question. The overarching answer from an academic, practitioner, and consulting perspective is: Yes, toward a "collective approach."[6] The "collaboration imperative"[7] triggers close alignment with many internal and external stakeholders in extended value chains and transforms the procurement function to be(come) equally "competitive, compliant and green."[8] After all, sustainability, defined as economic, environmental, and social performance, is "a license to do business in the twenty-first century."[9]

---

[5] Schneider, Lena; Wallenburg, Carl M.: "Implementing sustainable sourcing—Does purchasing need to change?", in: Journal of Purchasing and Supply Management, 18/4 (2012): 243–257.

[6] Villena, Verónica H.; Gioia, Dennis A.: A more sustainable supply chain, in: Harvard Business Review, 98/2 (2020): 84–93.

[7] Nidumolu, Ram; Ellison, Jib; Whalen, John; Billman, Erin: The Collaboration Imperative: New partnership models can protect the environment and create value for everyone, in: Harvard Business Review, 92/4 (2014): 76–84.

[8] Inverto: Beyond Compliance: Leveraging sustainability to drive procurement value, Cologne 2025.

[9] Carter, Craig R.; Easton, P. Liane: Sustainable supply chain management: evolution and future directions, in: International Journal of Physical Distribution & Logistics Management, 41/1 (2011): 46–62.

Carter, Craig R.; Jennings, Marianne M.: The role of purchasing in corporate social responsibility: a structural equation analysis, in: Journal of Business Logistics, 25/1 (2004): 145–186.

Carter, Craig R.; Rogers, Dale S.: A framework of sustainable supply chain management: moving toward new theory, in: International Journal of Physical Distribution & Logistics Management, 38/5 (2008): 360–387.

By applying the Three-Pillar-based approach and contributing to their respective *Sustainable Purpose*, *Traveling Organizations* have recognized the necessity for intensified collaboration, for *Connected Resources* within and across their value chain partners. Removing organizational boundaries, internally between divisions, regions, and functions, and externally with end-to-end business partners and within a wider value chain network, has led to new and emerging concepts: open sourcing,[10] open innovation,[11] open organization,[12] and open strategy.[13]

This chapter will demonstrate, through different use cases, how business organizations across a multitude of sectors and industries have joined forces to collaboratively leverage sustainable and ethical supply solutions for "economies *and* ecologies of scale." The various initiatives targeting the 17 UN SDGs are examples of cross-sectoral, multi-stakeholder partnerships (MSPs) between the private and public sectors, between for-profit, non-profit, governmental, and non-governmental organizations (NGOs).

As stated in the book's subtitle, the so-called Three-Pillar Model, a meta-level thinking and analysis concept developed and published in 2020,[14] based on Sustainable Purpose, Travelling Organization, and Connected Resources or Connectivity, will be applied (more details can be found in Chap. 2 of this book and in the conclusion of this chapter).

13.2 Sustainable Purpose: Service-Dominant Logic and Co-Creation

Considering the implications of TBL, GHG, and SDGs, of a constantly growing body of national and international legislation and regulation (e.g., the EU Omnibus, former Corporate Sustainability Reporting Directive, CSRD), of self-imposed ESG reporting, and/or of external pressure through NGOs (e.g., Greenpeace), business organizations have recognized the need to move from reactive compliance to proactive commitment.[15] Thus, the ambition level in many organizations has matured

[10] Peterson, Christine L.: How I coined the term "open source," www.opensource.com, 01.02.2018.

[11] Chesbrough, Henry W.: Open Innovation: How Companies actually do it, in: Harvard Business Review, 81/7 (2003): 12–14.

[12] Whitehurst, Jim: The Open Organization: Igniting Passion and Performance, Boston 2015.

[13] Stadler, Christian; Hautz, Julia; Matzler, Kurt; von den Eichen, Stephan F.: Open Strategy: Mastering Disruption from outside the C-Suite, Cambridge/London 2021.

[14] Wollmann, Peter; Kühn, Frank; Kempf, Michael (Eds): Three Pillars of Organization and Leadership in Disruptive Times—Navigating Your Company Successfully through the twenty-first Century Business World. Cham: Springer Nature, © 2020

[15] Nidumolu, Ram; Prahalad, Coimbatore K.; Rangaswami, Madhavan R.: Why sustainability is now the key driver of innovation, in: Harvard Business Review, 87/9 (2009): 56–64.

from "must do" and "expected to do" to "smart to do,"[16] from "what is right to competitive advantage."[17]

With a commitment to a *Sustainable Purpose*, the focus on sourcing and selling is also shifting from primarily physical goods, materials, components, and equipment, the so-called *operand* resources, to intangible services, capabilities, or know-how, the so-called *operant* resources.[18] Intangibles for sustainability, such as assurance services, carbon credits, eco-labels, audits, renewable energy certificates (RECs), and traceability data, have become table stakes and are preconditions for continuing business with tangible assets.

Instead of "value-in-exchange" (goods-dominant logic), intangibles are based on "value-in-use" (service-dominant logic). Technical parameters no longer drive specifications. External standards and accreditations are replacing internal testing protocols. Supplier selection and evaluation are less driven by "hard" (quantitative) performance criteria, such as costs, price, quality, and volumes. Instead, they are measured by "soft" (qualitative) perception criteria, including credibility, exclusivity, reliability, transferability, and validity.

Intangibles for sustainability are co-created by the shared investments of MSPs in the value chain and in cross-sectoral ecosystems. As a matter of fact, the co-creation, co-development, and co-promotion of sustainable and ethically responsible conditions in many industries through these partnerships, e.g., the Forest or Marine Stewardship Councils (FSC & MSC), are the "raison d'être" for their foundation. The inherent operating model is based on collaboration among its members to jointly mitigate negative externalities, such as hunger, poverty, inequality, or pollution, and promote positive externalities, including education, health, the circular economy, and renewable energy.

Sidebar 1 (see below) illustrates BASF's challenges and opportunities associated with external co-creation in the context of the Business-to-Business (B2B) in the Chemical Industry.

Foerstl, Kai; Azadegan, Arash; Leppelt, Thomas; Hartmann, Evi: Drivers of supplier sustainability: Moving beyond compliance to commitment, in: Journal of Supply Chain Management 51/1 (2015): 67–92.

Polman, Paul; Winston, Andrew: Net Positive: How courageous companies thrive by giving more than they take, Cambridge 2021.

van Weele, Arjan J.; van Tubergen, Kristine: Responsible Purchasing: Moving from compliance to value creation in supplier relationships, in: Bouchery, Yann; Corbett, Charles J., Fransoo, Jan C., Tan, Tarkan (Eds.): Sustainable Supply Chains, Cham 2024: 257–278.

[16] Aichbauer, Stefan; Buchhauser, Martina; Erben, Agnes; Steinert, Sven; Tietze, Detlef; Wiking, Emilia: Responsible Procurement: Leading the way to a sustainable tomorrow, Cham 2022.

[17] Aichbauer, Stefan et al.: ibid.

[18] Vargo, Stephen L.; Lusch, Robert F.: Evolving to a New Dominant Logic for Marketing, in: Journal of Marketing 68/1 (2004): 1–17.

Sidebar 1: BASF Creator Space—Co-Creating for a Sustainable Future!

BASF, the Ludwigshafen, Germany headquartered chemical company, launched a global initiative in connection with its 150th company anniversary in 2015, coined "Creator Space." The project, captured in a two-hour documentary film[19] as well as in a Harvard Business Case,[20] is anchored in BASF's corporate mission statement to "create chemistry for a sustainable future." As Elise Kissling, Head of Project 150, explains in the documentary, the initiative has been driven by the mantra to inspire and engage BASF employees with stakeholders outside the organization to co-create innovation.

This purposely required a new and different way of connecting with NGOs, communities, and governmental agencies, specifically targeting the megatrends around urban living, smart energy, and food. Katharina Felgenhauer, responsible for Public-Private Partnerships at BASF, describes the "empathic design" approach for innovative solutions as active listening to and learning from external stakeholders, based on an open-minded dialogue without preconceived ideas. It requires time, sensitivity, and the expressed will of all constituents for such experimental collaboration.

Overall, approximately 1000 ideas were generated by 50 teams worldwide, of which more than 100 were moved to implementation, including 11 long-term projects across various BASF divisions. The documentary offers further insights into several selected projects, including water distribution and sanitation in Mumbai, India (SDG 6); organic waste composting for agricultural products in Shandong Province, northwest of Shanghai, China (SDG 12); and the Red Hook urban community project in Brooklyn, New York, USA, which tackles ongoing gentrification (SDG 11). Long term, BASF decided to integrate the Creator Space initiative into the company's ongoing orientation toward innovation.

At the same time, the film openly addresses the inherent conflict between the desired co-creation for societal value in the long run and the short-term business paradigm to justify investments based on stage-gate processes and convincing return-on-investment cases. The presentation of the Creator Space projects to some of BASF's divisional business managers in a "Shark Tank or Dragon's Den" set-up, as shown in the film, amplifies a still-existing gap in many organizations.

[19] Experiment 150: Co-creating for a sustainable future, documentary film by Thomas Grube, 2016. https://www.youtube.com/watch?v=X-M9Vt-l0AE&ab_channel=BASF

[20] Rangan, V. Kasturi; Billaud, Emilie; Dessain, Vincent: BASF: Co-Creating Innovation, Harvard Business Cases 9-517-073 (A) & 9-517-074 (B), 23.04.2017.

While traditional product-driven innovation and business models are largely based on existing relationships in established markets with defined customers, new emergent service-driven innovation is built on awareness, education, and transparency toward rather unknown market and customer territories. As Hartmut Unger, Managing Director of the non-profit and charitable BASF Foundation, points out, community-building processes have not been the go-to-market approach of BASF in the past, but they certainly offer an opportunity for the future!

13.3 Traveling Organization: Navigation and Orchestration

The SDGs are addressing the most important and urgent, yet at the same time, the most wicked and complex environmental, social, and economic challenges. They require, almost by default, "Partnerships for the Goals" (SDG # 17). The internal collaboration warrants close relationships between the procurement and the sustainability function, likewise with finance (on risk management), legal (on compliance), research and development (on design for sustainability), and/or marketing and sales (on product and account management). The external collaboration relies on partnerships with many MSPs involving non-traditional suppliers, such as governmental agencies, audit firms, certification bodies (e.g., "B Corp"[21] or "EcoVadis"[22]), data providers, and traceability platforms.

Traveling Organizations[23] have started to shift their ESG focus increasingly from internal to external.[24] By doing so, the complex nature of MSPs and the changing regulatory landscape for the upstream supply markets call for an advanced set of skills from the procurement function, especially for navigation and orchestration (cap)abilities.

[21] Smith, John: The Rise of B Corporation: Can business be both profitable and sustainable?, www.planetflagship.com, 19.08.2025.

[22] Fröhlich, Elisabeth; Baid, Vaishali; Lenz, Jennifer; Peters, Silke: Developing Partnerships Across Global Supply Chains, in: Leal Filho, Walter; Marisa Azul, Anabela M.; Brandli, Luciana; Lange Salvia, Amanda; Wall, Tony (Eds.): Partnerships for the Goals. Encyclopedia of the UN Sustainable Development Goals, Cham 2021: 281–291.

[23] Defined in the Three-Pillar Model as "Travelling Organization refers to an organization accepting to be in a permanent state of flux and transformation while interacting with the external requirements of its journey by means of rapid learning and adaptivity. Ideally, this is the mindset of loving the unknown and being well able to manage it."

[24] Smith, Peter; Perera, Mark: Procurement with Purpose, 2021.

Navigation in business organizations is defined as proactive workplace behavior, characterized by a high level of orientation, coordination, and self-initiated leadership in changing context conditions.[25] Especially in boundary-spanning roles, as Purchasing & Supply Management (PSM), a combination of sensing (risks and opportunities), securing (internal and external stakeholders), and shifting (completing projects for transformation and change) capabilities is needed.[26] The Inner Development Goals (IDGs), a non-profit and open-source initiative, confirms that "without the development of [such] transformational skills at scale, external solutions to our global challenges may be limited, too slow or short-lived" (www.innerdevelopmentgoals.org).

Orchestration in same-industry value chain networks, value constellations,[27] or cross-industry ecosystems highlights the ability of the involved entities to explore new ways of working together, learn from each other, and jointly develop standards and protocols.[28] As organizational boundaries become blurred, the involved partners "open up" by aligning objectives, pooling resources, and sharing data, information, and knowledge.[29] This is often an iterative process of continued small steps, explorative and experimental in nature. It is particularly these *inter-firm*, as opposed to *intra-firm* collaborative capabilities, that make the difference[30] or as John Clarkeson, Emeritus CEO BCG, once phrased: "The winning organization of the future will look more like a collection of jazz ensembles (collaboration) than a symphony orchestra (cooperation)."[31]

The case of Allbirds, a US-American Direct-to-Consumer (D2C) footwear company, described in sidebar 2, highlights the open-source approach in MSP collaborations.

---

[25] Scherer, Jürgen: The Role of "Navigators" in Travelling Organizations, in: Kempf, Michael; Kühn, Frank (Eds.): Navigating a Travelling Organization, Cham 2022: 53–64.

[26] Scherer, Jürgen: Navigators in boundary-spanning roles: Leadership insights from 12 interviews, in: Leader to Leader, 116 (Spring 2025): 80–86.

[27] Normann, Richard; Ramírez, Rafael: From Value Chain to Value Constellation: Designing Interactive Strategy, in: Harvard Business Review, 71/4 (1993): 65–77.

[28] Schmelzle, Ulrich; Tate, Wendy L.: Purchasing orchestration practices—Introducing a purchasing-innovation framework, in: Journal of Purchasing and Supply Management, 28/2 (2022): 100756.

[29] Zarei, Hamid; Rasti-Barzoki, Morteza; Altmann, Jörn; Egger, Bernhard: Cooperation, coordination, or collaboration? A structured review of buyers' partnerships to support sustainable sourcing in supply chains, in: Environmental Science and Pollution Research, 30 (2023): 76491–76514.

[30] Luzzini, Davide; Brandon-Jones, Emma; Brandon-Jones, Alistair; Spina, Gianluca: From sustainability commitment to performance: The role of intra- and inter-firm collaborative capabilities in the upstream supply chain, in: International Journal of Production Economics, 165 (2015): 51–63.

[31] Pechstein, Arndt; Schwemmle, Martin: Future Skills Navigator: Ein neues Menschsein für die Welt von morgen, München 2023.

Sidebar 2: Allbirds—Sustainability and Profitability: Two Birds with One Stone?
Allbirds is a sustainable footwear (and partly apparel) company known for simplicity in design, confidence in comfort, and made from nature. The company, co-founded in 2015 by former New Zealand professional soccer player Tim Brown and San Francisco-based biotech engineer Joey Zwillinger, emphasizes environmental responsibility and transparency across its entire value chain. Driven by constantly innovating with new materials, such as merino wool (Wool Runners®), Allbirds is targeting comfortable and sustainable sneakers while meeting high-performance product standards.

With a strong focus on reducing the CFP of footwear (SDG 13)—traditionally a high-impact industry—Allbirds' life cycle assessment (LCA) involves examining materials, vetting supplier practices, factory energy use, packaging, shipping methods, and end-of-life disposal. By utilizing "waterfall charts," Allbirds measures the CFP of every product designed, produced, and sold from "brief-to-box." The company publicly shares its "Flight Plan," outlining carbon reduction goals and progress toward carbon neutrality, with aggressive targets such as Net Zero by 2030.[32]

Along this journey, in early 2025, the M0.0NSHOT shoe was successfully launched as the first zero-carbon shoe (0.0 kg of CO_2). Allbirds is one of the only companies to print a CO_2 emissions label on each shoe, helping consumers understand the product's environmental footprint. The company has adopted an open-sourcing approach for some materials, such as the plant-based EVA midsole made from sugarcane, co-developed with Brazil-based Braskem, a CO_2-negative contribution toward its Net Zero goals.[33] The methodology and process have been documented in their "Recipe B0.0k," an open-sourced toolkit inviting the footwear industry to follow their footsteps and emulate similar activities to accelerate sustainable innovation.

Prior, in recognition of Earth Day on April 22, 2021, Allbirds had announced to open source a version of its proprietary Carbon Footprint Calculator by sending a letter (published as a full-page advertisement in the New York Times) to 16 apparel and footwear companies, incl. Adidas, H&M, New Balance, Nike, Zara, raising awareness among their key competitors and seeking a broader industry leverage for responsible consumption and production (SDG 12) in the entire value chain.[34]

[32] Toffel, Michael W.; Pucker, Kenneth P.; Kuzucu, Eren: Allbirds: Decarbonizing Fashion (A), Harvard Business School case 9–622-024, Boston 2021.

[33] Toffel et al., 2021, ibid.

[34] Rašković, Matevž; Nagar, Swati: Driving Social Sustainability through Strorytelling, in: Šilenskytė, Aušrinė; Cordova, Miguel; Schmitz, Marina A.; Toh, Soo M. (Eds.): The Palgrave Handbook of Social Sustainability in Business Education, Cham 2024: 209–225.

At the same time, the young company has been struggling to perform financially, driven by a softening and derailing from its core business model. By adding apparel and accessories, expanding its footwear product portfolio into segments outside the comfort positioning, experiencing serious product quality issues in its manufacturing network, and trying to compete in the physical distribution via brick-and-mortar flagship stores, Allbirds "had lost its footing."[35] A refocus and turnaround plan led by a new management team is underway, trying to ensure sustainability and profitability at the same time.

13.4 Connected Resources: Multi-Stakeholder Partnerships and Ecosystems

Nearly all multi-stakeholder initiatives for sustainability incorporate not-for-profit and/or non-governmental organizations, like the WWF (World Wide Fund for Nature, formerly World Wildlife Fund) or the WEF (World Economic Forum). By involving third-party governance through independent NGOs and ensuring neutrality, MSPs provide a safe space to address systemic sourcing risks (e.g., child labor, deforestation, toxic chemicals, conflict minerals), a platform to share common pain points, jointly develop transparency and traceability, and enable the development of industry standards for scalability. The independent nature of NGOs in membership-based MSPs is actively facilitating many of the cross-company, cross-industry initiatives on SDGs, working as "meta governors."[36]

Connected Resources through industry-wide MSPs and in extended cross-industry ecosystems require a different set of performance metrics. Instead of functional or company-specific proprietary key performance indicators, typically output- or outcome-related, industry-wide initiatives, like most collaborative MSPs to meet the 17 SDGs, are moving toward impact-driven metrics. H&Z Consulting has created a sustainability radar, identifying key industry sectors, like chemicals, logistics, oil & gas, packaging, or textiles, that are characterized by a high category exposure and a high sustainability impact.[37] Such ESG impact metrics, beyond the scope of single organizations, can be defined as effects on root causes and sustained systemic change (e.g., a drop in poverty, obesity, or illiteracy) measured in terms of

---

[35] Bellezza, Silvia; Wilson, Annie: Allbirds: Can the sustainable shoe company reinvigorate the brand, Columbia Business School case CU380, New York 2024.

[36] Mariani, Laura, Trivellato, Benedetta, Martini, Mattia; Marafioti, Elisabetta: Achieving Sustainable Development Goals Through Collaborative Innovation: Evidence from Four European Initiatives, in: Journal of Business Ethics, 180 (2022): 1075–1095.

[37] Aichbauer et al. 2022. op cit.

communities, populations, or ecosystems.[38] This is achieved via intensified knowledge sharing, the leveraged development and application use of shared platforms, and jointly generated transparency. Hence, "the era of corporations integrating sustainable practices is being surpassed by a new age of corporations actively transforming the market to make it more sustainable."[39]

Tony's Chocolonely, the Dutch-based confectionery Business-to-Consumer (B2C) company, can be considered a trailblazer that has successfully transformed an entire industry toward meeting various of the 17 SDGs through MSPs (see sidebar 3).

Sidebar 3: Tony's Chocolonely—The Notorious SDGs: Go Sweet for Sustainability!

Tony's Chocolonely has revolutionized the entire cocoa value chain: from sourcing beans in West Africa (mainly Ghana and the Ivory Coast), grown by millions of farmers, through multiple tiers and layers of cooperatives, brokers, exporters, processors, fast-moving goods manufacturers, to selling confectionery products via retailers to millions of consumers. Founded in 2005 by Dutch journalist Teun (hence: Tony) van de Keuken, Tony's Chocolonely converts its sustainable purpose to "make chocolate 100% slave-free" (SDG 8) into a strategic roadmap defined as: "leading by example, creating awareness, and inspiring others."

Leading by example is based on their widely marketed five sourcing principles[40]: traceable beans ("bean tracker," a digital blockchain technology platform), higher prices (living income to farmers), strong farmers (training and development, educating and coaching of farmers and cooperatives), the long term (income commitment for 5 years), and better quality and productivity (professionalization). By fully segregating the material flow in its supply chain and ensuring the traceability of all cocoa components after processing (liquor, butter, powder), Tony's Chocolonely has positioned itself as the leading sustainable chocolate brand.

[38] Ebrahim, Alnoor; Rangan, V. Kasturi: What impact? A framework for measuring the scale and scope of social performance, in: California Management Review, 56/3 (2014): 118–141.

[39] Hoffman, Andrew J.: The next phase of business sustainability, in: Stanford Social Innovation Review, 16/2 (2018): 34–39.

[40] Pot, Viktor; van Wassenhove, Luk N.: Tony's Chocolonely: The road to 100% slave-free chocolate, INSEAD Business School case IN 1767, Fontainebleau 2021.

Creating awareness is mainly achieved through an unapologetically outspoken and unique form of information and communication with the demand chain, especially the consumers. The chocolate bars, unlike any other brand, are unequally divided, symbolizing the unequal distribution of income and profits in the cocoa value chain. The colorful and bright packaging wrappers with retro typography are the main communication tool to educate and update consumers, providing insightful information on major value chain milestones and achievements, e.g., through the child labor monitoring and remediation system (CLMRS).

Inspiring others is driven by Tony's "Open Chain" initiative of attracting all major stakeholders in the value chain to join the sourcing principles for an industry-wide impact. Besides suppliers and customers, there are meanwhile more than 20 competitors and complementors as actively participating "mission allies." As an example, Ben & Jerry (Unilever) joined the Open Chain and has marketed "Chocolatey Love A-Fair," co-branding ice cream and chocolate bars inspired by their new sustainability partnership.[41]

Tony's ability to create a unique selling proposition (USP) based on a unique buying proposition (UBP), driven by co-created tangible and intangible assets, and openly shared with all stakeholder groups in an extended value chain network, is a prime example of successful differentiation and competitive advantage through sustainability. By raising awareness of ongoing problems in the cocoa value chain, including deforestation (SDG 15), the company encourages and motivates consumers to demand more ethical and sustainable chocolate.

In this vein, Tony's Chocolonely views itself as an impact company that sells chocolate, not a chocolate company trying to make an impact.

13.5 Collaborative Sourcing

Joining forces for combined sourcing across multiple organizations is not a new phenomenon. Although buying cooperatives dates back as early as the eighteenth century, the movement gained significant momentum in the middle of the nineteenth century, both in the primary (agriculture, farming) and in the secondary (manufacturing) sector. Besides other reasons (e.g., credit unions, production and/or housing cooperatives), purchasing cooperatives were among the main driving forces for joint activities. In response to increasing price pressure in many industries

[41] Pannekoek, Frans; Breugem, Thomas; van Wassenhove, Luk N.: How Tony's Chocolonely created a purpose-driven (and profitable) supply chain, in: www.hbr.org, 11.09.2023. https://hbr.org/2023/09/how-tonys-chocolonely-created-a-purpose-driven-and-profitable-supply-chain

resulting from the first mass production, cooperatives offered small, single businesses the benefits of economies of scale, leveraging larger discounted volumes. Until today, purchasing cooperatives play a significant role in various sectors, e.g., food retail (Edeka, REWE in Germany, Coop in Switzerland), agriculture and farming (Dairy Farmers of America, Land O'Lakes in the USA), and financial services (Crédit Agricole in France).

During the 1980s/early 90s, *joint buying* activities became an important, yet tactical and operational measure to improve costs for materials and services and to ensure supply continuity in scarce supply markets by teaming up with other industry players. Mindful of legal anti-trust restrictions, e.g., the EU "safe harbor threshold" of a combined market share of less than 15%, the cooperation partners remained largely independent compared to membership-based associations and cooperatives, targeting proprietary benefits. For example, Premier, Inc., a US network of healthcare providers, negotiates contracts with medical suppliers, pharmaceutical firms, and equipment manufacturers to reduce procurement costs for participating hospitals.

By extending the scope across markets and industries and leveraging not only volumes, but also improving purchasing processes, *consortia (group) buying* activities evolved. Joint purchasing offices with a specific geographical scope, e.g., the Otto Group Asian Desk in Hongkong, founded in 1966 and later on expanded to provide sourcing services to other companies, or the third-party site management in industry parks, e.g., by Infraserv, a former Hoechst AG company, to consolidate the sourcing of services, like energy, facility, and/or waste management, are prime examples for such group purchasing activities.

By the mid-1990s, companies started to realize diminishing returns on their vertical, cooperative activities focused on cost and supply security from joint buying and group purchasing. Especially the lack of growth opportunities, the missing out on innovation from upstream markets, triggered a mindset shift toward a horizontal and collaborative sourcing approach. First, the IT industry successfully introduced *open sourcing* to innovate the development of software products and services. Linux, an important software kernel for many operating systems, introduced by the Raleigh, NC-based company Red Hat (acquired in 2018 by IBM), is freely shared and collaboratively developed.[42] Hence, Jim Whitehurst, former CEO of Red Hat, dedicated his bestselling book "Open Organization to the millions of open-source contributors and users out there who make what we do possible."[43] Other noticeable open sourcing projects, besides Allbird's CFP calculator (Sidebar 2) and Tony's Open Chain (Sidebar 3), are Starbucks' open-source agronomy approach in Costa

[42] Sandberg, Kirsten D.: Open Source for Sustainability. How projects under the Linux Foundation are helping to advance the UN Sustainable Development Goals, September 2023.

https://www.linuxfoundation.org/blog/open-source-for-sustainability-how-linux-foundation-projects-accelerate-progress-toward-the-united-nations-sustainable-development-goals-sdgs#

[43] Whitehurst, 2015, op cit.

shared benefit (horizontal, *growth*)	**open sourcing** (innovation-driven)	**collaborative sourcing** (value-driven)
proprietary benefit (vertical, *scarcity*)	**joint buying** (cost-driven)	**consortia (group) buying** (process-driven)
	limited scope (product & service)	**extended scope** (markets & industries)

Fig. 13.2 Collaborative sourcing, developed by the author, all rights reserved, used with permission

Rica,[44] Dow Chemicals and Ralph Lauren's jointly developed ECOFAST™ Pure sustainable textile treatment,[45] or the Green Fiber Bottle™ by Carlsberg,[46] all defined by shared, instead of proprietary benefits.

Extending the scope above and beyond innovation toward the overall growth and development of an entire market and/or industry, instead of a specific product or service category, defines *collaborative sourcing* as a horizontal value chain strategy driven by shared sourcing benefits.[47] Dedicated purchasing initiatives include the former Global Impact Sourcing Coalition (GISC) of the Business for Social Responsibility (BSR) non-profit network, or the recently formed UN Global Compact Coalition for Sustainable Procurement, bridging between Chief Procurement and Sustainability Officers (CPOs and CSOs) and the public sector, and aligning supply chain transformation projects toward the 17 SDGs.

Figure 13.2 Summarizes the various approaches to pooled purchasing concepts, building on and expanding a few already existing categorizations[48]:

---

[44] Starbucks: Why you should care about open-source agronomy, 14.01.2022. https://historias.starbucks.com/en-es/stories/2022/open-source-agronomy

[45] Ralph Lauren and Dow open-source new process to transform how the fashion industry dyes cotton, www.corporate.ralphlauren.com, 13.10.2021.

[46] Chesbrough, Henry W.; Bogers, Marcel; Strand, Robert: Sustainability through open innovation: Carlsberg and the green fiber bottle, Haas School of Business Case B5922, University of California, Berkeley 2018.

[47] Fawcett, Stanley E.; Magnan, Gregory, M., McCarter, Matthew W.: A three-stage implementation model for supply chain collaboration, in: Journal of Business Logistics, 29/1 (2008): 93–112.

Loseby, David L.: Collaboration: A value enabler for procurement and supply chain, CIPS report 2019.

[48] Eßig, Michael: Purchasing consortia as symbiotic relationships: developing the concept of "consortium sourcing", in: European Journal of Purchasing and Supply Management, 6/1 (2000): 13–22.

Eßig, Michael: Cooperative Sourcing, in: Hahn, Dietger; Kaufmann, Lutz (Eds.): Handbuch Industrielles Beschaffungsmanagement, 2nd ed., Wiesbaden 2002: 263–280.

Schotanus, Fredo; Telgen, Jan: Developing a typology of organisational forms of cooperative purchasing, in: Journal of Purchasing and Supply Management 13/1 (2007): 53–68.

Schotanus, Fredo; Telgen, Jan; de Boer, Luitzen: Critical success factors for managing purchasing groups, in: Journal of Purchasing and Supply Management 16/1 (2010): 51–60.

As Cammarano et al. show in their research study 2022[49] on SDG-oriented supply chains, procurement plays a significant role in almost any ESG-driven MSP through *collaborative sourcing* for sustainability, emission reduction, circular economy, human rights, natural resource management, and traceability and transparency. However, two independent literature reviews on sustainability in Purchasing and Supply Management (PSM) confirm that so far "only a few studies shed light on the direct linkages between collaboration and the SDGs,"[50] nor "touched specifically upon SDGs."[51]

Hence, Table 13.1 provides an overview of the scope and role of collaborative sourcing in MSP for meeting the 17 SDGs. As there are likely several hundreds, if not thousands, of active MSPs at various levels (local, regional, national, global) within and across all business sectors (primary, secondary, tertiary), the exemplary list serves as a first structured attempt.

Most collaborative sourcing initiatives are related to the secondary, manufacturing sector, contributing predominantly to improvements of Responsible Consumption and Production (SDG 12), Climate Action (SDG 13), and Partnerships for Goals (SDG 17). Nonetheless, the sample listing of alliances and coalitions highlights the fact that sourcing positively impacts across all SDGs as an important "supply chain collaboration mechanism,"[52] like collaborative advertising, forecasting, innovation, and transportation.

13.5.1 Pre-Competitive (Same Industry) Sourcing

Pre-competitive, same industry sourcing combines initiatives that are designed to improve the collaboration of buying organizations with their multi-tier supplier base. Instead of separate and often contradicting activities and measures, the collective effort across the industry ensures an aligned and structured approach to improving the combined supply chain performance toward economic, environmental, and social targets. It acknowledges the fact that in consolidated, oligopolistic markets, many suppliers work with similar customers, as many customers work with the same supplier base. Thus, an industry standard is more likely to be implemented and

[49] Cammarano, Antonello; Perano, Mirko; Michelino, Francesca; Del Regno, Claudio; Caputo, Mauro: SDG-oriented supply chains: business practices for procurement and distribution, in: Sustainability, 14/3 (2022): 1325.

[50] Chauhan, Chetna; Kaur, Puneet; Arrawatia, Rakesh; Ractham, Peter; Dhir, Amandeep: Supply chain collaboration and sustainable development goals (SDGs). Teamwork makes achieving SDGs dream work, in: Journal of Business Research, 147/8 (2022): 290–307.

[51] Miandar, Toloue; Johnsen, Thomas E.; Caniato, Federico: The role of purchasing and supply management in diffusing sustainability in supply networks: A systematic literature review, in: Business Ethics, the Environment & Responsibility, 33 (2024): 505–532.

[52] Chauhan et al., 2022, op cit.

Table 13.1 Scope and role of collaborative sourcing in MSP for meeting the 17 SDGs, developed by the author, all rights reserved, used with pieces

Collaborative sourcing/SDGs #	1	2	3	4	5	6	7	8	9	10	11	12
Pre-competitive (same industry) sourcing												
Together for Sustainability (TfS)												x
Sustainable Apparel Coalition (SAC)								X				x
Leather Working Group (LWG)						x						X
Responsible Minerals Initiative (RMI)								x				x
Global Cement and Concrete Association (GCCA)							X		x		x	
Ecosystem (cross-industry) sourcing												
Renewable Energy (RE100)							X					
Extractive Industry Transparency Initiative (EITI)												x
Open Charge Alliance (OCA)							x		x		X	
Sustainable Freight Buyers Alliance (SFBA)									x			
Getting to Zero Coalition (GtZ)												
Value chain (network) sourcing												
Unstereotype Alliance (UA)					X					x		
Cisco Network Academy (NetAcad)				X				x		x		
Better Cotton (BC) & Beyond Cotton	x	X										x
Women in Informal Employment (WIEGO)	X				x					x		
Global Alliance for Vaccines and Immunisation (GAVI)			X	x	x							
Regenerative (circular) sourcing												
Roundtable on Sustainable Palm Oil (RSPO)												X
Global Battery Alliance (GBA)							X					x
Forest Stewardship Council (FSC)												x
Marine Stewardship Council (MSC)												x
Alliance for Water Stewardship (AWS)						X						x

Collaborative sourcing/SDGs #	13	14	15	16	17	Industries	Websites	Founded	Founding members	Σ members	Annual report	Purpose	Collaborative sourcing (examples)
Pre-competitive (same industry) sourcing													
Together for Sustainability (TfS)	x				**X**	Chemicals	www.tfs-initiative.com	2011	BASF, Bayer, Evonik, Henkel, Lanxess, Solvay	58	x	Accelerate the dvlpmt. of sustainable chemical supply chains	ESG assessments/ audits, CFP calculations, supplier training
Sustainable Apparel Coalition (SAC)	x					Textiles	www.cascale.org	2009	Patagonia, Walmart	>300	x	Drive sustainable/ equitable apparel/footwear	ESG assessments (Higgs-index), CFP calculations, supplier dvlpmt.
Leather Working Group (LWG)			x			Leather	www.leatherworkinggroup.com	2005	Adidas, Nike, New Balance, a.o.	1000	x	Promote responsible leather sourcing, production, usage	Life cycle assessments, deforestation-free leather sourcing
Responsible Minerals Initiative (RMI)				**X**	x	Mining/ Extraction	www.responsiblemineralsinitiative.org	2008	RBA members, Global e-Sustainability Initiative	>500	x	Promote responsible, conflict-free minerals sourcing	Shared due diligence of conflict-free smelters (3TG) for sourcing
Global Cement and Concrete Association (GCCA)	x					Construction	www.gccassociation.org	1999	Cemex, Lafarge, Holcim, HeidelbergerCement, a.o.	40	x	Promote sustainable cement/concrete construction	Development of alternative CO2 free-/ low-materials and processes
Ecosystem (cross-industry) sourcing													

(continued)

Table 13.1 (continued)

Collaborative sourcing/SDGs #	13	14	15	16	17	Industries	Websites	Founded	Founding members	Σ members	Annual report	Purpose	Collaborative sourcing (examples)
Renewable Energy (RE100)	x				x	Energy	www.there100.org	2014	IKEA, Swiss RE, Climate Group, Carbon Disclosure Project (CDP)	415	x	Drive renewable electricity transition	Pooling demand for Power Purchase Agreements (PPA)
Extractive Industry Transparency Initiative (EITI)				**X**	x	Oil & Gas	www.EITI.org	2003	Companies, governments, civil organizations	55*	x	Promote transparency/ accountability in oil/gas/mining	Responsible sourcing standards based on LME and OECD guidelines
Open Charge Alliance (OCA)	x					Mobilty	www.openchargealliance.org	2014	ElaadNL, ESB, Irish Utility, Greelots	>400	x	Promote open EV charging standards	Joint development, certification, testing of open charge protocols
Sustainable Freight Buyers Alliance (SFBA)	**X**	x			x	Transportation	www.smartfreightcentre.org	2022	WEF, Smart Freight Centre, BSR	>200	x	Decarbonize freight through collective action	Environmental attribute certificates (EACs) on low-carbon logistics
Getting to Zero Coalition (GtZ)	x	**X**			x	Shipping	www.globalmaritimeforum.org	2019	WWF, Global Maritime Forum, Friends of Ocean Action	215	x	Shaping the future of sustainable maritime trade	Development and launch of Zero Emission Vessels (ZEVs)
Value chain (network) sourcing													
Unstereotype Alliance (UA)						FMCG/ Advertising	www.unstereotypealliance.org	2017	UN Women, AT&T, J&J, P&G, Unilever	240	x	Promote equality, remove harmful stereotypes	Global studies/research, sourcing of unstereotype media & comm.

Collaborative sourcing/SDGs #	13	14	15	16	17	Industries	Websites	Founded	Founding members	Σ members	Annual report	Purpose	Collaborative sourcing (examples)
Cisco Network Academy (NetAcad)						IT	www.netacad.com	1997	Cisco	>12.000**	(x)	Promote & train digital upskilling	Development of course material, recruitment of IT skilled personnel
Better Cotton (BC) & Beyond Cotton						Agriculture	www.bettercotton.org	2005	WWF, GAP, H&M, IKEA, a.o.	>2.700	x	Promote sustainable cotton production	Sourcing of BC-certified material through chain of custody standard
Women in Informal Employment (WIEGO)						Informal Economy	www.wiego.org	1997	Activists, researchers	153	x	Empower and protect informal workers	Collaborative market engagement activities for selling goods/services
Global Alliance for Vaccines and Immunisation (GAVI)						Pharmaceuticals	www.gavi.org	2000	WHO, UNICEF, World Bank, Bill & Melinda Gates Foundation	28*	x	Increase the equitable/sustainable use of vaccines	Pooling demand for vaccines/immunization in lower-income countries
Regenerative (circular) sourcing													
Roundtable on Sustainable Palm Oil (RSPO)	x		x			Agriculture	www.rspo.org	2004	WWF, Malaysia Palm Oil Ass., Unilever, AAK, Migros	~ 6.000	2x	Make palm oil production/consumption sustainable	Trading of credits (book & claim), sourcing of RSPO-certified material
Global Battery Alliance (GBA)	x					Automotive	www.globalbattery.org	2017	BASF, Audi, Tesla, LG Energy, Eurasian Resource Group, a.o.	>170	x	Establish a sustainable/responsible battery value chain	Use of battery passport (digital twin), sourcing of recycled Co/Li/Ni

(continued)

Table 13.1 (continued)

Collaborative sourcing/SDGs #	13	14	15	16	17	Industries	Websites	Founded	Founding members	Σ members	Annual report	Purpose	Collaborative sourcing (examples)
Forest Stewardship Council (FSC)	x		X			Constr./Packg./FMCG	www.fsc.org	1993	Businesses, environmentalists, communities	>1.100	x	Promote responsible forest management	Development of forest certification and sourcing of FSC-certified wood
Marine Stewardship Council (MSC)		X			x	Fishing/Seafood	www.msc.org	1997	WWF, Unilever	>700	x	Certify sustainable fisheries and protect oceans	Sourcing MSC-certified blue fish label (sustainable catch practices)
Alliance for Water Stewardship (AWS)					x	Water Utility	www.a4ws.org	2009	WWF, The Nature Conservancy, Ecolab	>200	x	Promote and certify responsible water usage	Treatment of responsible freshwater practices, co-funding of basins

X primary, x secondary SDG targets
* Countries; ** Educational institutions

accepted. Most of the same industry initiatives are centered on prosperity (SDGs 7–11), the planet (SDGs 12–15), and partnerships (SDG 17) by setting standards, developing tools (carbon accounting, life cycle analysis, databases, supplier scorecards, indices), and scaling for standardization.

Together for Sustainability (TfS) is a successful initiative of the Chemical Industry, founded in 2011 by six European-based companies, steadily grown into a global alliance with currently 58 firms, representing an annual turnover > € 500 bill. TfS members collaboratively assess (standardized tools and methodologies),[53] audit (EcoVadis partnership), build, train, and develop (shared academy) sustainability capabilities to improve the performance of the high-impact chemical supply chain. The overarching goal for sourcing is to create a mindset change (positive incentive) among suppliers to proactively partner for joint sustainability goals (SDG 17)[54] and implement tangible steps at their respective organizations.[55]

The **Sustainable Apparel Coalition (SAC)**, in 2024 renamed "Cascale," emphasizing collective action (Ca) and scale ambition, was originally formed in 2009 by outdoor apparel company Patagonia and Walmart. The non-profit comprises over 300 fashion brands, retailers, manufacturers, NGOs, and governments. Its Higgs Index (named after British physicist and Nobel laureate Peter Higgs), an apparel and footwear self-assessment tool to measure sustainability performance,[56] is actively used by 25,000 organizations. The coalition, often used as a channel of influence in the industry,[57] supports the Social & Labor Convergence Program (SLCP) to drive decent work and economic growth (SDG 8), and has set a goal of a 45% reduction in greenhouse gas emissions by 2030.

The **Leather Working Group (LWG)** is a not-for-profit MSP setting industry standard, ensuring, through its auditing process, that "leather is sourced, produced, and used sustainably, protecting people and the planet." Beyond the primary focus on responsible production practices in the tanneries (SDG 12) and the use and reuse of safe water free of pollutants and contaminants (SDG 6), some brands (e.g., Koio) have aligned their sourcing for animal hides and skins with farming projects focused on sustainable land and water protection practices. Track and trace functionality, supported by blockchain technology,[58] is vital for sustainable leather sourcing, creating a secure, fixed digital ledger that records the product's journey from farm to fashion.

---

[53] Together for Sustainability: Assessment process: Together for sustainability, 2016 https://tfs-initiative.com/assessment-process/

[54] Aichbauer et al., 2022, op cit.

[55] Fröhlich et al., 2021, op cit.

[56] Hoffman, Andrew J.: Building transparency within the Sustainable Apparel Coalition: The road to successful pre-competitive collaboration, Michigan Ross School of Business Case W93C99, Ann Arbor 2020

[57] O'Rourke, Dara; Strand, Robert: Patagonia: Driving sustainable innovation by embracing tensions, in: California Management Review, 60/1 (2017), 102–125.

[58] Heldt, Lisa; Pikuleva, Ekaterina: When upstream suppliers drive traceability: A process study on blockchain adoption for sustainability, in: International Journal of Physical Distribution & Logistics Management, 55/3 (2025): 196–222.

Part of the mining and extraction industry is arguably in conflict with numerous SDGs, related to environmental (land degradation, contaminated water), social (human rights abuses, armed conflicts), and economic (increased inequalities) targets. The **Responsible Minerals Initiative (RMI)**, created in 2008 by the Responsible Business Alliance (RBA), the Global e-Sustainability Initiative (GeSI), and US-based electronic companies (e.g., Cisco, HP, Apple, Dell), serves as an umbrella organization to tackle these critical ESG issues.[59] The Responsible Minerals Assurance Process (RMAP) identifies smelters and refiners for the conflict-free (SDG 16) sourcing of minerals (especially for "3TG"—tin, tantalum, tungsten, and gold), in alignment with the corresponding OECD goals.

The cement and concrete industry is responsible for 8% of the global CO_2 emissions (1.6 bill. metric tons per year), mainly through the high-temperature (1.500 °C) manufacturing process of converting limestone to clinker. To achieve a near-zero-emission cement production by 2030, the **Global Cement and Concrete Association (GCCA)** partners with the World Business Council for Sustainable Development (WBCSD) and collaborates with architects, engineers, and innovators. Pioneering concepts are currently being addressed by 40 industry-leading companies, for example, replacing limestone with carbon-free calcium silicate, sourcing low-carbon fuel alternatives, or developing carbon capture and storage technologies across the process (SDG 7).

13.5.2 Ecosystem (Cross-Industry) Sourcing

Collaborative cross-industry sourcing is formed through MSPs for ecosystem protection and improvements utilized by a multitude of industrial sectors, e.g., energy, oil and gas, mobility, shipping, and transportation. The partnering entities are often from very different private (primary, secondary, tertiary), public/civil, and community backgrounds, and focus on overall infrastructure, technology, new materials, and biodiversity mostly related to the planet (SDG 12–15).

The **Renewable Energy (RE100)** initiative is a global collective effort of more than 400 organizations worldwide to push for 100% alternative (green) electricity based on wind, solar, hydro, geothermal, and biomass energy sources. Initiated by IKEA, Swiss Re(insurance), the Climate Group, and the Carbon Disclosure Project (CDP) in 2014, the member organizations are collaborating for climate actions, increasingly relevant toward all key stakeholders, i.e., customers, suppliers, current and future employees, investors, and authorities. Their sourcing activities involve pooling electricity demand to sign large-scale Power Purchase Agreements (PPAs), to influence policy change, and to accelerate a transition to renewable energy (SDG 7).

[59] Aichbauer et al., 2022, op cit.

Many governmental organizations are involved in the **Extractive Industry Transparency Initiative (EITI)**, defining international standards and protocols for how the oil, gas, and mining industries operate their extraction business (e.g., coal, copper, iron ore), focused on peace and justice (SDG 16). There are currently 55 countries across all continents implementing the EITI standards, plus supporting nations (e.g., USA, Australia) that promote the initiative without domestic adaptation. The EITI standard builds on other initiatives, like the responsible sourcing requirements of the London Metal Exchange (LME), which in turn are based on OECD due diligence guidance, following a rigid assessment of potential sourcing risks and compliance measures.

The **Open Charge Alliance (OCA)** is a non-profit MSP involving a wide range of industries, from charging station producers, grid operators, electric vehicle manufacturers, software providers, to testing labs. The consortium focuses on developing and maintaining open standards for electric vehicle (EV) charging infrastructure, specifically the Open Charge Point Protocol (OCPP). Based on collaborative participation, rapid prototyping, meritocracy, and community-driven hard- and software development, the alliance ensures best interoperability, fostering a flexible, scalable, and competitive EV mobility market for sustainable cities and communities (SDG 11).

Next to electricity production, the transportation industry is the second-largest contributor, responsible for over 13% of global CO_2 emissions. Two of many MSPs are collaborating across industries to decarbonize through collective guidelines and actions: the **Sustainable Freight Buyers Alliance** (SFBA) aims to accelerate the shift to zero-emission freight across **all** modes of transport, including road, air, and ocean shipping (SDG 13). Meanwhile, the **Getting to Zero Coalition (GtZ)** concentrates on maritime logistics (SDG 14). The SFBA, launched in 2022, is uniting various existing partnerships under the auspices of the WEF, the Smart Freight Center, and the BSR network. Its focus is on collaborating on projects, leveraging the collective sourcing power of the alliance members, and working with policymakers to lower the entry barriers for green freight solutions. The overarching goal of the GtZ alliance, with more than 200 organizations (incl. 180 private companies like Amazon, Ford, Microsoft, Unilever) in the maritime, energy, infrastructure, and finance sectors, led by the WWF, Global Maritime Forum, and Friends of Ocean Action group, is to get commercially viable deep-sea vessels powered by zero-emission fuels into operation by 2030.

13.5.3 Value Chain (Network) Sourcing

Value chain, network sourcing is a strategic procurement approach toward co-creation and exchange of **tangible and intangible value**, beyond just the physical flow of goods and services. It focuses on building interconnected, collaborative, and dynamic relationships among all stakeholders to achieve a shared goal and generate mutual benefit. This includes, but is not limited to, **j**oint sourcing from smallholders, women-/minority-owned businesses, and/or social enterprises in developing countries. Ensuring inclusiveness and equitable value distribution, it has a particularly strong impact on the social SDGs (SDG 1–6).

The **Unstereotype Alliance (UA)** is an initiative convened by UN Women in collaboration with various global-scale marketing organizations (e.g., AT&T, J&J, P&G, Unilever) to eliminate harmful stereotypes and prejudices in media and advertising. The UA's Unstereotype Metric (UM) is an open-access tool to determine the respective progressive or regressive image of a female/male character in advertisements. By applying the 3P framework, i.e., presence (who is featured), perspective (what viewpoint is on display), and personality (authenticity of characters), the sourcing of marketing media and communications can be directed toward an increasingly diverse and inclusive society model (SDG 5).

Created in 1997, the **Cisco Network Academy (NetAcad)** was originally designed in response to the fast-developing computer network industry and the lack of qualified programming talent. Over time, it has been expanded to a global online curriculum for high school and college students, adopted by more than 12,000 educational institutions for training and upskilling (SDG 4). Powered by a vision for an inclusive future for all, meanwhile, the NetAcad open university offers a variety of free, self-paced courses in various areas like networking, cybersecurity, data science, and AI. Cisco NetAcad provides collaborative sourcing (recruitment and hiring) functionality through its partnerships with businesses and organizations, connecting employers with IT-trained and skilled individuals via the talent bridge tool.

Better Cotton (BC) is a Swiss-based, global initiative and governance group contributing toward more sustainable agriculture, particularly addressing inequity and hunger in farming communities (SDG 2). The **Beyond Cotton** project is a regional add-on that builds on the large-scale global BC initiative to strengthen family farming, advance food security and nutrition, promote climate-resilient livelihoods, and increase farmer incomes through improved agricultural practices and market access. Focused on (South-)East Africa, the program is tailored to meet the specific needs of the local smallholders by providing education and training on reducing chemical use, improving soil health, integrating other food crops (like maize and beans), increasing cotton and food yields, and marketing cotton by-products. The value generated upstream by the farmers—meanwhile, 23% of global production is BC-certified—is appreciated downstream by the manufacturers and retailers, like GAP, H&M, and IKEA.[60] BC facilitates collaborative sourcing through its **Chain of Custody (CoC) standard.** Brands and retailers can choose the model that best fits their respective ESG goals, and their sourcing directly funds training for farmers on more sustainable practices.

[60] van Hoeven, Helen: Sourcing and procurement as driving forces for sustainable business, Initiatief Duurzame Handel (IDH), Utrecht 2009.

Women in Informal Employment: Globalizing and Organizing (WIEGO) is a UK-based, global MSP network focused on improving conditions for workers in informal economies, started by social activists in 1997. WIEGO covers a wide array of human labor, from street vendors, waste pickers, and home-based work to unregistered small, non-regulated businesses. Workers in these economies lack formal labor protection, health care, insurance, and retirement benefits, and as a result, are often living below the poverty line (SDG 1). The collaboration is restricted to membership-based public organizations (e.g., trade unions), researchers, and (non-) governmental agencies, but purposely excludes any private sector influence. Although there is no dedicated collaborative sourcing program, the platform offers collaborative market engagement activities for selling goods and services.

The **Global Alliance for Vaccines and Immunization (GAVI)**, in short Vaccine Alliance, is a public-private healthcare partnership with the World Health Organization (WHO), UNICEF, the World Bank, and the Bill & Melinda Gates Foundation as founding members. Missioned to save the lives and protect children's health by increasing the equitable and sustainable use of vaccines (SDG 3), GAVI played an important role during the global COVID-19 pandemic. By co-leading and administering COVAX (COVID-19 Global Access), the alliance was able to provide close to 2.0 billion vaccine doses to 146 lower-income countries. GAVI's collaborative sourcing is a core element of its Advanced Market Commitment (AMC), pooling the demand for vaccines and immunization from least developed economies and providing the vaccine manufacturers with predictable demand forecasts.

13.5.4 Regenerative (Circular) Sourcing

William A. McDonough, the American architect, known for the green and sustainable design movement, once said: "Nature doesn't have resources, it has sources," and therefore, "it is our job as humans to turn them into re-sources that can be used over and over."[61] Regenerative Procurement[62] is embedding this manifesto, actively restoring and improving ecosystems with a strong impact on systemic (economic) transformation. Through circular sourcing decisions, focused on the planet (forests, oceans, water, land, and agriculture; SDG 12–15), it is shifting product life cycles from cradle-to-grave toward a cradle-to-cradle design.

---

[61] McDonough, William A.; Braungart, Michael: Cradle to Cradle: Remaking the way we make things, New York 2002.

[62] Regnier, Elvire: Regenerative Procurement: Transforming procurement and supplier ecosystems for more resilient companies, Paris 2025.

The **Roundtable on Sustainable Palm Oil (RSPO)** is a non-profit group established in 2004 to promote the production and consumption (SDG 12) of sustainable palm oil products through global standards and MSP governance. With more than 6,000 member companies across all industry sectors,[63] from small-scale farmers to consumer-packaged goods (CPG) manufacturers, it has become a best practice industry standard. Growers must comply with RSPO's principles and criteria (P&C), especially that no new plantings cause deforestation or replace "high conservation value" forests and peatlands.[64] Once certified and registered in the RSPO PalmTrace system, participants can start trading credits. CPG companies (e.g., Unilever, Nestlé), in return, guarantee with RSPO-certified sourcing, the conservation of biodiversity, fair labor practices, safe working conditions, and scalable, continuous improvement programs.

The **Global Battery Alliance (GBA)** is a public-private collaboration platform, originally formed by the WEF in 2017, and incorporated as a non-profit in 2022. The alliance is attempting to overcome the complex challenges of the entire automotive EV battery (EVB) value chain, from extraction through processing, manufacturing, and vehicle end-use, with a circular concept. As one important step in this direction, the GBA introduced the first Battery Passport, a digital twin of the physical battery that tracks and traces (T&T) key product information about sustainability and lifecycle requirements. Based on the T&T data, circular regional value chains can develop (SDG 8) mechanical, pyro-, and hydro-metallurgical processing for the sourcing of battery-grade recycled cobalt, lithium, nickel, and graphite.[65]

The **Forest Stewardship Council (FSC)** collaborates on regenerative practices in the lumber industry, mainly, reforestation of degraded lands, natural or assisted replantation after logging, and maintaining forest health and resilience (SDG 15). Already formed in 1993 as one of the first MSPs on ESG by downstream businesses (retailers), environmentalists, and communities, it addresses the vital role of woodlands in absorbing CO2 emissions from the atmosphere and converting them through photosynthesis **into biomass**. Through standardization, accreditation, and certification, the initiative has collectively increased the percentage of FSC-certified wood.[66] For example, IKEA is meanwhile sourcing 97% of its wood as FSC-certified or recycled and is now focused on new 2030 goals to promote responsible forestry practices beyond its own supply chain.[67]

[63] Aichbauer et al., 2022, op cit.

[64] Hoffman, Andrew J.: A sweet dilemma: Sourcing palm oil with Ferrero SpA and Nestlé, Michigan Ross School of Business Case W13C26, Ann Arbor 2022.

[65] World Economic Forum: How Chief Procurement Officers are leading the charge on sustainability, 06.01.2025.

[66] Austin, James E.; Reficco, Ezequiel: Forest Stewardship Council, Harvard Business School Case 9-303-47, Boston 2002.

[67] Carlsson, Magnus: Strategic Sourcing and Category Management: Lessons learned at IKEA, London 2015.

One of the most complex global regeneration challenges relates to the marine, particularly oceanic ecosystems. Both the **Marine Stewardship Council (MSC)** and the **Alliance for Water Stewardship (AWS)** are among other MSPs to preserve and improve the circularity of the aquatic life (SDG 14) and ensure the management of clean water and sanitation (SDG 6). With a founding role of the WWF for both non-profits, the MSC focuses on certifying **wild-caught marine and freshwater fisheries and** promoting sustainable catch practices (i.e., **ending over-fishing**), while the AWS initiative mainly targets the **responsible use and treatment of freshwater at specific sites** (e.g., factories, farms) within a local watershed, e.g., by sourcing co-funded water basins.

13.6 Conclusion

Sourcing, a key function in both private and public organizations, plays a crucial role in meeting the respective environmental, social, and economic targets of any for-profit and non-profit enterprise. By applying the Three-Pillar approach of organizational development, the need for a new operational model and capability shift in procurement and sourcing has surfaced.

1. **Sustainable Purpose**: By sourcing intangibles (besides tangibles) for sustainability, the nature of the procurement objects is reshaping the procurement function itself. A more service-dominant portfolio requires a transition from simply buying goods to curating proof (evidence) and crafting a story (narrative) in support of ESG claims. Co-creating and sharing sustainable value propositions in the upstream markets has become a source of competitive advantage in the downstream markets.[68]
2. **Traveling OIrganization**: Complex internal and external collaboration requires the blurring of organizational boundaries, proactive end-to-end value chain alignment, smart pooling of resources, and sharing of knowledge.[69] Individual navigation and organizational orchestration are essential skills; only successful intra- and inter-organizational teamwork will make achieving the SDGs dream work.[70]
3. **Connected Resources**: Working together for environmental, social, and economic impact is dependent on the interconnection and interplay of private, public, and (non) governmental organizations in multi-stakeholder partnerships and extended ecosystems. Proprietary interests and incentives are to be replaced by

[68] Scherer, Jürgen; Marell, Susanne; Wenzl, Jutta: Building a Sustainable Brand in Specialty Chemicals (Carve Out from a Big Corporate), in: Wollmann, Peter; Kühn, Frank; Kempf, Michael; Püringer, Reto (Eds.): Organization and Leadership in Disruptive Times, Cham 2021: 253–265.

[69] Zarei et al., 2023, op cit.

[70] Chauhan et al., 2021, op cit.

"creating shared value."[71] Achievements will be measured by independent benchmarking indices, like the Dow Jones Sustainability Index, ultimately pushing for "broader societal interests over narrow corporate interests."[72]

As "two-thirds of the average company's environment, social, and governance footprint lies with suppliers, procurement leaders (…) can make a decisive difference in sustainability."[73] Capitalizing on this opportunity is based on the successful implementation of a collaborative sourcing approach, as illustrated in Fig. 13.2. A first possible structure for executing collaborative sourcing to meet the 17 SDGs is outlined in Table 13.1. In many of the showcased multi-stakeholder alliances and initiatives, the sourcing and procurement function is the driving force for sustainable business development[74] and is "holding the key to sustainable supply chains."[75]

Jürgen Scherer holds an MBA and PhD from Cologne University in Germany. He has 30 years spanning broad industry experience in Executive positions for global corporations, namely FMCG company Henkel, Specialty Chemicals company Cognis (now part of BASF), and Industrial Packaging company Mauser. He has published various papers in academic journals and readers and has presented Business-to-Business Management topics at leading Business schools and international conferences. Since 2019, he is a Business Consultant, Coach, & Lecturer with www.bxb-exchange.com.

[71] Strand, Robert; Freeman, R. Edward: Scandinavian Cooperative Advantage: The theory and practice of stakeholder engagement in Scandinavia, in: Journal of Business Ethics 127 (2015): 65–85.

[72] Strand, Robert: Global Sustainability Frontrunners: Lessons from the Nordics, in: California Management Review, 66/3 (2024): 5–26.

[73] Cherel-Bonnemaison, Celine; Erlandsson, Gustav; Ibach, Ben; Spiller, Peter: Buying into a more sustainable value chain, www.mckinsey.com, 22.09.2021.

[74] van Hoeven, 2009, op cit.

[75] Hargreaves, Libby: WEF: CPOs hold the key to sustainable supply chains, in: Procurement Magazine, 07.01.2025.

Fashioning the Future: How Luxury Brands Champion Sustainability—An Interview with Laura Tinti

14

Mersida Ndrevataj and Laura Tinti

Abstract

This chapter examines in an interview with Laura Tinti how luxury brands integrate sustainability initiatives with traditional marketing strategies, using Gucci's transformation as a case study. The analysis introduces the concept of "arketing" (art + marketing), which is strongly built upon the so-called artification (a term that describes how non-artistic objects, practices, or brands are enhanced by incorporating artistic values): "Arketing" is linked with "societing" intended to create a new dream narrative and inspiration of higher values, beyond the simple purpose of profit.

This development has been driven by two principal forces. First, the widespread exposure of the fashion industry's reliance on exploitative labor practices and its substantial environmental impact, amplified by the expansion of the Internet and social media, has compelled fashion companies to integrate Corporate Social Responsibility (CSR) into their core communication strategies. Second, the adoption of the United Nations Sustainable Development Goals (UN SDGs, see https://sdgs.un.org/goals) has provided a standardized framework for corporate sustainability programs, fundamentally reshaping how luxury brands articulate their values and commitments.

The emergence of complex strategies—blending marketing, philanthropy, and visual arts—has been fostered by the following three phenomena, thus creating fertile ground for its affirmation:

M. Ndrevataj (✉)
IUAV University of Venice, Venice, Italy
e-mail: mndrevataj@iuav.it

L. Tinti
Polimoda, Firenze, Toscana, Italy
e-mail: dot.lauratinti@gmail.com

© The Author(s), under exclusive license to Springer Nature Switzerland AG 2026

P. Wollmann et al. (eds.), *The Sustainable Organization – Volume 3*, Future of Business and Finance, https://doi.org/10.1007/978-3-032-28949-0_14

- The proliferation of the world wide web and, more recently, the global scale of social media.
- The 2030 Agenda introduction which has gone on to further standardize the vocabulary and models used by companies to manage programs dedicated to preventive and constructive practices, as outlined in the United SDGs.
- The historical contingency of the first two with the pandemic, environmental protection (FFF), human rights movements (BLM), and the renewed demand for women's rights triggered by the #Metoo movement.

The interview will cover as a core a brief case study of the fashion house Gucci and its Gucci Equilibrium program, analyzing the activities and campaigns transporting values around sustainability, UN SDG coverage and Diversity & Inclusion values, associated with the brand that meets UN SDG No. 3 (Good Health and Well-being) and No. 10 (Reduced Inequalities).

14.1 Introduction: Some Fundamental Facts on the Integration of CSR Communication Strategies in Fashion—Trans-disciplinary Strategies to Cover Selected UN 17 SDGs

To prepare the stage for the following interview, it is important to first place the following questions on fundamental developments and to get a first link to the Gucci use case which will be covered later in detail:

- How has the aspirational approach[1] of luxury goods changed before and after the 2030 Agenda?
- How have the trajectories of social issues perceived as urgent, propagated through social media been intertwined, and what brand strategy has been adopted in fashion to align with these protocols and desires in the digital ecosystem?
- Which of the United Nations' Sustainable Development Goals (UN 17 SDG) are more realistically achievable, and which ones are more functional to a luxury brand's communication: sustainability in production or inclusion, diversity, and reduction of the gender gap?
- What are the risks for companies that do not align themselves with these new parameters? How do you build a reputation online and how much does it cost to rebuild or consolidate it?
- To renew, consolidate, and rebuild an image or reputation, and connect with new generations, brands have relied on visual arts for a decade (2001–2014). What has changed in the way companies encourage consumers to dream?
- Are consumers' dreams of the 2000s still the same in 2025? Are the arts and arketing still enough to make consumers dream?

[1] By "aspirational approach" we refer to the marketing and branding strategies in the fashion industry that appeal to consumers' dreams and ambitions for the future.

It was on 31 December 1987 that the UN Commission Chair, Gro Brundtland, declared the work on the Report of the World Commission on Environment and Development: *Our Common Future*;[2] the document, drawn up for the United Nations, defined in clear terms the severity of the world situation and who were the responsible actors in the countries of the industrial-based area with the highest per capita income at that time. The assumption made by the Commission in 1987, in the words of its chairman Gro Brundtland, for the concrete pursuit of sustainable development could be summed up in one word: limit.

Putting limits on the consumption of main resources; soil, water, air, limits to energy consumption, "*Sustainable global development requires that those who are more affluent adopt life-styles within the planet's ecological means—in their use of energy, for example.*"[3] And while the tone, albeit clear, evokes hope for economic growth that does not exclude sustainable development, "*But technology and social organization can be both managed and improved to make way for a new era of economic growth.*"[4] On the other hand, it is said that an increase in poverty and environmental disasters will inevitably hit the poorest countries with their populations expanding.

A few years earlier, in 1979, Archie B. Carroll forged the most widespread definition of socially responsible business (Corporate Social Responsibility = CSR): "*The social responsibility of business encompasses the economic and the legal ethical and discretionary expectations that society has of organization at a given point in time.*"[5] Central to this concept is a company's responsibility to contribute to the well-being of the community in which it operates and from which it profits. Companies that seek sustainability in the long term must make a positive contribution to the environment in which they operate, balancing economic objectives with social obligations.

Carroll's foundational work identifies four dimensions of CSR: legal, ethical, economic, and philanthropic (discretionary).[6] If in 1979 the philanthropic dimension was discretionary, it is no longer so in the twentieth century and even more in

---

[2] World Commission on Environment and Development (1987) *Our common future*. Oxford: Oxford University Press. Available at: https://digitallibrary.un.org/record/139811?v=pdf (Accessed 6 Feb 2026).

[3] World Commission, Our Common Future (1987), 25.

[4] World Commission, Our Common Future (1987), 24.

[5] Carroll, A. B. (1979). A Three-Dimensional Conceptual Model of Corporate Performance. *Academy of Management Review*, *4*(4), 497–505.

Carroll, A. B. (1991). The pyramid of corporate social responsibility: Toward the moral management of organizational stakeholders. *Business Horizons*, 34(4), 39–48.

Carroll, A. B. (1999). Corporate social responsibility. *Business & Society*, 38(3), 268–295.

Carroll, A. B. (2015). Corporate social responsibility: The centerpiece of competing and complementary frameworks. *Organizational Dynamics*, 44(2), 87–96.

[6] Carroll (1991), 42.

Fig. 14.1 UN SDG (source: https://www.un.org/sustainabledevelopment/), used with friendly permission of UN. The content of this publication has not been approved by the United Nations and does not reflect the views

the twenty-first and it is absolutely not discretionary for the Fashion industry sector. Priorities have been redefined and many assumptions and previous balances have dramatically changed, as have the desires and dreams of consumers. However, the objectives of the actors of the capitalistic system always stay the same: the containment of expenses and the constant increase in profits.

This tension between profit imperatives and social responsibility has intensified following the enactment of the UN 17 SDG protocol.[7] In the fashion sector specifically, there has been a progressive adaptation and greater dissemination of information once confidential in the official communication of many companies that declare to prepare measures supporting the achievement of harm reduction objectives.[8]

Since the adoption of the UN SDGs (details see Fig 14.1) in 2015, the trend in fashion—both general and luxury—has been to communicate with a constant commitment to innovative and sustainable processes in the supply chain.[9] Despite the huge and constant attention that the fashion industry receives and in parallel the

[7] United Nations (2015), Transforming Our World: The 2030 Agenda for Sustainable Development, Resolution A/RES/70/1, adopted September 25, 2015.

[8] Amatulli, C., Angeli, F., Costabile, M., & Guido, G. (2017). Sustainable Luxury Brands: Evidence from Research and Implications for Managers. Palgrave Macmillan, p. 2.

[9] For analysis of sustainability trends since 2015, see Choi, T. M., & Cheng, T. (Eds.) (2015), Sustainable Fashion Supply Chain Management: From Sourcing to Retailing (Springer); McKinsey & Company (2016), Style that's sustainable: A new fast-fashion formula; Fashion Pact (2019), The Fashion Pact, presented at G7 Summit, Biarritz.

attention given to environmental issues in recent years, clothing consumption doesn't seem to stop, and fashion initiatives to legitimize the industry multiply.

The environmental stakes are particularly relevant in this sector. It is widely known that the production process of the clothing and accessories industry ranks among the most polluting on the planet, contributing significantly to water pollution and greenhouse gas emissions.[10]

The above considerations on CSR are of course valid for companies operating in any type of sector.[11] However, when it comes to CSR in the fashion industry, with its extremely polluting production system and high pace production schedule, the inconsistencies increase and become more evident.

Researchers have been questioning in numerous commentaries for more than a decade now, the paradox of sustainability in fashion.[12] This paradox becomes even more acute for historical luxury brands, which must navigate a dual challenge: remaining true to their heritage and identity while launching new, desirable products five times a year that must be—a statement that resonates with the words of Gucci CEO Marco Bizzarri when he took over the management of Gucci perceived as both exclusive and appealing to an ever-larger mass of global consumers. As industry specialists note, the old brands' association is no longer enough; it need to be proven, renewed, and refreshed for today's customers,[13] criticizing the previous period for relying too much on heritage skills and craftsmanship.[14]

The alarms launched by the UN since the 1982 World Charter for Nature,[15] denouncing the consequences of hyper-production and environmental damage that have only increased exponentially since, have required a formal response from the players of the fashion industry.

---

[10] United Nations Environment Programme (2019), UN launches drive to highlight environmental cost of staying fashionable. Available at: https://news.un.org/en/story/2019/03/1035161 (Accessed: 25 January 2026).

[11] Carroll (1991).

[12] Pucker, K. P. (2022). The myth of sustainable fashion. Harvard Business Review. Available at: https://hbr.org/2022/01/the-myth-of-sustainable-fashion (Accessed: 25 January 2026).

[13] See: Todesco, F. (2021). A big bang theory about Gucci success. Bocconi University News. Available at: https://www.unibocconi.it/en/news/big-bang-theory-about-gucci-success (Accessed: 25 January 2026).

[14] Marco Bizzarri, President and CEO of Gucci, speaking at the 21st Fashion & Luxury Summit organized by Pambianco and Deutsche Bank in Milan on November 10, stated: "In my view, between 2008 and 2014 the brand positioning path was too focused on a discourse of tradition and heritage. The objective was to bring Gucci back to being one of the most desirable brands in the fashion system" (FashionNetwork, 2016). Citation retrieved from: https://it.fashionnetwork.com/news/Marco-bizzarri-velocita-e-flessibilita-sono-le-chiavi-del-successo-di-gucci-,751880.html#tory-burch (Accessed: 12 February 2026).

[15] United Nations (1982), World Charter for Nature, General Assembly Resolution 37/7, adopted October 28, 1982.

14.2 The Interview with Laura Tinti

Questions by Mersida Ndrevataj

First Question: What have the marketing focuses for luxury brands in the last century?

Answer:

I would like to start with the so-called Arketing (sometimes written also "art-keting" or art-keting)—a made up world combining art and marketing—which is a strategy born in the context of the French luxury market to vary the product of a historical brand with strong heritage and a conservative and traditional image like Louis Vuitton.[16]

The term "art-keting," "artketing," or "arketing" was used for the first time in 2010 when the French brand strategist Guy Boche[17] referred to the mix between luxury brands and the arts with the term "Art-keting," highlighting the heterogeneous approaches by which luxury brands incorporate art within their communication activities. These co-operations would allow luxury brands to "reconnect with their time" and continually recreate the idea of "newness."[18]

Before the term arketing was introduced, its concept was already being applied by French and Italian luxury goods manufacturers, initially specialized in suitcases and leather goods for travel (see Louis Vuitton's first logo patented in 1855, and Guccio Gucci's in 1927). These houses were admired as models of excellence and *savoir-faire*, initiators and custodians of a tradition of local production and family values, and careful and slow manufactures—a process similar to art creation.

Over the years, both France and Italy became a global model (with a large clientele in the USA, Japan, Russia, and today also the Asian Pacific area) with a subsequently growing scale, and immense economic power. A well-founded image (heritage) and reputation have always kept essential core values to preserve, intact, for many decades. The primary catalyst for transformation in heritage luxury brands has been the urgent need to attract younger consumers while maintaining brand prestige.

So, getting back to the initial question of our discussion: luxury brand marketing strategies have evolved in three main stages over the last century. Initially, they emphasized tradition, craftsmanship, and exclusivity (first half of the twentieth century), targeting an elite clientele. This approach then evolved into ambitious marketing strategies, in which brands became status symbols and luxury was made accessible to a wider community of consumers (1970–1990, through entry-level

[16] Examples include Louis Vuitton's collaborations with artist Stephen Sprouse for the Graffiti line in 2001 and with Takashi Murakami for the Japanese market.

[17] Boche, G. (2010), The artketing age of luxury. Available at: http://fblog.futurebrand.com/the-artketing-age-of-luxury/ (Accessed: 23rd December 2012).

[18] Boche, G. (2010).

products). The marketing strategies of arketing were introduced in 2000, shifting the focus from material status to cultural capital and artistic expression.

Second Question: And how did this marketing movement develop so far in the new millennium?

Answer:

After 2000, the "*little for few*" luxury paradigm became a "*lot for many*" (masstige).[19] Niche production was increasingly put aside to diversify the production across three levels—affordable, medium, and inaccessible luxury—on a large-scale and with multiple consequences, including the expansion of the supply chain (while maintaining style offices in Europe). Also, as the economy of scale increases, the production requires delegating to contractors in order to expand and eventually delocalize the supply chain.

Since 2001, Louis Vuitton has been a pioneer in arketing strategies, offering on the market a re-interpretation of the bestseller items of the house: the shape of the product remains identical and recognizable, while the becomes more conspicuous and colorful to attract a younger clientele, without the need to renew the production chain or materials (accessories, leather goods, ultimately from 1999 including also ready-to-wear).[20]

Following this trend, many fashion houses, with mixed fortunes, began to include contemporary artistic references in their collections, showing a preference for "street art" or "graffiti art." French sociologist Roberta Shapiro named this process "artification"—a term that describes how non-artistic objects, practices, or brands are enhanced by incorporating artistic values.[21] The artketing strategies have been mixed with societing strategies lately, and the intersection is intended to create a new dream narrative and inspiration of higher values, beyond the simple purpose of profit.[22] Historical and conservative brands such as Hermès have achieved sometimes unforgettable results, through such strategies, with great sales successes, and others less memorable, but nevertheless become collectible like art (e.g., *Hermès for Congo*).

As a countermeasure to current trade trends, art collaborations helped luxury brands regain lost distance from competing brands (masstige) and *untargeted*

[19] Silverstein, M. J., & Fiske, N. (2003). Luxury for the Masses. Harvard Business Review, 81, 48–57, 121.

[20] Dion, D., & Arnould, E. (2011), Retail luxury strategy: Assembling charisma through art and magic, in: Journal of Retailing, Vol. 87, No. 4, pp. 502–520.

[21] Shapiro R (2004) Qu'est-ce que l'artification? XVII Congrès de l'Association française de sociologie, HalSHS.Cnrs; Shapiro, R. (2019), Artification as process, in: Cultural Sociology, Vol. 13, No. 3, pp. 401–415.

[22] See: Badot, O., Bucci, A., & Cova, B. (2007). Beyond marketing panaceas: In praise of societing. European Journal of Marketing, 41(1/2), 85–98.

"average" customers,[23] based on the *strong social demarcation function of art.*[24] Though, after about 10 years of overexposure to limited or arty editions, the so-called wear-out effect has occurred in the market, that means a customer feedback defined as "wearing."[25] Wear-out-effect describes a negative perception that consumers express about this sort of collaboration or co-design or co-branding: they may find it irrelevant or, boring, anyhow not special. The large number of arty editions launched on the luxury market between 2001 and 2010 or limited editions has generated, in the higher end clientele, a certain boredom and, in the larger clientele, it has led consumers to question the actual exclusivity of such products.

At the same time, with the global spread of the World Wide Web and the ever-faster circulation of information and images, the obscure aspects of production such as the extraction of valuable materials, the conditions of workers, and the exploitation of animals, also in the market for luxury goods or masstige, have begun to become a variable to be considered in the formation of consumer purchasing decisions.

Third Question: How is the connection between Arketing and Artification? And how were in marketing art references, societing and musealization of the shopping experience connected with each other?

Answer:

According to the classic definition by Kotler,[26] marketing consists in identifying and satisfying human and social needs through four main areas: product, price, point of sale, and communication (Product, Price, Placement, and Promotion). It is obvious that Arketing might play a significant role in this context. As a result, LBACs[27] would give rise to a set of strategic issues such as an increased visibility in the marketplace, an improved likeability, and a substantial gain in brand value. It appeared

[23] Silverstein, M. J., Fiske, N., & Butman, J. (2005). *Trading up: Why consumers want new luxury goods—and how companies create them* (2nd ed.). Penguin Books.

[24] Petersen, A. (2008). *Art-Kommerz: Kunst und Mode sind eine gute Partie. Die Welt.* www.welt.de/lifestyle/article2869221/Kunst-und-Mode-sind-eine-gute-Partie.html (Accessed on June 29th, 2013).

[25] In order to counter the impending wear-out effect, luxury brands should be willing to truly engage with their artistic counterpart and to further explore unconventional opportunities for collaboration (See: Kastner, Olga Louisa:, *When Luxury Meets Arts*, p.73, 2014). Also, while the fact that ever more luxury brands decide to join forces with the arts today highlights indeed the current popularity of LBACs (Label Based Access Controls), it does not necessarily hint at the lasting appeal of these collaborations. Rather, the more luxury brands team up with the arts, the more likely it is that a certain wear-out effect sets in, in the course of which LBACs are no longer perceived as disruptive branding initiatives, but rather as a regular and hence predictable element of the luxury brand's universe.

[26] Kotler, P., Roberto, N., & Lee, N. (2002). Social marketing: Improving the quality of life (2nd ed.). Sage Publications.

[27] Label-based Access Controls.

1 year later in 2011 in a contribution of C. Rioux[28] which makes (without reference to Boche) an extremely comprehensive treatment of the "collaborations" at corporate and product levels, and in a contribution of Kastner,[29] but there is no mention of *"artification."*

The connection between artification and arketing that has been treated separately, while they seem (even chronologically) one propaedeutic to the other. In particular, the dimension evoked by Shapiro (2004)[30] of social world—worlds and forms of social aggregation, excluded from the marketing sector until 2008, but social aggregation, it seems to me, is today central in the process of creating a positive brand image for a top-level player. We also add societing (or societal marketing)[31] that has proven to be central for Gucci from 2019 to 2022 for the reconstruction of a positive brand image using the definition of Giampaolo Fabris based on that of social marketing by Kotler:[32] *"a third way of being of so-called ethical marketing and so-called social marketing, defined as the use of marketing principles and techniques to influence a target to accept refuse to modify abandon, voluntary behavior to the benefit of groups or people or society as a whole."*[33] Our hypothesis is that two aspects that have been excluded deserve renewed interest: the artification theory and arketing strategies in conjunction with societing strategies (new social network creations Shapiro, 2004) are the prerequisites that made Gucci 2019–2022 arketing strategy's success possible.

Fourth Question: You mentioned Gucci. What are core elements of Gucci's marketing in the context of arketing, artification, and societing?

Answer:

The Gucci marketing approach has chosen to mix social content aggregation inclusion and visual arts. Thanks to the contingency of historical events that provided the brand with a fertile ground in order to approach its most important target audience (Gen Z and Millennials), showing itself aligned with the youth audience needs and desires deemed urgent through social interactions (twitter and IG).

In the Gucci hybrid-arketing strategies, common luxury goods are ennobled by intangible values, reaching a higher status, similar to that of an artistic work.

Why is this happening? To differentiate them from luxury goods, making them more desirable for the consumer; to raise, as a result, the role and an elevated

[28] Rioux, C. (2011). Le luxe et l'art : Du marketing à l'artketing. In O. Assouly (Ed.), Le luxe : Essais sur la fabrique de l'ostentation (pp. 331–351). Institut Français de la Mode & Éditions du Regard, p. 331.

[29] Kastner, O. L. (2014). *When luxury meets art.* Springer Fachmedien Wiesbaden, p.8.

[30] Shapiro (2004).

[31] Societal marketing is a marketing approach that goes beyond satisfying customer needs and also takes into account the long-term interests of society. It aims to bring about positive social change by encouraging companies to address social issues and develop sustainable solutions.

[32] Kotler, Roberto, & Lee, 2002, as cited in Fabris, *Societing. Il marketing nella società postmoderna,* p. 480, 2008.

[33] Fabris, G. P. (2008). Societing. Il marketing nella società postmoderna. Milano, Italia: Egea.

mission or the company from mere profit to a more noble goal, such as patronage or philanthropy. By the way, it is considered that the benefits are multiple: joining forces with the arts offers different opportunities, and so CSR.[34]

Thus, since the art world is not commonly easily accessible to the average consumer, *luxury brands, entering the artistic sphere, affirm their distinctive connotation and emphasize their* membership in elite networks.[35] In this way, luxury brands *renourish their exclusive image and alleged (since only artificially) scarce character.*[36]

In the societing strategies, the aim is to ennoble the ethical dimension of the company and make it associated with the product as much as possible; the societing neologism (from G.P. Fabris, 2008) originates from a definition of *ethical marketing* by Kotler (2002): "a *third way of being of so-called ethical marketing and so-called social marketing, defined as the use of marketing principles and techniques to influence a target to accept to refuse to change or abandon, voluntary behavior to the benefit of groups or people society as a whole.*"[37]

Fifth Question: How was this marketing profiling embedded in the Gucci Organization?

Answer:

Gucci is part of the Kering Group which is a luxury conglomerate born from the PPR Group (Pinault Printemps Redoute) named after its founder, Francois Henri Pinault, son of the entrepreneur of the timber Francois Pinault. The group changed its name from PPR to Kering in 2013, and acquired and controlled a number of historic European luxury brands in the leather goods, clothing, luxury watches, and eyewear sector: Gucci (acquired and integrated 1999–2013),[38] Saint Laurent, Bottega Veneta, Balenciaga, Alexander McQueen, Brioni, Boucheron, Pomellato, DoDo, Qeelin, Ulysse Nardin, Girard-Perregaux, and Kering Eyewear.

Between 1999 and 2001, PPR undertook a corporate escalation to the equity of Gucci SpA to accumulate company shares, in direct contrast with the group LVHM

---

[34] CSR=Corporate Social Responsibility.

[35] Meffert, H., & Lasslop, I. (2004). Luxusmarkenstrategie. In M. Bruhn (Ed.), Handbuch Markenführung: Kompendium zum erfolgreichen Markenmanagement. Strategien - Instrumente - Erfahrungen (2nd ed., pp. 927–947). Springer Gabler, p. 940.

[36] Catry, B. (2003), The great pretenders. The magic of luxury goods, in: Business Strategy Review, Vol. 14, No. 3, pp. 10–17.

[37] See: Fabris, G. P. (2008); Kotler, P., Roberto, N., & Lee, N. (2002).

[38] Gucci was founded in 1921, in Florence by the founder Guccio Gucci, and until 1993 it was a brand symbol of the birth of made in Italy, born in Florence with Emilio Pucci (1947) and Ferragamo, (1927) founded in Florence and with models that for quality and slow production rates represent the archetype of the excellence brands of made in Italy.

In 1993, the family crisis and a sale of a substantial share of the stock package, previously managed by the Gucci family, to a group of Saudi investors, Investcorp, changed the brand's history forever.

The company, then literally on the verge of bankruptcy, thanks to the direction of Gucci America CEO Domenico de Sole, and the new Creative Director Tom Ford, from 1994 to 2004 becomes the most influential brand in terms of image, and over time the most profitable brand of the parent company.

of Bernard Arnault, and, passing through the courts, Pinault finally was successful in 2001, when it became 99% of the total shareholder body.

This moment constituted a fundamental step in the more recent history of what was made in Italy, but not only; the balance of European finance as a scenario has redesigned, and former family founded and owned luxury brands are absorbed by conglomerates, which can decide abruptly either to acquire them or to sell depending on their profitability.

The Kering Group (like the other conglomerates) includes well-known brands, which guarantees stability, due to the reputation of the companies that form it, and for the balance that in case of loss of one player is offset by the good performance of the other. The advantages for subsidiaries of being part of a group are concrete: the logistics network, an integrated network, the supply chain (supply chain), the human resources interchangeable, experienced and specialized management (in the shared sector of leather goods or textiles).[39]

From 2017 onward, Gucci has boasted the achievement of record profits, a "desirable and countercurrent image," (the dream); which has been paired from 2019, with a program dedicated to sustainability and to goals (on people and planet included in the UN17 SDG goals (number 3,5,10,12,13, see above)). Not only the Gucci company launched this very specific program, but the founder and owner of the entire Kering company, Francoise Pinault, was appointed by Emmanuel Macron himself, the coordinator of the G7 for Fashion Pact Biarritz, in 2017: "*With the aim of mitigating the damage and finding a common solution to this problem, therefore, was born the Fashion Pact. This is a pact strongly wanted by the French president Macron who, on the eve of the G7, has entrusted to François-Henri Pinault, president and CEO of the Kering group, always at the forefront of sustainability, the task of bringing together the biggest fashion companies and involving them in a project for the protection of the plane*t."[40]

In 2 years 2015–2017 of record, only a few isolated false steps overshadowed the adamantine reputation of the Group. However, Gucci has responded promptly and has structured initiatives that involve the company at all levels (CSR) for the pursuit of universal values: "*Gucci is not a company where you must leave your values at the door, but one where they are enhanced, challenged and amplified. Gucci Equilibrium is about us (Gucci management-employees), spreading that energy and the positive intent to everyone who loves our brand ... These are critical times where we can all play our part in helping to deliver on the UN global goals and the Paris Climate Agreement. The only way to do that is by bringing people together, sharing ideas, innovation and experiences. This is the objective we have set for Gucci Equilibrium.*"[41]

[39] Cabigiosu, A. (2020). Digitalization in the Luxury Fashion Industry. Strategic Branding for Millennial Consumers. Palgrave Macmillan, pp. 178–179.

[40] G 7 Fashion Pact; Biarritz. Global Fashion Agenda and The Boston Consulting Group (2017). Pulse of the Fashion Industry 2017.

[41] Official press conference presenting the Gucci Equilibrium Program 2018. Full interview available at: https://wwd.com/fashion-news/designer-luxury/gucci-kicks-off-new-web-destination-1202688838/

Sixth Question: What gives Gucci the profile as a Sustainable Organization in the Kering Group

Answer:

For Gucci (from 2015 to 2022), the image is based on being inclusive, in adv and official model casting and testimonials, inside the company, in human resource and management, with the highest number of women employees in the management level, as the best workplace for people with disabilities (inside to outside CSR).

According to the management, the secret of Gucci's success is "*An inclusive approach to building customer relationships, guided by the brand's mantra of promoting self-expression; engaging customers across all age groups, demographics and geographic locations*"[42] (Kering Financial Document 2019). The Gucci 2015–2022 case is the ultimate example of the ability to associate a luxury product intermediate-masstige values of solidarity equal inclusion and celebration of diversity. The brand has become the most popular especially for the part of public, Gen Z and Millennials (born between 1981 and 1996),[43] and digital natives among the most sensitive to certain social issues.

In fact, the brand has another primacy: when it comes to reputation in the years 2017–2022, it was perceived as the most inclusive of the landscape, since the models and symbols associated with the brand are aligned and consistent with the idea of inclusiveness.[44]

First for the human resources that operate within the company;[45] then for the customer, then outside the corporate reality, with philanthropic activities, program Chime for Change (from 2013)[46] then Change-makers (2019): that means impact, funding, education/scholarships.

The economic and image success of the operation has attracted great attention in the market and among analysts, so much so that the Gucci/Big Bang operation was defined from a managerial point of view, to describe the scope« *For dozens of different communities—from businessmen to punk rockers, and from art lovers to rappers in the Bronx—Gucci created dedicated events and built a narrative around its*

[42] Kering Financial report (2019).

[43] Millennials, born between 1981 and 1996, were labeled the "Me Me Me Generation" by Time magazine (May, 2013).

[44] Cabigiosu, 2020.

[45] In 2004, Gucci voluntarily implemented a program aimed at protecting and respecting workers; it obtained certification SA8000 for this effort. See: Bonacchi, Massimiliano & Perego, Paolo & Ravagli, Rossella. (2012). The Evolution of Corporate Social Responsibility in Gucci: From Risk Management to Stakeholder Engagement.

[46] CHIME FOR CHANGE, the girls' and women's empowerment campaign founded by Gucci in 2013, has now funded more than 390 projects in 86 countries through 132 non-profit partners. Among the founding partners, we find Kering Foundation, the Bill & Melinda Gates Foundation, Facebook, and Hearst Magazines. Retrieved from: https://www.gucci.com/us/en/st/stories/article/10-years-gucci-chime?srsltid=AfmBOoouqc3fj1M749pyJVjmsuM9S_zUmEYB50rzjT9k9o6H-6SAn2DR

new values in a way that resonated with authenticity for each specific group.»[47] This type of operational marketing, in the sector of high symbolic value goods, can include many activities: event design exhibitions, talk, etc. Gucci has refunded the brand image and relaunched heritage in a contemporary key.

Seventh Question: What about special Gucci CSR-linked programs?

Answer:

I would first name the Gucci program "Gucci Equilibrium Milano," announced publicly on 5 June 2018 in Milano. It is divided into two sub-programs: people and planet. Planet includes activities for the environment (SDG 12 and 13) and part with "*three objectives environment, fur free from 2018, carbon neutral while the people section concerns «people diversity*."[48] This project is forged to give substance to Gucci as a sustainable organization "as a brand with a strong vocation for improving the living conditions of people and the environment, in the words of CEO Marco Bizzarri and on social « *Supporting @gucci ongoing commitment to generate positive change for people and planet.*"[49]

Gucci Equilibrium "committed to generate positive change for People and Planet. Prioritizing inclusivity and Respect, everyone in the Gucci community is free to express their authentic diverse selves." The Equilibrium project was officially created in 2018 and is linked specifically to the product and image of the Gucci brand, not other brands within the Kering group, but reflects some of the overall goals that the entire group pursues in its activities and subsidiaries.

The 2018 Kering report lists among the UN SDGs pursued by the Kering Group: SDG 3 (Health and Well-being), SDG 5 (Gender Equality), SDG 6 (Clean Water and Sanitation), SDG 8 (Decent Work and Economic Growth), SDG 12 (Responsible Consumption and Production), and SDG 13 (Climate Action). Among the Kering tools included in the Gucci equilibrium sector Planet program (corresponding to SDG goals number 12 and 13 of the 17 UN SDG Goals), there is an instrument called EP&L[50] report (environmental profit and loss annual report).

Kering had developed and applied since 2011 the mentioned internal EPL tool, which brings together all the environmental costs of the production chain: the environmental costs are calculated on the basis of a formula developed by the company for each area (couture, leather goods, silk, accessories, etc.) with the environmental cost of the operation is calculated in monetary terms. In 2012, Kering announced that the group is *committed to roll out the EP&L across its brands by 2016* » and in 2015, coverage was declared achieved; since 2017, Kering's sustainability targets for 2025 have been made public.

---

[47] Todesco (2021).

[48] Gucci. (2018). *Official press conference: Presentation of the Gucci Equilibrium program.* Press conference.

[49] Gucci. (2018).

[50] EP&L= Environmental Profit & Loss.

The EP&L, which is on a digital platform available from Gucci as of 2019, also contains the methodology for the environmental income statement and the aggregated data sets that characterize it. These downloadable datasets include the Environmental Performance Indicators (EOI) with the environmental impacts of the Group's activities and assessment coefficients.

Eighth Question: Is there another program you would like to mention?

Answer:

Second, the **Changemakers Scholarship Program,** announced in 2009, allocated $1.5 million distributed over a multi-year plan for the provision of scholarships to students residing in North American states and who are eager to study, mainly design, to allow the industry access of the African-American community and other ethnic groups currently under-represented. "The Changemakers Council will also help select eligible students from schools across North America to participate in the Gucci Changemakers Scholarship Fund. Over the course of four years, each student shall receive a US$ 20,000 grant toward completing their college education."[51]

This opened up new opportunities for the youth to be involved in the creative process in the global fashion industry, and to avoid them feeling unrepresentative in the future by facilitating their access. The Changemakers Fund allocates $5 million in a multi-year plan to associations active on the territory in various North American states and more specifically in 12 cities: Atlanta, Chicago, Detroit, Houston, Los Angeles, Miami, New York, San Francisco, Toronto, and Washington DC. The associations eligible for funding are selected on the basis of their commitment identified in specific terms "*within the African-American community and communities of color at-large, while bringing positive change and inspiring solutions for a better future.*"

Initially announced to employees in 2018, the Gucci Changemakers volunteering program empowers all 18,000 Gucci employees worldwide to dedicate up to 4 paid days off for volunteering activities in their local communities. This represents up to 8000 days of volunteer supporting North America, the first region in the global roll-out, where volunteering opportunities cover four main pillars: equality; support for refugees and the homeless; protection of the environment; and education.

Ninth Question: Something you would like to add to describe the Gucci Use Case?

Answer:

I would first like to quickly cover Gucci's art-related narratives. Besides philanthropic activities, the main character of the whole Gucci image from 2015 to 2022 is the A-temporality, an example of retro-marketing. The narrative of the Gucci

[51] Kering Press release from 05.06.2019. Available at: https://www.kering.com/it/news/kering-lancia-la-piattaforma-digitale-per-l-ep-l-e-hackathon-per-favorire-piu-trasparenza-e-innovazione-nella-sostenibilita/ (Accessed: 12 February 2026).

product is based on a series of ancient archetypes (Pompeii, ancient Rome, Florence and the Renaissance, Manierismo of the Bronzino) and postmodern (pop icons, glam rock, television superheroes or heroines, and video games characters).

The images of the campaign show a very different humanity, specific for age and glamour, but this is not new for the fashion recent history (see Benetton Oliviero Toscani, United Colors of the World, Diesel for Steven Meisel). The beauty mode chosen by Gucci includes gender fluid models, unusual looks, elderly professional models, next to ordinary people,[52] but the key element of the sense of novelty formulated by the company to sell the product is the body in transformation.

The models presented or illustrations are neither purely male nor female, in line with an emerging model of a person who does not want to be identified precisely with either sex, such as has happened in history, based on social definitions or traditional and customary representations.

The psychedelic rock star look is mixed with mannerist and pre-raffaellite painting, graffiti and American cartoons; in SM posts the product and logo are introduced, but not always in a central position, and the product is integrated into the visual commissioned from artists who are sometimes unknown, emerging, or outsider of the art system but young people active at a local level.
They receive legitimacy thanks to the collaboration with this global brand and an instant visibility unimaginable before the spread of social media. The centrality of the online brand's product experience is under the lens of scholars since the first decade of the 2000s; Mosca wrote "... *the differentiating factor that will enable a competitive advantage to be obtained is bound to become increasingly real.*"[53]

Second, it is important to have a look at the values of the twenty-first century in the context of which Gucci built a new image of CSR, "a trade between money-oriented stakeholders and more general oriented to well-being stakeholders."[54]

In 2017, Goldman and Sachs and Altagamma commissioned market research that reveals how in 2020 the most determining consumer classes will be the Gen Z and Millennials; the Millennials were born between 1980 and 1995 and the people born between 1995 and 2010 were Gen Z.

According to this research, 22% of GEN Z are willing to pay more for sustainable products in the process, while only 13% of Millennials are willing to pay more for a sustainable product. The result shows that the question posed by the research is to find out what and how much profit can be made, in the transition to a slower production sustainable supply chain, to compensate for the same product created in a sustainable way with a higher price that the customer will pay.[55]

---

[52] For its 2018 Cruise Collection, Roman Rapsody, Gucci has chosen ordinary Roman residents as models, regardless of age. The aim was to represent the true Roman society. Sonia Zhou was among them, a Chinese lady owner of the famous Chinese restaurant Hangzhou by Sonia. Retrieved from: https://luxurysociety.com/en/how-guccis-storytelling-winning-millennials-china/

[53] Mosca, F. (2017). Heritage e prodotto di marca: Modelli teorici e strumenti operativi di marketing per le imprese nei modelli globali del lusso. FrancoAngeli: Milano.

[54] Glavas, A. (2016). Corporate social responsibility and organizational psychology: An integrative review. *Frontiers in Psychology, 7*, Article 144.

[55] According to the PwC Global Consumer Insights Observatory (2018), based on a panel of 2424 Italian consumers, 39% of Millennials (born 1980–1994) and 63% of Generation Z respondents (born 1995–2010) were represented in the sample.

Based on the data of this historical moment, if in the consumer target group most important for the market only the 13% of Millennials, and among Gen Z only the 22% are oriented to pay more for the product obtained from a more sustainable supply chain, it can be assumed that the producer who will face costs for adapting to sustainable protocols in order to make the chain sustainable, may not be motivated to do so because he does not see the economic advantage of making the change of pace.

But the mission of the Kering favorite subsidiary, understood as support to achieve the UN 17 Goals as declared is Sustainable purpose, is perceived as coherent and achieved. At all levels of the holding, in the subsidiaries, by management and employees and on the other side customers or prospective customers are involved in a process of "*empowering imagination*." The corporate brand Gucci, strategy under the control of parent company Kering, is consistent, invites to dream and to be active and ready for change.

14.3 A Linkage to the Three-Pillar Model and Its Application on the Gucci Case

The relevance of the Three-Pillar Model[56] as a preferred thinking and mental style for challenging transformations is already described in detail in the introductory chapters of the book. In brief, the model is based on three key building blocks (pillars) for organization design and leadership for the VUCA[57] world:

(1) The first Pillar: **Sustainable Purpose**

Sustainable Purpose is the raison d'être of an organization, having a guiding objective means bringing orientation and certainty to the people for their joint endeavor and success. Contributing to sustainability is for sure a convincing sustainable purpose of an organization.

(2) The second Pillar: **Traveling Organization**

Traveling Organization means for an organization accepting to be in a permanent state of flux and transformation, while interacting with the external requirements of its journey with rapid learning and adaptivity. Ideally, this is the mindset of loving the unknown and being well able to manage it. The realization of the Paris Agreement is an especially challenging journey—in order to secure nothing less than the world's long-term survival—and these endeavors

[56] Wollmann, P. et al. (Eds.): Transforming Public and Private Sector Organizations – Implementing Sustainable Purpose, Travelling Organization and Connectivity for Resilience. Cham: Springer Nature, © 2022.

Wollmann, P. et al. (Eds.): Organization and Leadership in Disruptive Times – Design and Implementation of the 3-P-Model. Cham: Springer Nature, © 2021.

Wollmann, P. et al. (Eds.): Three Pillars of Organization and Leadership in Disruptive Times – Navigating Your Company Successfully through the 21st Century Business World. Cham: Springer Nature, © 2020.

Wollmann, P. et al. (Eds.): Reading an Unknown Organization - Leaders in New Roles Thriving in Challenging Environments with the 3-P-Model Mindset. Cham Springer Nature, © 2024.

[57] VUCA = Volatile, Uncertain, Complex, Ambiguous.

which often lead through unknown territories require the mindset and the capabilities of a real Traveling Organization.

(3) The third Pillar: **Connectivity**

Connectivity is all about the ability to interconnect on the journey with all needed parties and resources inside and outside the silos, therefore creating high efficacy and consistency and during this process building also a shared learning network. It is obvious that an enormous degree of connectivity is needed for organizations like the Greenhouse Protocol to cover all stakeholders' demands and interests and to give the needed orientation.

Applied to Gucci and its development shows that the company is a real role model especially regarding its change journey over the last couple of year. The strong and diverse sustainable purposes (and their use in the stakeholder and market communication) were described in detail in the previous sections (including the interesting links to the terms artification, arketing, and societing and the strong commitment to sustainability and the coverage of selected UN SDGs and in general CRS requirements). The intensive transformation journey the company took built on a mindset as the mentioned "Traveling Organization" and the diverse programs all need a high level of connectivity or a well-connected organization internally and with the whole contemporary society.

14.4 Conclusions

The chapter shows that fashion in general and Gucci as a special use case are running a large number of activities in the context of Sustainability, UN SDG Coverage, and CSR—together with Diversity & Inclusion projects, often connected with positive connections to old and new art, history, etc. This offers large opportunities to underlie the experience of purchasing and owning luxury goods with positive values related to sustainable attitudes to connections and thus actually supporting sustainability. In line with the complexity of fashion in the luxury goods sector, the corresponding marketing concepts have many facets and require—particularly for reasons of credibility—a very stable, easily comprehensible foundation—but they are definitely possible and successful, as Gucci proves.

Mersida Ndrevataj is an urban scholar with a background in architecture, urban planning, and environmental psychology. She holds a PhD in Urban Planning and Public Policy from the IUAV University of Venice, where her research focused on the implementation of public policies in cultural heritage management. She is currently a lecturer at the IUAV University of Venice and collaborates with several COST Action research networks focused on urban studies. Her work brings together academic research and field-based inquiry by adopting audiovisual and sensory approaches to investigate urban spaces, everyday socio-spatial practices, and lived experiences.

Laura Tinti has 18 years of experience in contemporary art production and communication. Her interest goes across several fields: public space design projects, from art production for public spaces to education, and in education especially the invention of educational tools to learn basic elements of contemporary visual arts and fashion design and to introduce comprehension of contemporary artwork to a wider audience. Since 2017, Laura has been regularly teaching as a professor at Polimoda Institute in Florence, Italy, and since 2022 as guest lecturer at Ca' Foscari University in Venice.

Laura who also has a solid experience in cultural mediation thanks to years of designing and developing solutions for the La Biennale di Venezia Foundation.

Part IV

Conclusions and Take-Aways

Desirable Impact of Our Sustainability Books on the Consulting Industry and for Academic Education and Research

15

Ovais Sarmad, Claudia Lemke, and Peter Wollmann

Abstract

In this chapter, we want to shed light on a special desirable impact from both books on "The Sustainable Organization—Volume 1 and 2," namely the urgent call for action in two special fields: academic education and research and the consulting industry. Both fields are crucial for the long-term success of the activity landscape to reach Sustainability and UN SDG Coverage. The academic field educates and trains the next generation which will finally have to reach the ambitious global sustainability transformation targets, and also trains the consulting industry, therefore academia is involved in many of the single activities and gives advice to global and local institutions, countries, enterprises, and NGOs. Thus, it is obvious that it makes sense to devote a special chapter to these two areas before the general conclusion.

15.1 Introduction

Consultancies operate under the same mantra as any other business: "Grow, compete, win." The mutually amplifying pursuit of money and power has emerged as the dominant force in our world. Fearing to lose, we keep feeding the

O. Sarmad (✉)
Independent Adviser on Sustainability, Benissa, Alicante, Spain
e-mail: osarmad@gmail.com

C. Lemke
Berlin School of Economics and Law (Hochschule für Wirtschaft und Recht Berlin), Berlin, Germany
e-mail: claudia.lemke@hwr-berlin.de

P. Wollmann
Bonn, Germany
e-mail: pw@peterwollmann.com

© The Author(s), under exclusive license to Springer Nature Switzerland AG 2026
P. Wollmann et al. (eds.), *The Sustainable Organization – Volume 3*, Future of Business and Finance, https://doi.org/10.1007/978-3-032-28949-0_15

machine—even when we feel something is fundamentally wrong. Decision-makers worry about the global crisis' impact on their businesses and on their children. Only a fundamental shift of values and market rules can change that, and only when the business community itself pushes for that shift, politicians and policy makers will act. To support them, consultancies need to re-invent themselves as well: from competition to collaboration, from self-promotion to promoting the cause, and from self-optimization to bringing actual progress to humankind. It is necessary to foster a respective significant change in consultancy and academia.

Nevertheless, a "final" or better more advanced evaluation of the current situation of the world and advancement of humanity, needs more ongoing discussion because of partly contradictory developments, see results of Hans Rosling's[1] book *Factfulness*[2] and doomsday-like analysis from climate scientists. It might make sense to take a perspective like Hannah Arendt, one of the famous political and philosophical thinkers of the last century, who had after the Second World War and the holocaust nevertheless the optimism that people coming together with a sustainable purpose have "the power to change the world." It is important in this context that "power springs up between men when they act together and vanishes the moment they disperse," according to Arendt.[3] Freedom is according to Arendt the freedom to participate in shaping society.

In the first chapter of this book, we described the power of the numerous existing activities in the nearly infinite activity landscape to reach Sustainability and UN SDG Coverage—and that the biases in public perception of the real progress achieved so far. The respective experience and recommendations of the corresponding author of this chapter supporting these statements were summarized in Sect. 1.3.2.

And it is obvious from our perspective that one work stream for the shaping refers to a change in education for consultancy and academia.

15.2 Resulting Demand for Action: Academia and Consulting Transformations

Our consideration in the previous sections might be (again) summarized in our following key statement:

"Transformation of education is a precondition for the transformation of consultants and business leaders in general, both is a precondition for future transformational leadership."

What are the impacts for the academic area? We want to place two fundamental remarks:

Remark 1: The thoughts described mean for universities the urgent demand to transform academic staff education/to start further development.

[1] https://en.wikipedia.org/wiki/Hans_Rosling

[2] https://en.wikipedia.org/wiki/Factfulness

[3] See https://monoskop.org/images/e/e2/Arendt_Hannah_The_Human_Condition_2nd_1998.pdf

Remark 2: These transformations are rather long-term endeavors but—as a sort of good news—they have been already started in many places as we will show below in Sect. 15.4 introducing some few use cases.

We will not offer final solutions for a comprehensive action plan in the chapter but draft some ideas for future key activities to be started or continued (and describe how a group of Three-Pillar authors intends to support). The proposed future activities or outcome will mid-term contain:

- The formulation of a practical codified framework for (future) education (at universities), something like a sort of "SDGs for education"[4] and a broad range of use cases covering specific problems.
- The formulation of a practical codified framework for (future) consultancy incl. key messages and shared values. This should include a sort of SDGs for consultancy and a broad range of use cases covering specific problems.

15.3 The Earth Charter (EC) and EC Organization: Comments from Ovais Sarmad and Claudia Lemke on a Platform Representing the Required Values for the Transformation

We will focus in this section on the following questions:

- Why did we choose Earth Charter to build on or to cooperate with respectively?
- What does Earth Charter strategically/operationally mean for you and what could it be for others and in general?
- What did you do with Earth Charter and what were your experience and your lessons learned?
- What you recommend for future use/cooperation?

15.3.1 Introduction of the Earth Charter[5]

The Earth Charter (EC) is a document[6] with 16 principles powering a global movement including an Earth Charter organization. The EC document enunciates the values and principles by which sustainable development can be achieved. The underlying thought is that when one does apply it to one's business, school, or community, one can begin turning conscience into action to make all life on Earth thrive.

The Earth Charter, adopted in 2000, is an inspiring framework grounded in values and a shared understanding of how we can create a sustainable global society. It

[4] The Earth Charter (see following paragraph) calls that "Education for Sustainable Development."

[5] Partly cited from the website: https://earthcharter.org/about-the-earth-charter/

[6] The Earth Charter document starts with the Preamble, followed by the four pillars: Respect and Care for the Community of Life, Ecological Integrity, Social and Economic Justice, and Democracy, Nonviolence, and Peace. It concludes with The Way Forward.

outlines an ethical vision for human and ecological well-being, where the principles encourage us all to envisage and create just and sustainable futures rooted in respect and care for the community of life, ecological integrity, social and economic justice, and democracy, nonviolence, and peace (Earth Charter International, 2020).

The mentioned EC organization is an ethical foundation for actions with the purpose to build a more just, sustainable, and peaceful global society in the twenty-first century. The organization articulates a mindset of global interdependence and shared responsibility. It offers a vision of hope and a call to action.

The EC movement is a product of a decade-long, worldwide, cross-cultural dialogue on common goals and shared values. It is in our days used as an educational instrument getting more and more important—and also as a reference and tool in the development of policies, planning, and decisions in the context of Sustainability and UN SDG Coverage and in many other ways.

15.3.2 The Earth Charter's Significance in More Detail

As already stressed, the Earth Charter provides a moral foundation and a compass for shaping a sustainable and peaceful global society. The Charter's guiding principles of ecological integrity, social justice, democracy, and global responsibility offer practical pathways for aligning these with long-term human and planetary well-being.

This foundational idea of the Earth Charter can be traced back to some profoundly spiritual and moral principles enshrined in various well-known authors and declarations. Some of those are:

> The principles of **Al-Mizan (balance)** are based on an Islamic ethical framework for environmental stewardship, emphasizing harmony with nature, justice, and moderation. Key principles include protecting the environment as a divine trust, using resources equitably and wisely, and calling for community action to address environmental challenges through public good and forbidding wrong action.
>
> **Laudito si**—Pope Francis argues that environmental degradation and social injustice are intertwined; therefore, solutions must address both ecological and human well-being. The encyclical's core message is that caring for our environment is not just an ethical duty but a spiritual one.

Mahatma Gandhi's philosophy embraced both **Equality** as a fundamental moral value and **Equity** as a practical approach to ensure everyone's needs are met. He championed equal opportunity for all individuals, while acknowledging natural differences in capacity, and advocated for an equitable distribution of resources based on need.

15.3.3 Practical Aspects of the Earth Charter in Relation to Technological Developments: A Focus on AI and Renewable Energy Transition

As emerging technologies—especially artificial intelligence (AI) and renewable energy systems—rapidly transform the world, the need for an ethical compass becomes urgent. By applying these principles to the development and deployment of AI and renewable energy, societies can ensure that technological progress contributes meaningfully to sustainability and equity.

15.3.3.1 Ecological Integrity: AI and Renewable Energy for Environmental Protection

The Earth Charter underscores the need to "protect and restore the integrity of Earth's ecological systems." Both AI and renewable energy technologies have direct roles to play in realizing this vision.

AI for Ecological Stewardship

Practical applications include:

- Precision environmental monitoring: AI-powered remote sensing systems can detect deforestation, track biodiversity, monitor coastal erosion, and predict natural hazards with high precision. These applications enhance early intervention and reduce ecological harm.[7]
- Optimization of resource use: AI-driven models can optimize water management, agricultural inputs, and energy consumption, reducing waste and environmental impact.

Renewable Energy for Climate Stability

Renewable energy systems—solar, wind, hydro, geothermal—are central to stabilizing the climate. Practical actions aligned with Earth Charter principles include:

- Lifecycle sustainability: Designing renewable infrastructure with recyclable materials, and responsibly managing battery waste, supports the Charter's call for responsible consumption.
- Smart renewable grids: AI enables dynamic load balancing, predictive maintenance, and optimized energy distribution in smart grids—maximizing the efficiency and resilience of clean energy systems.
- Community microgrids: Small-scale renewable systems empower local communities, reducing dependence on fossil fuels while enhancing energy security.

---

[7] We have nevertheless to emphasize here that the use of AI also carries the risk of excessive energy consumption and future energy shortages. In addition, the first AI adoption projects in companies are currently showing that business cases often do not deliver what was hoped for.

We should also address (see the Stanford study "Canaries in the Coal Mine") the fact that entry-level jobs in particular are already being replaced and that "only" experts are benefiting from AI augmentation. This will cause fundamental changes in the labor market when the baby boomers all retire and the subsequent generations are "automated."

Together, AI and renewable energy form a powerful partnership for ecological integrity, enabling cleaner production, smarter resource stewardship, and a reduced carbon footprint.

15.3.3.2 Social and Economic Justice: Ensuring Equity in AI and Renewable Energy Adoption

The Earth Charter emphasizes social justice, human rights, and the eradication of poverty. Technological advancements must uplift all communities rather than widen existing inequalities.

AI and Inclusive Development

Practical measures include:

- Reducing algorithmic bias: Ensuring diverse datasets and transparent AI development processes prevents discrimination and promotes fairness in areas like employment, credit scoring, and public services.
- AI for public good: Deploying AI in healthcare diagnostics, climate adaptation planning, and educational support helps extend essential services to underserved communities.

Equitable Renewable Energy Access

To align renewable energy with the Charter's social principles:

- Decentralized energy systems: Deploying solar home systems and mini-grids in rural or marginalized areas provides affordable, reliable power where traditional infrastructure is lacking.
- Local capacity building: Training local technicians and entrepreneurs to install, maintain, and manage renewable energy projects ensures job creation and economic participation.
- Fair transitions: As fossil fuel industries decline, just transition policies—including retraining and social protections—ensure that workers and communities are not left behind.

In both domains, equity-focused strategies help ensure that the benefits of AI and renewables are shared widely, fostering social cohesion and economic justice.

15.3.3.3 Democracy, Nonviolence, and Peace: Ethical Governance of AI and Energy Systems

The Earth Charter calls for democratic participation, transparency, and nonviolence. These values are vital as AI and energy technologies increasingly influence politics, social relations, and security.

AI Governance and Democratic Values

Practical applications include:

- Transparent- and explainable AI systems with "clear" accountability, developed based on human-centered approaches regarding the UN Charta of Human

RightsAI systems: Open algorithms, explainable models, and public oversight mechanisms help maintain trust and accountability.

- Protection from digital harms: Ethical safeguards against misinformation, surveillance abuse, and autonomous weapons reflect the Charter's commitment to peace and human dignity.
- Community-inclusive decision-making: Engaging diverse stakeholders in AI policy development ensures that technological decisions reflect collective values rather than corporate or governmental interests alone.

Renewable Energy and Peacebuilding

Energy systems directly impact geopolitical stability. Practical actions aligned with the Charter include:

- Reducing conflict over resources: Renewables reduce dependence on oil and gas reserves, which are often linked to geopolitical tensions and conflict.
- Participatory energy planning: Community engagement in energy projects promotes transparency, reduces opposition, and fosters peaceful local development.
- Resilient infrastructure: Renewable systems fortified with AI-driven predictive maintenance reduce disruptions that could fuel social instability.

Together, democratic AI governance and participatory renewable energy systems strengthen peace, fairness, and human dignity.

15.3.3.4 Interconnectedness and Global Responsibility: A Holistic Vision for AI and Clean Energy

A core idea of the Earth Charter is the recognition of interdependence—ecological, technological, and social. Applying this worldview to AI and renewables involves global cooperation and ethical responsibility (this also covers solutions for the energy hunger of AI).

Collaborative Innovation

- Open-source AI for sustainability: Sharing AI tools for climate modeling, renewable integration, and environmental monitoring accelerates collective efforts to address global challenges.
- International renewable energy partnerships: Cross-border grids, technology transfer, and capacity development help democratize access to clean energy.
- Sustainable Production and Circular Systems
- Circular economy models for tech: Designing solar panels, wind turbines, and AI-enabled electronics with repairability and recyclability reduces waste and resource pressures.
- Responsible mineral sourcing: Ethical supply chains for critical minerals—lithium, cobalt, rare earths—honor the Charter's call for respect for life, justice, and human rights.

Shared Ethical Frameworks

International alignment on AI ethics, renewable energy standards, and climate responsibilities reflects the Earth Charter's call for a global ethic of care for people and the planet.

A Potential for Practical Application of The Earth Charter

Artificial intelligence and renewable energy movement and transitions are among the most influential technological and ecological developments shaping humanity's future. When guided by the Earth Charter's principles of sustainability, justice, democracy, and global responsibility, these transitions become powerful leverages for ecological restoration, equitable development, and lasting peace.

The Earth Charter is a moral compass and not an autopilot. Therefore, its practical integration needs to be driven by individuals, society-at-large, governments, and corporations—supported by consultants and experts who specialize in the implementation of The Earth Charter, to ensure that innovation does not merely advance human capability but strengthens our ethical commitment to one another and to the living Earth. In doing so, AI and renewable energy transition can form the backbone of a truly sustainable and compassionate global society.

15.3.4 Mainstreaming the Earth Charter into Education and Learning

We live in an increasingly complex world marked by significant geopolitical shifts, new forms of injustice, and inequality driven by rising economic and social polarization, often compounded by the effects of man-made climate change. We also live in a world that is becoming more technologized, where the participation and sovereignty of individuals and states depend on their access to digital and networked technologies, which have become deeply woven into our lives and work.

The Earth Charter offers us a compass—developed over more than 10 years through international collaboration and dialogue with deep respect for Mother Earth and recognition of all life upon her—to navigate this world and find solutions to current and future challenges (see diverse papers from Steven Rockefeller as for example https://cdn.prod.website-files.com/5b61560c6d61a0848e9d8359/5d9e5457e8c37c9029d2f98b_Rockefeller_A_Short_History_of_the_Earth_Charter_Initiative.pdf). We cultivate these through empathetic and peaceful coexistence for all, as it is a value-based framework that honors cultural differences worldwide and considers the needs of all social groups.

Following John Dewey (1897, see: https://en.wikipedia.org/wiki/My_Pedagogic_Creed), who argued that education itself is not merely preparation for life but is life itself, our cultural and social belonging shapes education to enable us to share in the intellectual and moral resources humanity has accumulated (Dewey, 1897, see above). The Earth Charter provides a fundamental and powerful framework for creating shared understanding and helps teachers and learners to clarify their own and others' worldviews through dialogue, thereby defining pathways toward cultural and social harmony and belonging. By reflecting on how they and others relate to the world, learners identify common actionable pathways toward just and sustainable futures. In today's world, education must equip learners to navigate uncertainty, complexity, and ambiguity. It needs to be transformative, empowering learners to critically examine the economic, social, and ecological roots of inequality, injustice, and non-sustainability.

Our current relationship with technology requires transformative education grounded in the principles of the Earth Charter more than ever. We recognize the dual nature of technology—it allows us to create new things but can also lead to destruction. We must reflect more carefully on how digital and networked technologies are changing economic, social, and increasingly, ecological systems. Therefore, understanding the choices involved is more important than ever.

By educating about the four central pillars and their principles, previously vague perceptions of technology can be clarified, reflected upon in relation to one's own values and those of others, and thus sustainable practical solutions can be developed. In this way, education about technology has a transformative effect and genuinely provides approaches for guiding technological progress in a sustainable and peaceful manner.

For students, the Earth Charter is often entirely new, so the framework itself needs to be introduced first. Reflection activities, categorizing one's own worldview, exploring other perspectives, and applying the principles to specific technological solutions (e.g., smartphone usage duration and intensity) quickly lead to increased knowledge and create emotional spaces for a mindset shift.

Therefore, the Earth Charter should be recognized by everyone as a powerful value-driven framework for the sustainable development of our world. It should be promoted more broadly, understood, and communicated within organizations as a voluntary commitment, similar to the legal document of the UN Convention on Human Rights.

15.4 First Concrete Sustainability Activities in Academia and Consulting (Use Cases)

15.4.1 Introduction into Academia Transformation

"Sustainability" is not a defined framework, but a broad umbrella covering many initiatives and concepts aspiring to have transformative impact on domains like economy, engineering, society, and others. Many of these concepts are so far neither matured nor aligned, resulting in various inconsistencies and open questions. This is not only a challenge to society and economy in pursuit of sustainability, but also to academic institutions that want to offer education and orientation around sustainability. Nevertheless, many universities have already started numerous activities to tackle at least some selected topics around sustainability and UN SDG coverage.

As we have many representatives of academic institutions in our author community, we are able to show a remarkable use case from some of them in this book to give a bit of a taste of what is already in place. One of the noble aims of this book is to establish a lively community of authors from the academic field who, guided by the findings of the book, will engage in a lively exchange about the activities initiated and the people responsible for them, ensure a certain degree of systematization, influence curricula in the long term, etc.

In this context, some fundamental questions may arise for the respective universities or university departments, for example (all questions are meant in the context of sustainability and UN SDG coverage):

- What is your (institution's/department's) educational mission, i.e. what impact do you want to have on your students, what impact on the principles and practices in the above-mentioned domains?
- What areas and concepts are your institution/department focusing on? Based on what criteria have you selected them?
- Are you seeing possibilities to align and integrate the concepts that your department is covering?

In order to also stress the Three-Pillar Management Concept, based on the 3-P-Model, and to introduce it also at universities as a systemic approach to manage complex organizations with its three pillars addressing fundamental aspects of complexity, an additional question is obvious:

- "Traveling Organization" addresses the need for adaptability in a dynamic and complex world, "Sustainable Purpose" promotes to prioritize long-term benefit for the overall system over pure short-term economic considerations, and "Connectivity" acknowledges that only what is connected and integrated can effectively and efficiently collaborate. As practitioners of the Three Pillars: do we see opportunities to integrate and promote these principles in your educational concepts?

The following—intentionally very condensed presentation of—three activities, concepts, and use cases from different universities from three different countries give a first taste of the start of a respective development journey and partly already reflect some of the above stressed questions (and give preliminary answers on partial aspects of them).

15.4.2 Activities, Concepts, and Use Cases at School of Economics and Law (HWR) Berlin, Germany (Claudia Lemke)

HWR Berlin has a large number of activities, concepts, and programs on sustainability in place, strongly supported by the HWR's vice president for Sustainability Prof. Dr. Silke Bustamante.[8] For reasons of space, we will limit ourselves here to following on from the previous explanations in this chapter and mentioning the following activity in more detail: Transformative learning is a concept that can change

[8] See https://www.hwr-berlin.de/en/hwr-berlin/university-management/vice-president-3

the way people see themselves and their world, thereby creating a changed worldview, often triggered by their own powerful experiences.[9] In her role as professor in Business Information Systems and Academic Director of the Master's Program Digital Transformation at the Berlin School of Economics and Law (HWR Berlin), Claudia Lemke has also been intensively involved with The Earth Charter and Transformative Education[10] in recent years and started to integrate elements of this into Academic Education and Research in close connection with her global academic network.[11]

15.4.3 Activities, Concepts, and Use Cases at ZHAW Zurich, Switzerland (Albena Björck)

Drawing on more than a decade of research, teaching, and executive advisory work, Dr. Albena Björck[12] has established herself as a leading voice in purpose-driven strategic management. At ZHAW, she heads the Global Business Lab—the Center for Purpose, Resilience and International Strategic Management—where she advances cutting-edge research on purpose-driven organizations, strategic renewal, and crisis readiness, and leads long-term collaborations with global companies such as Johnson & Johnson, Logitech, Clariant, and ASMPT. Her academic work spans conceptual scholarship, systematic reviews, and empirical studies on how corporate purpose shapes strategy, leadership, and organizational transformation. She is also the Co-Founder of the Chief Purpose Officer Network, contributes internationally as Chair of the EURAM SIG Strategic Management, and serves as an expert for Innosuisse (the Swiss federal innovation agency) on purpose-led innovation and strategy evaluation. In collaboration with Stanford University, she leads teaching programs in purpose-driven innovation.

Complementing these roles, she has established the Purpose Inside Lab, a platform dedicated to purpose-led strategies, organizational transformation, and performance measurement—enabling organizations to translate purpose into strategic decisions, operating models, and measurable impact. In this effort, she is

[9] Clark, M. C. (1993). Transformational learning. New directions for adult and continuing education, 1993(57), 47–56.; Mezirow, J. (1997). Transformation theory out of context. Adult education quarterly, 48(1), 60–62.

[10] "Transformative Education ... empowers learners of all ages with the knowledge, skills, values and attitudes to address the interconnected global challenges we are facing, including climate change, environmental degradation, loss of biodiversity, poverty and inequality. It builds a sense of belonging to a common humanity and help them become responsible and active global citizens in building inclusive, peaceful and sustainable societies."

[11] Details on the theoretical and conceptual background of her work, see Chap. 5 of this book.

[12] See also: Björck, A.; Pugnetti, C.; Aebischer, A.; Steller, N. and Vogel, B.: "Purpose in Business Models: Strategic Principles and Key Decisions" in the previous book: Wollmann, P., Pemler, D., Ndrevataj, M.: The Sustainable Organization - How Organizations Address the 17 UN SDGs Using the 3-P-Model; Cham 2025, SpringerNature (https://link.springer.com/book/10.1007/978-3-031-89549-4).

collaborating with Dr. Carlo Pugnetti from the Lucerne School of Business and Prof. Dr. Henk Volberda from Amsterdam Business School.

15.4.4 Activities, Concepts, and Use Cases at Canberra University, Australia (Ehssan Sakhaee)

Ehssan Sakhaee from the University of Canberra (formerly the University of Sydney) developed the AAUL (Awareness, Acceptance, Understanding, Leadership) framework for sustainable leadership in a VUCA[13] environment and has used it for many years in his lectures. The framework—shown in Fig. 15.1—addresses organizational resilience by aligning with the three common components of resilience identified by Diane Coutu in the resilience literature,[14] which Sakhaee incorporated into his teaching.

The AAUL framework is particularly well suited to crisis situations, as the self-awareness component enables rapid self-leadership, allowing individuals to move from panic to a calm, grounded state that supports more informed decision-making. The framework aims to cultivate a holistic mindset that begins with the self and extends outward to the broader VUCA environment. The attitudes, mindsets, and skills developed at the individual level naturally extend to one's immediate

Fig. 15.1 AAUL leadership framework introduced in Three-Pillar Book 2 (Wollmann, P.; Kühn, F.; Kempf, M.: Püringer, R. (Eds.): Organization and Leadership in Disruptive Times—Design and Implementation of the 3-P-Model. Cham: Springer Nature, © 2021) by Ehssan Sakhaee, used with permission. All rights reserved

[13] Volatile, Uncertain, Complex, Ambiguous.

[14] Review, Harvard Business et al. Resilience (HBR Emotional Intelligence Series). La Vergne: Harvard Business Review Press, 2017. Print.

environment and relationships, and then further expand to a global level encompassing our interconnected world.

At its core, AAUL fosters holistic awareness and understanding by recognizing the interconnectedness of all things—from self to others, and from organizations to society and the natural environment, both locally and globally. This mindset is non-judgmental, open, and curious, and is grounded in a deep sense of care for collective wellbeing and sustainability. It integrates concern for the self, others, the systems and organizations to which one contributes, as well as wider society and the environment.

The AAUL framework has been integrated into several units of study at the University of Sydney, including *Building a Sustainable World*, which was open to students across the university and fostered an interdisciplinary cohort. It has also been developed and applied at the University of Canberra within *Technology and Engineering Management* and *Professional Practice in Engineering*, equipping Engineering and IT students with a leadership framework that supports deeper awareness and understanding of themselves, others, and the complex world they inhabit.

The Three-Pillar model[15] has likewise been adopted both implicitly and explicitly. Skills and knowledge related to adaptability at individual and collective levels, sustainable purpose, and meaningful connection across academia and industry have been embedded within curricula and extra-curricular activities. This includes cross-disciplinary collaboration between students and staff on projects that contribute positively to society and the environment at both the University of Sydney and the University of Canberra.

15.4.5 Summary Academia Area

The use cases shown are, of course, only the tip of the iceberg. The topic of sustainability is now firmly established in most universities—with responsible persons directly in university management, with dedicated institutes and departments and research projects, with integration of the topic into traditional degree programs, etc. In several countries, there are specific responsibilities for the topic in overarching university associations.

One could say that the transformation journey—perhaps predominantly as a grassroots movement—has begun. It will now be important to gradually systematize this movement without taking away its momentum. This means that, based on a meta-level concept such as the aforementioned "SDGs for Academia" and on a uniform value system such as the Earth Charter, degree programs that exert a particular influence on the world through their graduates will also be gradually transformed.

In this context, we would like to refer to the current discussion on the so-called Bermuda Triangle of Talent, for example, described in a book from Simon van Teutum (originally in Dutch: De Bermuda-driehoek van talent, April 2025, De

[15] Details see below in Sect. 15.4.6.2.

Correspondent)[16] which questions why the best talents from university who are motivated to change the world and make it a really better place end up after their university degree in the triangle of (big) finance industry, (big) consulting, and (big) law firms.[17] On the one hand, they would be urgently needed in other places and on the other hand they often lack their organizations' and their own personal Sustainable Purpose.[18] This means that especially MBA and Law studies at universities might need a bit of transformation to extend the scope through integration of sustainability and UN SDG coverage aspects.

15.4.6 The Consulting Area: Some First Thoughts

15.4.6.1 Introduction

In the consulting area, we find the topic Sustainability and UN SDG coverage being at least formally addressed by all large international and national consultancies and we additionally find many smaller consultancies specialized on aspects often connected with support for respective general reporting requirements and linked governance topics.

A typical specialized consultancy was described in the first volume[19] in Chap. 20: "Strategic Sustaina-bility Implementation in SMEs: Leveraging the SDGs for Context-based Impact" by Patricia Moock,[20] founder and CEO of the consulting company 4L Impact Strategies.

Another typical specialized consultancy is Sustayn,[21] an organization with a focus on health management and motivation, learning and personal development, leadership and cultural management as building blocks for Sustainability and for resilience of organizations.

---

[16] See also: https://www.linkedin.com/posts/simon-van-teutem-74061a74_the-bermuda-triangle-of-talent-27-year-old-activity-7388547685571768320-tgLr/ and https://fortune.com/2025/10/26/gen-z-college-selling-out-elite-universities-2025-admissions-mckinsey-harvard-oxford/

[17] https://ideas.lukemac3000.com/2025/05/31/simon-van-teutem-the-bermuda-triangle-of-talent/

[18] Cited from ibid: "When Morgan Stanley intern Simon van Teutem's hair falls onto his keyboard at the end of a long working day, he asks himself: what am I doing here? Like many other ambitious students, he wants to change the world. And yet he works late into the night for a bank. No world peace, cultured meat and climate breakthroughs, but Excel sheets, conference calls and business class flights. How did he end up here? In conversations with dozens of fellow students, bankers, lawyers and consultants, Simon van Teutem reconstructs how idealistic young people swim into the trap of big capital. The result is a candid insight into the world of big money, a mirror for everyone who is looking for meaning in work, and an urgent call to education, NGOs and the government to prevent the waste of talent."

[19] Wollmann, P., Pemler, D., Ndrevataj, M.: The Sustainable Organization - How Organizations Address the 17 UN SDGs Using the 3-P-Model; Cham 2025, SpringerNature (https://link.springer.com/book/10.1007/978-3-031-89549-4).

[20] https://4l-impact-strategies.de/team/patricia-moock/

[21] https://sustayn.de

15.4.6.2 The Three-Pillar Model as Base for Consulting Direction

In general, we propose on a meta-level the implementation of a more purpose-driven direction combined with journey thinking and accompanied by the respective narratives for a new consulting: that means a pragmatic orientation toward the Three-Pillar (Mindset or Thinking) Model[22] for organization design and leadership for the VUCA[23] world:

(1) The first Pillar: **Sustainable Purpose**

Sustainable Purpose is the *raison d'être* of an organization. Having a clear guiding objective provides orientation and certainty to people, fostering a sense of shared effort and success. This is especially important in times of crisis and transformation, as it becomes the key source for the narrative of transformation.[24]

For consultancies, a Sustainable Purpose has two facets: on the one hand, the consultancy itself needs an own convincing Sustainable Purpose (e.g., to attract and develop talents)—and that has to be a lot more than "making money," but it has to be connected to make the world—and the organization of the client—a sustainably better place. This immediately establishes a potential connection to the UN Agenda 2030 (it can at least not be ignored). And on the other hand, there has to be a Sustainable Purpose for nearly each consultancy project for a client which has also to be transparently defined—especially in the case of large transformations to show its deeper sense which has to be carefully and convincingly communicated to all stakeholders to make the transformation successful.

(2) The second Pillar: **Traveling Organization**

Traveling Organization refers to an organization accepting to be in a permanent state of flux and transformation while interacting with the external requirements of its journey by means of rapid learning and adaptivity. Ideally, this is the mindset of loving the unknown and being well able to manage it.[25]

For consultancies, it is crucial to underline that each (especially larger) client project and/or transformation is a journey which potentially leads through terra incognita which means that it cannot be defined as a waterfall proceeding where everything is clarified from the outset and there is no longer any risk (of necessary modifications). It is widely recognized that especially transformational

[22] First described in: Wollmann, P.; Kühn, F.; Kempf, M. (Eds.): Three Pillars of Organization and Leadership in Disruptive Times – Navigating Your Company Successfully through the 21st Century Business World. Cham: Springer Nature, © 2020.

[23] VUCA= Volatile, Uncertain, Complex, Ambiguous.

[24] Everybody in the organization, but also its key stakeholders in the entire ecosystem and organization environment, must know what the organization stands for and what value and societal contribution it creates.

[25] One has an intuitive idea of a Traveling Organization from the metaphor of an expedition to Africa in former times. The participants of the Traveling Organization must be curious, open to learn, courageous, keen to experiment, and they deal well with uncertainty, stress, unforeseen incidents in a totally new world. This mindset of the Traveling Organization is the core of a Learning Organization. It is impossible in and for the Traveling Organization not to learn. Learning is a question of survival.

journeys of organizations are explorative in nature and to be handled in an agile way. This is a challenge especially for those consultancies very strongly basing on methods and tools—but in a VUCA world with unknown territories there are always situations without precedence where traditional methods and tools—and also data—are no longer applicable.

(3) The third Pillar: **Connectivity**
Connectivity focuses on the ability to interconnect during the journey with all necessary parties and resources inside and outside the silos, thus creating high effectiveness and consistency and also building shared learning networks[26] during this process.

It is obvious and well-known that an enormous degree of connectivity is needed for consultancies to understand the markets of their clients, the dynamics, and the respective preconditions and ecosystems, etc., to successfully define their client projects and the transformational journeys of their clients. This requests a very open mindset and a cooperative personality.[27]

To exemplify the narrative of the Three Pillar Model in general, one should perhaps start with the Traveling Organization. In our imagination, this term can evoke the image of a very agile, flexible team embarking on an unforeseeable journey to explore unknown territories. In former times, this meant, as already mentioned above, for example, a scientific expedition to Africa or Australia to explore territories previously unknown to us. It is intuitively clear—and supported by various travel reports and even video footage—what kind of team dynamics, personal attributes, and collective capabilities are essential for the successful execution of such an endeavor.[28]

We regard the described mindset and thinking style shift of consultancies toward the Three-Pillar Model as a precondition to successfully integrate elements of the UN Agenda 2030 and UN SDG Coverage into the consultancy day-to-day practice in the future.

15.4.6.3 Concrete Integration of the UN Agenda 2030 and the UN SDG into Future Consultancy

Innovative and effective consulting strategies in **SDG** go beyond box-ticking. The most impactful approaches integrate strategy, data, governance, and value creation. It is important and necessary to move from **traditional to innovative to frontier practices**, with a focus on what actually delivers results for clients. Here are seven

[26] This means that organization has to be thought of within a network perspective, including interfaces to adjacent organizations and across organizational borders. In many cases, it might be impossible to learn effectively without collaborating closely with people from outside your silo.

[27] See Schmidt, H.: "Connectivity and Personality" in Wollmann, P.; Kühn, F.; Kempf, M. (Eds.): Three Pillars of Organization and Leadership in Disruptive Times – Navigating Your Company Successfully through the 21st Century Business World. Cham: Springer Nature, © 2020, page 141–146.

[28] See also Chap. 2 of this book.

points to reflect on and elaborate in (3P) Progressive-Pragmatic-Practical consulting approaches and practices.

1. **Strategic Integration (From Reporting to Value Creation)—SDG-Anchored Business Strategy: Move from "Mapping Activities to SDGs" to Using SDGs as a Strategic Lens**
   Effective consulting practices include identifying material SDGs using double materiality by combining finance and impact. This can be achieved by translating SDGs into business-relevant objectives (e.g., SDG 13 in the context of carbon pricing and supply-chain resilience). Furthermore, embed SDGs into, inter alia, corporate strategy, capital allocation, innovation pipelines, and M&A and risk assessments. By doing this, clients see SDGs as drivers of competitiveness, not compliance costs.
2. **Advanced Materiality and Stakeholder Intelligence**
   Advance Data-Driven Double Materiality through Innovation by applying AI-supported materiality assessments combining ESG risk exposure, value-chain impacts, regulatory foresight (CSRD, ISSB, SEC), and Stakeholder sentiment analysis. Such an approach will provide faster, defensible, audit-ready materiality results, aligned with EU CSRD, GRI 2021, and ISSB standards.
3. **Smart SDG Metrics and Impact Measurement—Move Away from Outputs to Outcomes and Impact**
   Shift from activity-based indicators to outcome metrics, science-based targets, and context-based KPIs. Apply Digital Impact Measurement Systems such as ESG data platforms integrating IoT (energy, water, emissions), satellite data (land use, deforestation), supplier ESG data via APIs, and blockchain for traceability and auditability (select use cases). This will reduce reporting friction and increase credibility.
4. **Regulatory-Ready SDG Reporting—It is Critical to Have SDG Alignment with Mandatory Frameworks**
   An innovative consulting approach will require a "Report once, comply many times" architecture that will include SDGs ↔ GRI; SDGs ↔ CSRD/ESRS; SDGs ↔ ISSB; SDGs ↔ TCFD/TNF.
5. **Effective Strategies for SDG Governance and Accountability Models at the Board-Level**
   This will include: SDG dashboards for Executives; Executive incentives tied to SDG KPIs and SDG risk heatmaps integrated into ERM. This will promote Internal Capacity Building and train "SDG translators" inside the organizations and promote micro-learning scenario simulations which in turn will reduce long-term dependence on external help.
6. **Integrity by Walking the Talk**
   It will be absolutely critical for consultancy services to uphold the highest level of integrity by walking the talk on Anti-Greenwashing and Credibility Safeguards. High-value consulting services will ensure SDG claim verification through

impact boundary setting, vouching for third-party assurance readiness and establishing a mechanism for red-flag analysis of false or misleading SDG marketing claims.

7. **Development of Frontier Mindset Practices and Back-Casting Approaches** In taking consultancy to the next generation of businesses, frontier mindset practices and back-casting approaches need to be developed and integrated in the solutions toolbox. This will include systems-thinking SDG consulting (cross-SDG trade-offs), nature-positive accounting (TNFD which stands for Taskforce on Nature-related Financial Disclosures + SDGs), migration-aware SDG impact metrics (SDGs 8, 10, 13 in climate-vulnerable regions), and scenario-based SDG transition planning (climate, biodiversity, social instability).

15.5 Outlook

It will be important to discuss in this book's author community—or even in the Three-Pillar Author Community as a whole—how to proceed in detail and with whom for which work stream. We have many potential candidates—e.g., from academia and from consultancy—in the community who might want to participate. As already mentioned above, the future activities will mid-term contain:

- The formulation of a practical codified framework for (future) education (at universities), some-thing like a sort of "SDGs for education" and the collection of a broad range of use cases covering specific problems. This includes the identification of already existing activities in this area.
- The formulation of a practical codified framework for (future) consultancy incl. key messages and shared values. This should include a sort of SDGs for consultancy and the collection of a broad range of use cases covering specific problems. This includes the identification of already existing activities in this area.
- Integration of the Earth Charter values and concepts into the mentioned activities.

Ovais Sarmad served the United Nations Climate Change secretariat (UNFCCC)—COP23 to COP27 (2017–2023). Prior to UNFCCC, he served as Director of Cabinet & Chief of Staff to the Director General at the International Organization for Migration (IOM). At IOM, he worked in several management and policy development capacities over a period of 27 years. Currently, he serves as the Vice Chair of GHG-Protocol Steering Committee, and his other areas of focus include engaging with and supporting activities and organizations around the world (a) that promise to innovate and accelerate sustainable development and growth without damaging the environment; (b) to contribute to public policy discourse and development; (c) to empower youth through quality education at all levels; and (d) to advise, mentor, and coach individuals and organizations in multilateralism, diplomacy, geopolitical issues on climate change and migration, and organizational change and development.

Claudia Lemke holds a Professorship in Business Information Systems at the Berlin School of Economics and Law (HWR Berlin). She is also the Academic Director of the Master's Program Digital Transformation at the Berlin Professional School, the Professional Institute at the HWR Berlin, and a member of the HWR Berlin Harriet Taylor-Mill Institute (HTMI) of Economics and Gender Studies. She has over 25 years of research, teaching, and consulting experience in different countries. Her most recent work focuses on the impact of digitalization on the economy and society driven by emerging technologies.

Peter Wollmann has been a responsible manager, initiator, mentor, and facilitator in large, predominantly global transformations and strategic developments for almost 40 years in large organizations and since 2017 as independent consultant. He is also a member of an advisory board of a R&D oriented company. Peter holds a degree in mathematics and physics from the University of Bonn. He is the author and publisher of numerous books and articles on strategy, leadership, organization design and project and project portfolio management.

16 Toward a Reconstructive Foundation for Sustainability—A Philosophical Deepening of Sustainability Beyond Perception and Performance

Ashkan Tashvir

Abstract

This concluding chapter reframes sustainability from a predominantly technical, regulatory, and operational challenge into a fundamentally ontological and reconstructive question. While contemporary sustainability frameworks—most notably Elkington's Triple Bottom Line Model (s Elkington, J.: Cannibals with Forks: The Triple Bottom Line of 21st-Century Business. 1997, Capstone. https://doi.org/10.1002/tqem.3310080106) and the United Nations World Summit Outcome 2005 (see: https://www.un.org/en/conferences/environment/newyork2005)—have generated important metrics, policies, and institutional responses, persistent patterns of fragility, short-termism, fragmentation, and implementation fatigue suggest the presence of a deeper explanatory gap. This chapter addresses that gap by asking why sustainability stabilizes in some systems yet repeatedly collapses in others, even where intentions, governance mechanisms, and performance indicators appear robust.

Building on the empirical, institutional, and sectoral contributions of the volume, the chapter situates these insights within a deeper philosophical foundation. It argues that sustainability is not merely an objective to be managed, measured, and enforced, but an emergent property of systemic integrity—the coherence between purpose, values, structures, incentives, decision-making, and long-term consequences. Where this coherence is absent, sustainability initiatives tend to become performative, fragmented, and increasingly dependent on continuous external intervention.

The chapter then turns to the human ontological dimension of sustainability. It demonstrates that enduring sustainability originates not only in what systems *do*, but in how humans are *being* within those systems. Through an examination

A. Tashvir (✉)
Engenesis, Sydney, NSW, Australia
e-mail: Ashkan@engenesis.com

© The Author(s), under exclusive license to Springer Nature Switzerland AG 2026
P. Wollmann et al. (eds.), *The Sustainable Organization – Volume 3*, Future of Business and Finance, https://doi.org/10.1007/978-3-032-28949-0_16

of Being (Tashvir, A. (2021) *BEING*. Sydney, NSW: Engenesis Publications), meaning, intention, and sense-making, the chapter explains why behavioral change programs, compliance regimes, and awareness campaigns often fail to produce lasting outcomes when deeper architectures of responsibility, perception, and value remain misaligned.

To explain why sustainability discourse often remains fragmented and unresolved—despite apparent agreement on objectives—the chapter introduces the concept of "metacontent"—the underlying assumptions, narratives, perspectives, and cognitive frameworks that shape interpretation beneath surface-level discussion. By foregrounding layered sense-making, the chapter clarifies how sustainability debates can remain stalled at the level of competing positions, limiting integration and long-term coherence.

Crucially, the chapter moves beyond critique toward reconstruction. It argues that deconstructive analysis alone is insufficient for addressing systemic sustainability failures. What is required is a reconstructive ontology capable of redesigning economic, institutional, organizational, and cultural systems around first principles of coherence, integrity, regeneration, and long-term responsibility.

From this foundation, the chapter advances the concept of Authentic Sustainability as a regenerative design paradigm. Sustainability is reframed not as preservation, optimization, or crisis response, but as the regeneration of coherence, trust, meaning, and systemic continuity. The chapter concludes by integrating this ontological foundation with the Three-Pillar Model, showing how Sustainable Purpose, organizational movement (Traveling Organization), and Connectivity acquire durability only when grounded in intrapersonal, interpersonal, structural, and systemic integrity. In doing so, it positions sustainability not as an external constraint on development, but as the architectural logic of systems—including entire civilizations—capable of enduring themselves.

16.1 Foundational Context

16.1.1 The Sustainability Paradox

At no point in human history have we spoken so extensively about sustainability while living so unsustainably. Governments design policies, corporations publish ESG reports, universities establish sustainability centers, and entire professional fields now exist around impact measurement, compliance, and transition management. Sustainability has become embedded in the language of progress, governance, and responsibility across public, private, and civic domains.

And yet, ecological systems continue to degrade, trust in institutions continues to erode, inequality widens, social cohesion fractures, political polarization intensifies, and economic systems grow more extractive rather than regenerative. Despite unprecedented coordination, expertise, and investment, the global condition remains fragile.[1]

This coexistence of expanding sustainability effort and deepening systemic instability is no longer a peripheral paradox. It has become a defining feature of contemporary civilization.

Common explanations typically point to insufficient political will, inadequate technology, regulatory gaps, funding constraints, or public resistance. Each of these factors contributes to the problem, and many are addressed directly within the contributions of this volume. Yet none of them, on their own, explains the persistence of sustainability failure across otherwise well-intentioned systems.

What remains insufficiently examined is whether sustainability failure originates not only in policies, structures, or implementation, but in the deeper assumptions through which reality, value, progress, and responsibility are understood and enacted by human beings and institutions.

Much of contemporary sustainability discourse operates at the level of optimization and transition. It specifies what should be measured, reduced, governed, or transformed. However, it less frequently addresses the underlying architecture of meaning, intention, and interpretation that continues to shape decision-making beneath the surface.

Consequently, sustainability is often added onto systems whose foundational logic remains unchanged. Growth imperatives persist. Short-term performance dominates. Political survivability outweighs generational responsibility. Organizational incentives reward speed over coherence, while cultural norms continue to privilege consumption, competition, and immediacy.

Under such conditions, sustainability becomes structurally constrained. It functions as an accessory rather than an organizing principle. Progress is reported, yet the forces producing unsustainability remain largely intact. This generates a distinctive form of dissonance. Sustainability is publicly affirmed and institutionally codified, yet weakly embodied. It is visible across strategies, reports, and declarations, while remaining inconsistently internalized within the deeper logic of systems.

The resulting crisis is not only ecological or economic. It is ontological, insofar as it concerns how reality is interpreted, how value is constructed, how responsibility is distributed, and how continuity across time is understood within the systems we design and inhabit. In this sense, sustainability is not failing because it is misguided, but because it remains incomplete at the level of first principles. This chapter begins by examining those deeper conditions.

Taken together, these patterns point to a persistent tension running through contemporary sustainability efforts. While activity, investment, and institutional attention have increased, coherence has not. What is missing is not effort, goodwill, or intelligence, but an adequate way of understanding why well-intentioned

[1] **IPBES (2019)** *Global Assessment Report on Biodiversity and Ecosystem Services*. Bonn: IPBES Secretariat. **OECD (2021)** *Trust in Government*. Paris: OECD Publishing.

Fukuyama, F. (2022) *Liberalism and Its Discontents*. London: Profile Books.

World Economic Forum (2024) *Global Risks Report 2024*. Geneva: WEF.

sustainability initiatives so often fragment, stall, or require continuous external enforcement to remain viable.

To address this gap, it is necessary to step briefly outside conventional sustainability discourse and examine the deeper conditions that determine whether systems can—or should—endure at all. The following section introduces a condensed ontological perspective developed in *Sustainabilism*,[2] not as a competing framework, but as a diagnostic and reconstructive lens that helps clarify why sustainability so often fails upstream before it fails in practice.

16.1.2 An Ontological Lens on Sustainability

The patterns identified here point to a condition that recurs across domains, sectors, and scales: sustainability initiatives tend to fragment, stall, or require continuous external enforcement when the systems within which they operate lack internal coherence. This condition is best understood not as a failure of effort or intent, but as a state of systemic disintegration.

Disintegration, in this context, refers to the erosion of coherence and agency across systems—from individuals and collectives to organizations, institutions, communities, and societies. It is not synonymous with dysfunction, nor is it presented as a moral judgment. Rather, it describes an observable condition in which systems lose their capacity to hold purpose, responsibility, and continuity together over time. Although disintegration can create space for renewal, it does not automatically do so. When underlying fractures remain unexamined, it more often produces downward spirals than meaningful recalibration.

From this perspective, integrity and disintegration are not binary opposites but dynamic states on a continuum. Systems rarely collapse suddenly. More commonly, they enter prolonged periods of flux in which some forces stabilize coherence while others quietly erode it. The instability visible across contemporary sustainability efforts is therefore not random. It follows patterns. When those patterns are acknowledged and understood, they can be interrupted, modulated, and reconstructed.

Many sustainability initiatives that struggle to endure share a common structural characteristic: they operate primarily through external prescriptions, targets, and compliance mechanisms applied to systems whose underlying logic remains unchanged. This dominant mode of engagement is referred to here as "sustainabilism." Sustainabilism encompasses doctrine-based sustainability approaches that prioritize measurement, reporting, optimization, and enforcement, often with considerable sophistication and goodwill. However, when pursued in isolation from deeper questions of coherence, meaning, and responsibility, sustainabilism remains dependent on continuous intervention and vulnerable to fragmentation.

The recurring limitations of sustainabilism do not indicate that sustainability itself is misguided. Rather, they reveal the ontological constraints of pursuing

[2] Tashvir, A. (2025) *Sustainabilism: Exposing the Sustainability Illusion*. Engenesis Publications, Sydney.

sustainability as an external corrective rather than as a property of system design. When sustainability is treated primarily as something to be imposed, audited or incentivized from the outside, it struggles to endure once attention, enforcement, or favorable conditions recede.

It is in response to these recurring failure patterns that the concept of "authentic sustainability" becomes necessary. Authentic sustainability does not position sustainability as a prescriptive objective layered onto existing systems. Instead, it understands sustainability as an emergent property of systems grounded in systemic integrity. From this perspective, durability arises when coherence is maintained between purpose, value, responsibility, incentive, and structure across time.

Central to this shift is the recognition that human beings are not external observers of the systems they seek to sustain. They design them, legitimize them, inhabit them, and reproduce them through everyday decisions, interpretations, and incentives. Sustainability therefore cannot be separated from the inner architecture of human participation, nor can systemic integrity be reduced to technical performance, governance mechanisms, or regulatory compliance alone.

The ontological lens developed in *Sustainabilism*[3] provides a way of making these dynamics visible without displacing the empirical, institutional, and sectoral contributions that precede it. Ontology is used here not as abstract philosophy, but as a means of clarifying the foundational assumptions about reality, value, agency, and responsibility that quietly shape how systems are designed and sustained. When these assumptions are misaligned, sustainability efforts tend to become performative, characteristic of sustainabilism. In contrast, when they are coherent, sustainability can emerge as a native property of the system. This is authentic sustainability.

Within this lens, systemic integrity and disintegration are understood across a small number of interacting domains that describe how coherence is stabilized, distorted, recalibrated, or lost within human systems. The purpose of this framing is diagnostic rather than exhaustive. It provides sufficient structure to locate where sustainability failure originates before it becomes visible in outcomes, without requiring wholesale adoption of a new theoretical architecture.

Crucially, this perspective also clarifies why critique alone has proven insufficient. While deconstructive analysis can expose contradictions, inefficiencies, and unintended consequences, it does not by itself rebuild the conditions required for endurance. What is required instead is a reconstructive orientation: an approach capable of reconfiguring existing systems around coherent first principles without discarding the accumulated knowledge, capacity, and intent they already contain.

Within the arc of this volume, this ontological lens functions as a philosophical deepening rather than a competing framework. It offers a language for understanding why sustainability initiatives that are technically robust and institutionally supported may still struggle to endure when coherence at the level of human participation and system design is absent.

[3] Tashvir, A. (2025) *Sustainabilism: Exposing the Sustainability Illusion.* Engenesis Publications, Sydney.

With this foundation established, the chapter now turns to its central task. It examines how sustainability can be understood as an emergent property of systemic integrity, how reconstruction operates as a design logic rather than a crisis response, and how this perspective can be integrated with the Three-Pillar Model to enhance durability across the contexts explored throughout this volume.

16.2 Introduction: The Need for an Ontological Foundation of Sustainability

This concluding chapter steps beyond technical, regulatory, and sector-specific perspectives to examine the deeper conditions that determine whether sustainability efforts endure. It does not seek to replace existing frameworks, case studies, or institutional approaches, but to deepen the interpretive ground upon which they rest.

The analysis proceeds from the recognition that sustainability initiatives often stall not because of insufficient tools or intent, but because unresolved questions of meaning, integrity, and systemic coherence remain unaddressed. The inquiry is therefore foundational rather than procedural. It asks what must be structurally true for sustainability to persist across time rather than only what must be operationally improved within existing models.

The preceding chapters of this volume demonstrate, with clarity and practical rigor, that sustainability is no longer a marginal concern. Across policy, organizational design, leadership practice, and sector-specific applications, the contributions reflect a field that has matured significantly in both sophistication and intent. This chapter does not seek to contest or displace those contributions. Rather, it proceeds from the recognition that such work, while necessary, is not always sufficient to explain why sustainability initiatives often struggle to endure once initial momentum, leadership attention, or external pressure diminishes.

The breadth of contemporary sustainability practice reveals a deeper challenge. Sustainability is pursued across multiple domains with differing assumptions, values, incentive structures, and interpretive frameworks. This pluralism is not inherently problematic. However, when there is no shared foundational understanding of what sustainability fundamentally is, fragmentation becomes structurally unavoidable.

Most sustainability models operate at the level of capability, governance, policy, metrics, innovation, and organizational transformation. These dimensions are essential. Yet they presuppose deeper assumptions about reality, value, responsibility, human agency, and systemic causality that are rarely examined explicitly. When these assumptions are misaligned or internally contradictory, sustainability efforts become vulnerable to misalignment, performativity, and eventual erosion.

This chapter begins from the premise that sustainability cannot be secured purely through technical mastery or institutional coordination. It requires coherence at a more fundamental level. It requires an ontological grounding that clarifies what kinds of systems can be sustained, what kinds of human participation are required to sustain them, and what conditions of integrity must be present for sustainability to persist across time rather than across reporting cycles.

Ontology, in this context, does not refer to abstract metaphysics removed from practice. It refers to the most basic assumptions about reality that organize perception, value judgments, economic priorities, political decisions, organizational design, and individual behavior. Whether explicitly acknowledged or not, every sustainability strategy is already grounded in an implicit ontology. The difficulty is that these ontologies are often fragmentary, inherited, or left unexamined.

When this foundation is incoherent, sustainability becomes vulnerable to three recurring distortions. It may be reduced to a technical problem to be optimized, appropriated as an ideological instrument, or confined to a compliance exercise detached from deeper questions of meaning, responsibility, and systemic integrity. None of these distortions negate the value of existing frameworks, but they limit their depth, endurance, and transformative capacity.

The purpose of this chapter is therefore not to introduce another operational model, nor to compete with the Three-Pillar Model that underpins this volume. Its purpose is to articulate the ontological conditions that must be present for any such model to function in a coherent, regenerative, and enduring way.

Drawing on the philosophical foundations developed in *Sustainabilism*[4] and the broader authentic sustainability discourse, the chapter advances the view that sustainability is not primarily an objective to be achieved, but a systemic condition to be embodied. It emerges where integrity, coherence, intentionality, and meaning are aligned across human, institutional, and ecological domains.

From this perspective, the central question of sustainability shifts. It is no longer only concerned with how organizations should operate or how societies should regulate impact. It becomes a question of what kinds of beings we must become, what kinds of institutions we must design, and what kinds of systemic logics we must inhabit for sustainable life to endure at all.

The sections that follow develop this ontological foundation progressively. They move from the consequences of first-principle ambiguity, to sustainability as a systemic property, to integrity as a structural prerequisite, to the human dimensions of Being and sense-making, and finally toward a reconstructive paradigm capable of integrating coherence, regeneration, and systemic continuity.

In doing so, this conclusion seeks to provide a conceptual horizon within which the empirical, institutional, and sectoral insights of this volume can be interpreted with greater depth, coherence, and long-term clarity.

[4] Tashvir, A. (2025) *Sustainabilism: Exposing the Sustainability Illusion.* Engenesis Publications, Sydney.

16.3 The Core Problem: Sustainability Without First Principles

Modern sustainability frameworks demonstrate impressive breadth and methodological sophistication. They draw from environmental science, economics, engineering, organizational behavior, ethics, digital transformation, psychology, and public policy. They map interdependencies, quantify impacts, model future scenarios, and identify risks and levers for transition. This pluralistic knowledge base is one of the field's core strengths.

Yet beneath this diversity lies a systemic vulnerability that is rarely addressed directly. Contemporary sustainability discourse still lacks a shared set of first principles that define the nature of sustainability itself. Without such grounding, even the most well-designed strategies rest on conceptual foundations that remain unsettled.

In this context, first principles refer to the fundamental assumptions about reality that determine how systems operate. They are not policy goals or strategic priorities, but the often-unexamined beliefs about human nature, value, responsibility, power, growth, and causality that quietly shape design choices, incentives, and decision-making across systems. These assumptions are rarely articulated, yet they influence everything from economic design and organizational culture to political judgment and technological ambition.

The absence of first principles manifests in several recurring patterns.

First, sustainability is often defined by its outcomes rather than its nature. It is described in terms of reduced emissions, responsible resource use, social equity, institutional resilience, or long-term economic viability. These are valuable objectives, but they do not explain what sustainability fundamentally *is*. They describe indicators and aspirations, not the underlying condition from which those outcomes would reliably arise.

Second, sustainability is frequently approached as a technical challenge to be managed through innovation, regulation, or organizational optimization. This instrumental framing assumes that sustainability is achieved when systems function efficiently within environmental limits. However, efficiency alone does not resolve questions of meaning, value, integrity, or human motivation. A technically efficient system may still be fundamentally unsustainable if the forces shaping it remain misaligned.

Third, sustainability is often treated as a policy domain rather than a systemic condition. It becomes one concern among many rather than the organizing logic of the entire system. This compartmentalization fragments effort and creates internal contradictions. An organization may pursue sustainability initiatives while simultaneously operating according to economic, cultural, or political logics that undermine them.

These challenges arise not because sustainability frameworks are inadequate, but because they attempt to operate without a clear ontological center. When a concept becomes widely used without grounding in first principles, it expands in meaning until it risks describing everything and nothing. Sustainability increasingly occupies this paradoxical position. It is invoked across almost every organizational, institutional, and governmental context, yet interpreted differently in each.

The central question therefore emerges: how can we sustain what we cannot coherently define?

In the absence of first principles, sustainability defaults to the dominant logic of the systems it seeks to influence. Economic systems interpret it as cost optimization and risk management. Political systems interpret it through the lens of public approval and electoral viability. Corporate systems interpret it as brand value, compliance, or strategic differentiation. Academic systems interpret it as conceptual expansion or methodological refinement. None of these interpretations are inherently incorrect. Yet none, on their own, provide the coherence required for sustainability to endure.

This helps explain why, despite decades of frameworks, standards, summits, and reporting mechanisms, the global condition remains fragile. We have become increasingly skilled at describing sustainability and more adept at performing it. But performing sustainability is not the same as embodying it. Embodiment requires structural transformation at the level of first principles.

If these underlying assumptions remain unchanged, sustainability becomes structurally constrained. Even well-intentioned efforts are absorbed into pre-existing logics and redirected by incentive structures that privilege short-term performance, competitive advantage, or political stability. The result is not failure in a conventional sense, but gradual dilution. Sustainability becomes integrated without becoming transformative.

This is the core problem this chapter addresses. Sustainability requires not only new strategies, but a clearer understanding of the systems within which those strategies operate. It requires first principles that allow sustainability to function as a systemic property rather than a collection of objectives. It requires an ontological foundation capable of bringing coherence to the dispersed and often competing interpretations that currently shape the field.

This challenge is examined in depth in the book, *Sustainabilism*. The argument developed there is not that sustainability should be abandoned, but that sustainability cannot be understood purely through external indicators or organizational behaviors. Instead, it must be understood as a condition arising from the alignment of systemic forces across human, institutional, ecological, and economic domains. In this sense, sustainability is not an intervention to be imposed, but an emergent property of systemic integrity.

The next section develops this claim by examining sustainability as a specific kind of systemic phenomenon, one that cannot be produced through technical optimization alone.

16.4 Sustainability as a Systemic Property Rather Than an Objective

Sustainability is most often spoken of as a goal to be achieved. Organizations "pursue" sustainability. Governments "commit" to sustainability. Industries "transition" toward sustainability. In each case, sustainability is positioned as a future state toward which effort is directed, measured, and evaluated. This framing is understandable. Yet it subtly misrepresents the nature of what sustainability actually is.

When sustainability is reduced to an objective, it becomes comparable to performance targets such as efficiency, profitability, compliance, or growth. It enters the familiar logic of input, output, optimization, and reporting. This logic is not inherently flawed. However, it belongs to a different category of phenomena. Objectives can be pursued and achieved within existing system architectures. Sustainability cannot.

Sustainability is not an outcome that can be produced on demand. It is a property of systems that exhibit internal coherence, regenerative capacity, and temporal continuity. A system does not become sustainable because it adopts sustainability initiatives. It becomes sustainable when the way it is structured, governed, valued, and enacted supports endurance without self-erosion.

This distinction is critical. A system can meet sustainability objectives while remaining fundamentally unsustainable. It can reduce emissions, publish transparent reports, invest in renewable technologies, and still operate on incentive structures, cultural logics, and power arrangements that undermine its own long-term coherence. In such cases, sustainability functions as an overlay rather than an organizing principle, applied to systems whose deeper architecture remains unchanged.

To understand sustainability as a systemic property requires a shift in causal logic. Rather than asking what actions produce sustainability, the more foundational question becomes:

What conditions must be true of a system for sustainability to naturally emerge and persist?

This shift requires attention to the internal conditions that determine whether systems hold together (remain coherent) or come apart (fragment) over time. Systems do not fail solely because of external shocks or insufficient performance. They fail when coherence between purpose, responsibility, incentives, and consequence erodes. When this fragmentation reaches a critical threshold, sustainability initiatives become increasingly performative, reliant on enforcement rather than internal alignment.

From this perspective, sustainability failure is better understood as a condition of systemic disintegration rather than a lack of activity or intent. The Authentic Sustainability Framework[5] articulates these conditions not as aspirational ideals, but

[5] Introduced in the book *Sustainabilism—Exposing the Sustainability Illusion*, the Authentic Sustainability Framework (ASF) is a unifying architecture developed to diagnose and reconstruct the conditions under which systems endure or deteriorate. It integrates a set of interrelated models that examine systemic coherence, distortion, and regeneration across human and institutional sys-

as design requirements for systems capable of sustaining themselves without reliance on constant enforcement or correction. Disintegration does not imply collapse or dysfunction alone, but the gradual loss of coherence that allows systems to endure without constant correction. When this loss remains unaddressed, even sophisticated sustainability efforts struggle to stabilize.

Consequently, sustainability is not something a system does. It is something a system is when certain conditions hold. These conditions include coherence between internal parts, alignment between purpose and action, integrity across relational boundaries, and continuity between short-term adaptation and long-term identity.

In natural systems, sustainability is never pursued as a goal. It is an emergent consequence of dynamic balance, regeneration, and responsiveness to real limits. Ecosystems do not manage themselves through targets. They maintain continuity through structural coherence.[6] When coherence is disrupted, collapse follows regardless of how well individual components perform.

Human systems differ in complexity, intentionality, and abstraction. However, they are not exempt from this principle. Organizations, economies, and institutions that fragment their internal logic, separate responsibility from consequence, or reward short-term optimization at the expense of long-term coherence may exhibit temporary success while accelerating their own unsustainability.

This is why sustainability cannot be secured through optimization alone. Optimization sharpens performance inside a given architecture. Sustainability requires that the architecture itself be coherent.

Seen in this light, many sustainability efforts inadvertently reproduce the very patterns they seek to correct. They operate within worldviews that continue to prioritize growth without limit, acceleration without resting cycles, extraction without regeneration, and competition without systemic reciprocity. The sustainability initiative becomes a correction mechanism within a structure that continuously generates the conditions it must then try to correct.

This explains a familiar experience across organizations and institutions. Sustainability programs multiply, yet systemic tensions deepen. More indicators are tracked while coherence declines. Reporting improves while trust erodes. Innovation accelerates while long-term stability weakens. These are not failures of effort. They are signals that the logic of sustainability as an emergent systemic property has not yet been fully integrated.

---

tems, including a systemic ontology of integrity, a diagnostic account of recurrent instability, and a reconstructive methodology that translates critique into coherent redesign. These elements are supported by applied tools designed to make integrity and fracture visible and actionable in practice.

[6] Holling, C.S. (1973) 'Resilience and stability of ecological systems', Annual Review of Ecology and Systematics, 4, pp. 1–23.

Folke, C. et al. (2010) 'Resilience thinking: Integrating resilience, adaptability and transformability', Ecology and Society, 15(4), Article 20.

In *Sustainabilism*,[7] this distinction is clearly made. Sustainability does not arise from what a system attempts to achieve. It arises from what a system structurally is. When coherence, integrity, and regenerative relationships are built into the identity of a system, sustainability follows as a consequence. When they are not, sustainability must be constantly imposed, defended, or simulated.

At this point, a further distinction becomes necessary.

There is a difference between systems that seek to appear sustainable and systems that are structurally capable of sustaining themselves. The first relies on visibility, signaling, and compliance. The second relies on coherence, integrity, and regeneration. Both may coexist within the same institution. However, only one can endure.

This reframing does not invalidate sustainability objectives. It situates them correctly. Objectives can only be fulfilled sustainably when the system within which they are pursued already exhibits the structural conditions that allow fulfillment to persist without continuous external correction.

The implication is direct. Sustainability strategies that do not address underlying conditions of coherence and integrity remain vulnerable to dilution, reversal, and unintended contradiction. In contrast, sustainability strategies that do address those conditions shift from being corrective to being generative.

It is at this deeper level that the question of integrity must now be introduced. Because coherence alone is not sufficient. A system may be coherent and still coherently destructive. Sustainability requires not only structural coherence, but a specific quality of coherence. It requires integrity.

The next section develops this distinction and introduces systemic integrity as the ontological precondition for sustainability.

16.5 Systemic Integrity as the Ontological Precondition for Sustainability

If sustainability is a systemic property rather than an objective, then the central question becomes what conditions enable that property to arise and endure. Numerous sustainability frameworks address governance, capability, resources, and coordination. Although these are essential elements of practice, none of them identify the deeper organizing principle that determines whether a system can—or should—sustain itself in the first place. That organizing principle is systemic integrity.

Integrity, in this context, does not refer to moral virtue or compliance with predetermined ethical norms. Systemic integrity describes the condition in which the internal architecture of a system is coherent with its purpose, consistent with its values, aligned across its parts, and capable of maintaining continuity over time without eroding its own foundations.

[7] Tashvir, A. (2025) *Sustainabilism: Exposing the Sustainability Illusion*. Engenesis Publications, Sydney.

Integrity is what allows a system to remain itself while adapting. It is the structure that connects identity with action, short-term behavior with long-term consequence, and internal processes with external realities. When integrity is present, systems exhibit resilience, clarity, and regenerative capacity. When it erodes, systems may continue functioning on the surface, but the conditions for their sustainability are already compromised.

In human systems, integrity functions as the foundational logic that determines not only what is possible, but what is durable. In this sense, integrity also provides a basis for discernment, enabling evaluation not only of whether systems can endure, but of whether their endurance remains coherent with the conditions of life, responsibility, and continuity they claim to support.

This question of discernment is particularly relevant when examining dominant sustainability instruments such as ESG reporting, the Sustainable Development Goals (SDGs), and related measurement-based frameworks. These initiatives have played a significant role in elevating sustainability into organizational, financial, and policy consciousness. They have created shared language, improved visibility and introduced mechanisms of accountability across systems that previously operated without common reference points.[8]

However, these frameworks operate primarily at the level of indicators, targets, and disclosures. They are designed to describe and compare performance rather than to reconfigure the internal logic of systems themselves. As a result, they often assume that sustainability can be achieved by improving what is measured, governed, or reported, while leaving underlying incentive structures, cultural norms, and ontological assumptions largely intact.

From the perspective of systemic integrity, this creates a structural limitation on what such frameworks can achieve on their own. When sustainability is mediated primarily through reporting regimes and external metrics, systems can learn to *perform* sustainability without becoming sustainable. Progress may be demonstrated through indicators while everyday decision-making remains driven by short-term cycles, competitive pressure, reputational risk, or political survivability.

This does not invalidate ESG, the SDGs, or similar frameworks. It clarifies their function. They operate effectively as instruments of coordination and visibility. However, on their own, they do not generate systemic integrity. Where integrity is absent, such instruments require increasing levels of enforcement, signaling, and compliance to sustain momentum. Where integrity is present, they can amplify coherence rather than compensate for its absence.

The limitations of instrumental sustainability become most visible through recurring failure patterns observed across sectors and geographies. One such pattern

---

[8] **United Nations (2015)** *Transforming our world: the 2030 Agenda for Sustainable Development.* New York: United Nations.

Eccles, R.G., Ioannou, I. and Serafeim, G. (2014) 'The impact of corporate sustainability on organizational processes and performance', *Management Science*, 60(11), pp. 2835–2857.

Berg, F., Kölbel, J.F. and Rigobon, R. (2022) 'Aggregate confusion: The divergence of ESG ratings', *Review of Finance*, 26(6), pp. 1315–1344.

is the decoupling of sustainability metrics from real incentive structures. Indicators improve while rewards, promotions, funding decisions, and leadership evaluations continue to privilege speed, growth, or short-term advantage.

A second pattern is the reduction of sustainability to branding or compliance layers. Sustainability becomes something organizations demonstrate rather than something they structurally embody, producing the appearance of progress without altering underlying operating logic.

A third pattern is fragmented ownership. Responsibility for sustainability is distributed across departments, committees, or functions without coherent integration into the system's core identity and decision architecture. Environmental, social, economic, and governance priorities operate in parallel rather than as a unified whole.

A fourth pattern is the dominance of short political and organizational time horizons.[9] Electoral cycles, quarterly reporting, and leadership tenure constrain decision-making, consistently subordinating long-term systemic consequences to near-term stability.

A fifth pattern is moral or activist polarization replacing shared sense-making. Sustainability becomes framed as an identity position rather than a systemic necessity, eroding trust and preventing the formation of a shared interpretive ground for durable action.

These patterns do not arise because sustainability frameworks are misguided. They arise because such frameworks are applied within systems whose deeper structures lack integrity. Without alignment at the level of purpose, incentives, meaning, and responsibility, sustainability initiatives remain vulnerable to dilution, fatigue, and reversal.

Organizations with low systemic integrity can achieve temporary performance but cannot sustain it without disproportionate cost. Societies with low integrity can maintain the appearance of stability while becoming increasingly fragile beneath the surface. Economies with low integrity can generate prosperity in the short term while creating structural risks that eventually manifest as crisis. Integrity is therefore not an optional attribute of sustainable systems. It is the ontological precondition that makes sustainability possible at all.

The common intuition that sustainability requires alignment, transparency, responsibility, reciprocity, and long-term thinking is accurate. However, these are consequences of integrity, not substitutes for it. They arise naturally when integrity is present. They cannot be consistently produced when integrity is absent.

Seen through the lens of systemic integrity, these patterns reveal a recurring challenge in sustainability practice. Many organizations invest in sustainability initiatives while their internal cultures reward behaviors that undermine trust, fragment decision-making, or prioritize short-term gains over long-term coherence. The

[9] **Laverty, K.J. (1996)** 'Economic "short-termism"', *Academy of Management Review*, 21(3), pp. 825–850.

Barton, D. and Wiseman, M. (2014) *Focusing Capital on the Long Term*. Harvard Business Review.

sustainability strategy becomes an external addition to a system whose internal operating logic remains misaligned. In such environments, sustainability requires increasing effort to maintain because the system continuously generates conditions that erode it.

Integrity shifts this dynamic entirely. When integrity forms the primary organizing principle of a system, sustainability does not require constant correction. It emerges as a natural consequence of a system whose internal and external relationships are aligned.

This is evident in natural systems. Forests, rivers, and ecosystems maintain continuity through internal integrity rather than external management. Each component functions in alignment with the larger whole, and the whole responds coherently to disturbances without losing its identity. Sustainability arises from the integrity of relationships rather than the pursuit of targets.

Human systems differ because they involve intentionality, abstraction, and competing interpretations of value. However, they remain subject to the same foundational principle. Without integrity, no amount of technical sophistication can secure sustainability.

The work presented in *Sustainabilism*[10] develops this insight by analyzing integrity not merely as an ethical preference but as a structural condition that governs the behavior of systems. It identifies specific dimensions of integrity across human, organizational, and institutional domains and shows how their presence or absence determines whether sustainability is possible.

At its core, sustainability requires coherence between what a system claims, what it does, what it rewards, what it values, and how it affects the world around it. When these dimensions are misaligned, integrity fractures and sustainability weakens. When they are aligned, sustainability becomes feasible without requiring continuous external enforcement.

This does not imply that integrity guarantees sustainability. Rather, it asserts that sustainability cannot exist where integrity is absent because it creates the environment within which sustainability can arise, take root, and persist.

The implications of this are profound. If sustainability depends on systemic integrity, then the central work of sustainability is not only environmental management, technical transition, or governance reform, although these remain essential. It is the reconstruction of systems around a coherent ontological center where human values, institutional behaviors, organizational structures, and ecological realities are aligned rather than competing.

This reconstruction cannot occur without addressing the human dimension directly, because integrity ultimately begins with the ways in which human beings perceive, interpret, and enact their participation within systems.

The next section turns to the question of Being and the ontological architecture that shapes meaning, motivation, and decision-making. Without coherence at this human level, integrity at the systemic level remains incomplete.

[10] Tashvir, A. (2025) *Sustainabilism: Exposing the Sustainability Illusion.* Engenesis Publications, Sydney.

16.6 Being: The Human Dimension

Systemic integrity cannot be understood purely at the level of structures, incentives, or governance arrangements. Every system is animated by human participation. Policies are drafted by people. Institutions are governed by people. Organizations are led by people. Technologies are designed, deployed, and directed by people. The deeper question therefore becomes not only how systems are structured, but how human beings are *being* within those systems. This is where sustainability confronts its most neglected dimension.

Contemporary sustainability discourse devotes enormous attention to behavior, awareness, and skills. Behavior is treated as something to be influenced. Awareness is treated as something to be raised. Skills are treated as something to be trained. Yet the deeper ontological structure of the human being as the originating agent of systemic action remains largely unexamined. This omission is not trivial. It is causal.

Being (how individuals and collectives are being) governs what is perceived as real, what is experienced as valuable, what is pursued as meaningful, and what is avoided as threatening. It shapes how responsibility is interpreted, how power is exercised, how risk is perceived, and how time is understood. All sustainability-related decision-making flows through these inner architectures long before it appears in policy, investment, or institutional design.

Within the Authentic Sustainability Framework developed in *Sustainabilism*,[11] Being is used as an analytic category to describe the underlying orientation from which human behavior emerges. It does not refer to personality traits or individual psychology alone. It refers to the deeper configuration of attentional patterns, emotional dispositions, ethical orientation, and existential stance through which individuals and leaders interpret the world. This orientation influences whether sustainability challenges are approached through fear, control, and competition, or through responsibility, care, and generative participation.

This distinction helps explain a phenomenon that is widely observed but rarely examined. Two leaders may operate within the same regulatory framework, possess comparable expertise, and face identical sustainability constraints. One may orient toward long-term regeneration, collective responsibility, and systemic coherence. The other may orient toward short-term optimization, image management, and personal advantage. The difference is not primarily one of knowledge or policy. It is a difference in underlying orientation—a difference of Being.

Human beings are the primary demand system of every economic, institutional, and organizational structure. However, sustainability debates often focus on supply side solutions: cleaner energy, greener products, better infrastructure, improved reporting, smarter technologies. These interventions are necessary, but they operate downstream of the more fundamental driver that is rarely named explicitly—human beings.

[11] Tashvir, A. (2025) *Sustainabilism: Exposing the Sustainability Illusion*. Engenesis Publications, Sydney.

Systems do not demand growth, speed, consumption, extraction, or short-term optimization by themselves. These demands originate in human priorities, fears, incentives, aspirations, and unresolved tensions. Markets respond to what people reward. Organizations evolve around what leaders tolerate. Institutions stabilize around what societies normalize.

Sustainability therefore cannot be sustained without addressing the quality of demand that human beings continuously generate. As long as demand remains shaped by insecurity, status competition, short-term survival logic, or distorted meaning, systems will reorganize themselves to satisfy those forces, regardless of how many sustainability initiatives are layered on top.

This explains a recurring contradiction. Sustainability strategies attempt to redesign supply while leaving demand structurally untouched. Cleaner production is pursued while consumption logic accelerates. Responsible governance is promoted while political incentives reward immediacy and polarization. Long-term resilience is discussed while individuals and institutions remain trapped in short-term psychological and economic cycles.

From an ontological perspective, demand is not merely economic. It is existential. It reflects how human beings relate to time, security, meaning, responsibility, and identity. Where meaning is shallow, demand becomes excessive. Where responsibility is fragmented, demand becomes externalized. Where Being is oriented toward fear and scarcity, systems are pressured toward extraction and control.

Authentic sustainability therefore requires a transformation not only of systems, but of the human architectures that continuously generate demand within those systems. This does not imply moral perfection or behavioral purity. It implies coherence between inner orientation and outer consequence.

Without this coherence, sustainability remains corrective. However, with it, sustainability becomes generative.

16.7 Meaning and the Architecture of Sense-Making

Meaning and intention function as central structural forces within this human architecture. Introduced in *Sustainabilism,* the concept of "Minalogy"[12] refers to the way meaning is constructed and assigned across individuals and institutions. It addresses how significance, value, and purpose are attributed to actions, outcomes, and trade-offs. Sustainability efforts tend to fail where meaning is reduced to abstract ideals without lived embodiment, or where intention is fragmented, conflicted, or subordinated to unexamined incentive systems.

This is why sustainability cannot be sustained through compliance alone. Compliance can produce behavioral alignment without ontological alignment.

[12] Minalogy is introduced in *Sustainabilism* as a new philosophical discipline that treats meaning-making as a distinct field. It integrates linguistic, ontological, phenomenological, psychological, and cultural dimensions within a coherent ontology, positioning meaning-making as a primary human act essential for discernment and sustainable action.

While it may generate short-term conformity, it cannot generate long-term regenerative capacity.

The architecture of sense-making further complicates this picture. Human beings do not encounter sustainability as a neutral set of facts. They interpret it through layered structures of cognition, narrative, identity, ideology, experience, and aspiration.[13] These layers shape what is perceived as threatening, what is perceived as possible, and what is perceived as desirable.

For some, sustainability appears primarily as a moral obligation. For others, it appears as an economic constraint or a technological opportunity. For others still, it appears as ideological overreach. These divergent interpretations do not arise from the facts themselves. They arise from deeper architectures of meaning and sense-making.

Without engaging these architectures directly, sustainability discourse remains confined to surface-level disagreement. It debates policies, targets, and mechanisms while the deeper sources of divergence remain untouched.

This also explains why sustainability initiatives so often encounter resistance that cannot be accounted for by ignorance or misinformation alone. Resistance frequently emerges from perceived threats to identity, power, autonomy, security, or belonging. These are ontological concerns rather than technical ones. They reflect how individuals and groups understand their place in the world and their relationship to authority, risk, and the future.

Systemic integrity therefore cannot be realized without intrapersonal integrity. Institutions cannot maintain coherence if the agents within them operate from fragmented inner architectures. Organizational alignment cannot be sustained if individuals are internally divided between stated values and lived incentives. Societal trust cannot emerge where fear, resentment, or meaninglessness dominate the interpretive landscape.

At this level, sustainability becomes inseparable from fundamental questions of human existence—who we are, how we make sense of reality, what we consider valuable, and what we are willing to sacrifice for continuity across generations. Therefore, sustainability ceases to be merely a technical or managerial challenge. It becomes an existential one.

The next section extends this analysis by introducing the role of metacontent and the layered nature of sense-making in shaping how sustainability is perceived, contested, and enacted across individuals, organizations, and societies. It is at this level that perception bias, narrative fragmentation, and ideological capture acquire their deeper systemic explanation.

[13] **Weick, K.E. (1995)** *Sensemaking in Organizations*. Thousand Oaks: Sage.
Johnson-Laird, P.N. (1983) *Mental Models*. Cambridge: Cambridge University Press.
Senge, P.M. (1990) *The Fifth Discipline*. New York: Doubleday.

16.8 Metacontent and the Architecture of Perception

Sustainability is not encountered as a neutral object in the world. It is encountered through layers of interpretation that shape what is noticed, what is emphasized, what is dismissed, and what is resisted. Data does not speak for itself. It is always mediated through meaning. Policy does not implement itself. It is carried through interpretation. Even ecological reality enters human systems through the filter of perception.

This distinction between *content* and *metacontent* is central. Content refers to what is happening in the world. Examples include emissions levels, biodiversity loss, climate data, policy instruments, technologies, and targets. Metacontent refers to how that content is interpreted: the assumptions, narratives, cognitive frames, and paradigms through which meaning is assigned, and action becomes possible or impossible.

Figure 16.1 illustrates this distinction by showing sustainability not only as a set of objectives, indicators, and interventions (content), but as something always already filtered through interpretive architectures (metacontent). This distinction helps explain why sustainability initiatives that are technically robust often fail to generate durable change. When sustainability is addressed only at the level of content through policies, metrics, targets, and reporting frameworks, it remains

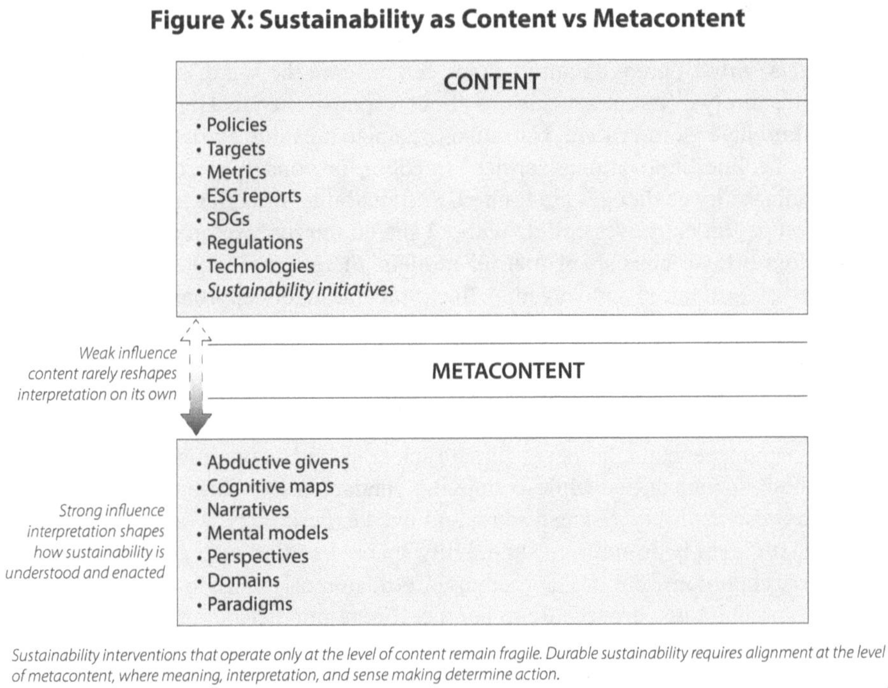

Fig. 16.1 Sustainability as content vs. sustainability as metacontent. Developed by the author, used with permission, all rights reserved

vulnerable to distortion by unexamined interpretive structures. Meaning, perception, and sense-making operate beneath visible interventions, shaping how sustainability is understood, prioritized, resisted, or performed.

Unless coherence is established at the level of metacontent, sustainability remains dependent on continuous external enforcement rather than internal regeneration.

In the Nested Theory of Sense-making—a model introduced in *Metacontent—The Intellectual Substrates for Sense-making*[14]—metacontent refers to the underlying structures through which content is interpreted. Every sustainability discourse operates on both levels simultaneously. However, most contemporary approaches focus almost exclusively on content. Consequently, emissions, indicators, technologies, and policies are treated as the primary drivers of change, while the deeper structures that determine how these elements are perceived, evaluated, and acted upon remain largely implicit.

These interpretive structures operate across multiple nested layers, each shaping how sustainability is perceived, contested, and enacted.

At the most immediate level is the **abductive given** or **initial insight**. This refers to what is unconsciously taken as real, normal, inevitable, or impossible before deliberate reasoning begins. For some, perpetual economic growth is an unquestioned assumption. For others, ecological limits are the primary given. These pre-rational assumptions silently organize perception long before sustainability policies enter the conversation.

The second layer is the **cognitive map**: the internal representation of how the world works. Cognitive maps shape how causes and consequences are linked. Individuals with different cognitive maps can observe the same sustainability data and reach entirely different conclusions about responsibility, feasibility, and priority.

The third layer is **narrative**. Narratives organize meaning across time. Stories of progress, decline, innovation, sacrifice, freedom, or control profoundly influence how sustainability challenges are framed. Sustainability struggles to gain coherence where competing narratives collide without shared interpretive ground.

The fourth layer consists of **mental models**, the internal logics through which problems are structured and solved. A linear mental model approaches sustainability through prediction and control. A systemic mental model approaches it through feedback, emergence, and non-linearity. A mechanistic model seeks optimization. A regenerative model seeks continuity and renewal.

The fifth layer is **perspective**, the situated viewpoint of the observer. A policy-maker, an entrepreneur, a farmer, a climate scientist, and a community elder may all speak about sustainability while occupying fundamentally different perspectives shaped by responsibility, risk exposure, and lived experience.

The sixth layer is **domain**. Sustainability spans domains such as ecology, economy, law, technology, culture, and ethics. Confusion arises when the logic of one domain is applied inappropriately to another. Economic reasoning is used to settle

[14] Tashvir, A. (2024) *Metacontent – The Intellectual Substrates for Sense-making*. Engenesis Publications, Sydney.

moral questions. Legal frameworks are invoked to resolve existential dilemmas. Technological optimism is employed to avoid ethical responsibility.

The seventh layer is **paradigm**. Paradigms define what counts as real, what counts as knowledge, what counts as success, and what counts as human progress. Sustainability is ultimately a paradigmatic challenge because it questions the foundational assumptions upon which modern civilization has been built.[15]

These layers operate simultaneously in every sustainability debate. They explain why consensus remains elusive even when scientific evidence is widely accepted. They explain why sustainability becomes polarized along cultural, ideological, economic, and political lines. They explain why good faith disagreement often degrades into adversarial positioning.

It is important to note that all of these layers rest upon a deeper and often overlooked foundation: context. Sense-making does not occur in abstraction. It is always situated within specific historical, cultural, environmental, institutional, subjective, and intersubjective conditions that shape what is available to perception, what feels plausible, and what appears negotiable or fixed.

Context includes factors such as the era in which sense-making occurs, prevailing cultural narratives, material and ecological conditions, institutional arrangements, and shared social norms. These contextual variables exert powerful influence over how paradigms form, how narratives stabilize, and how meaning is assigned. They quietly frame interpretation long before conscious reasoning begins.

For this reason, sense-making cannot rely on immediate insight or raw information alone. Understanding is cultivated through cognitive and relational practices such as analysis, comparison, reflection, dialogue, and the negotiation of shared meaning. What appears as disagreement at the level of policy, targets, or values often originates in unexamined contextual differences rather than in the content itself.

Without awareness of these layers and their underlying contextual variables, sustainability discourse remains trapped in surface-level disagreement. Content is debated while incompatible metacontent remains unexamined. Sustainability becomes shouted rather than understood, imposed rather than embodied, and defended rather than integrated.

Perception bias, therefore, is not a peripheral issue. It is a central structural feature of sustainability failure. Bias is not merely a cognitive error. It is the expression of deeper metacontent conflicts that remain unresolved.

This is also why information alone rarely changes behavior. Data enters interpretation through existing cognitive maps, narratives, and paradigms. Where those structures are threatened, information is distorted or rejected. Where they remain unaltered, sustainability reforms are either resisted or superficially adopted.

Therefore, metacontent represents a missing architecture in sustainability transformation. Without structured engagement at this level, sustainability initiatives

[15] **Kuhn, T.S. (1962)** *The Structure of Scientific Revolutions*. Chicago: University of Chicago Press.
Wilber, K. (2000) *A Theory of Everything*. Boston: Shambhala.

remain corrective rather than generative. They attempt to repair outcomes without transforming the interpretive engines that generate those outcomes.

Restoring coherence to sustainability discourse requires that institutions, leaders, and societies develop the capacity to recognize and work with these deeper layers of meaning. This includes the ability to discern domain boundaries, examine cognitive maps, question narratives, update paradigms, and realign abductive givens with lived ecological and social realities.

This is not a purely academic exercise. It is a structural prerequisite for systemic integrity. Integrity cannot be sustained where perception is systematically distorted and meaning remains structurally fragmented.

The next section builds directly from this insight. If sustainability fails at the level of metacontent and meaning, then it cannot be secured through corrective technical programs alone. It must be reconstructed as a regenerative design paradigm. That is the essence of authentic sustainability.

16.9 Authentic Sustainability as a Regenerative Design Paradigm

If sustainability is to move beyond correction, compliance, and ideological contestation, it must be reconceived not as a policy agenda but as a design paradigm for living systems. This is the central proposition of authentic sustainability as developed in *Sustainabilism*.[16]

Authentic sustainability begins from a simple but far-reaching distinction. Systems can be preserved, optimized, or extended without becoming sustainable. Preservation seeks to maintain what already exists. Optimization improves performance within existing structures. Extension increases capacity or scale. In contrast, authentic sustainability refers to something more fundamental: the capacity of a system to regenerate the very conditions of its own continuity.

This shift from preservation to regeneration marks a civilizational turning point. While preservation resists collapse, regeneration restores coherence. A system that merely avoids failure remains vulnerable. However, a system that regenerates its internal integrity becomes capable of enduring change without erosion.

Authentic sustainability therefore does not ask only how to limit damage or reduce harm. It asks how systems can be designed so that coherence, responsibility, reciprocity, and continuity are reproduced from within rather than imposed externally.

This distinction explains why many sustainability efforts feel exhausting. They operate as constant acts of resistance within architectures that continuously generate the conditions they must then oppose. The system runs in one direction while the sustainability function pulls in another. Increasing effort is required merely to

[16] Tashvir, A. (2025) *Sustainabilism: Exposing the Sustainability Illusion*. Engenesis Publications, Sydney.

prevent regression. This is the structural signature of a system that has not integrated sustainability into its own generative logic.

Authentic sustainability reverses this dynamic. It seeks to embed regenerative logic into the identity of systems themselves. In such systems, sustainability ceases to be a corrective force. Instead, it becomes a native property of how the system operates.

This regenerative logic expresses itself through several interdependent principles.

First, coherence becomes the central design criterion. Systems must be internally aligned across values, incentives, structures, and behaviors. Fragmentation is no longer tolerated as a normal operating condition. Organizational, institutional, and social systems are assessed by the degree to which their components reinforce rather than undermine one another.

Second, integrity becomes the organizing force of continuity. Integrity binds identity to action across time. Authentic sustainability requires that what a system claims, rewards, tolerates, and produces remains aligned, not only statically but dynamically as conditions change.

Third, regeneration replaces extraction as the guiding economic and institutional logic. Regenerative systems restore capacity rather than merely consume it. They restore ecological viability, social trust, institutional credibility, and human meaning as part of their normal functioning rather than through external repair.

Fourth, long-term systemic continuity replaces short-term performance dominance. Time is no longer treated as a constraint to be exploited but as the medium through which sustainability is realized. Decisions are evaluated not only by immediate returns but by their capacity to preserve coherence across generations.

Fifth, meaning and human purpose are integrated into system design. Authentic sustainability recognizes that systems do not remain viable through structure alone. They remain viable because people find dignity, significance, and coherence in their participation. When meaning collapses, sustainability collapses regardless of technical sophistication.

Under this paradigm, sustainability ceases to be primarily a regulatory or reporting function. It becomes a civilizational design discipline, shaping how economies are structured, how institutions govern, how organizations operate, how leadership is exercised, how education forms perception, and how culture sustains value.

For this reason, authentic sustainability cannot be implemented as an add-on. It emerges only where systems undergo reconstructive transformation at the level of first principles, integrity, meaning, and generative logic. Systems must be designed not only for efficiency and growth, but for coherence, responsibility, and continuity.

This paradigm also clarifies the distinction between sustainability as moral aspiration and sustainability as an ontological condition. Moral aspiration without structural embodiment produces perpetual tension between intention and outcome. Ontological embodiment aligns intention with systemic reality.

Authentic sustainability therefore represents not the next iteration of sustainability policy, but a shift in how sustainability itself is understood. It replaces the question of how to manage impact with the deeper question of how to design systems that naturally generate coherence, regeneration, and continuity as expressions of their basic identity. This regenerative paradigm leads directly to the next step.

Frameworks such as the Three-Pillar Model have played a vital role in structuring sustainability practice by clarifying environmental, social, and economic domains and enabling coordination across them. Their value is not diminished here. Rather, this analysis suggests that such models achieve their full potential only when embedded within systems whose underlying design, integrity, and ontological foundations are coherent. If sustainability requires reconstruction at the level of system design and first principles, then the prevailing mode of sustainability discourse—still largely deconstructive and corrective—must itself be examined.

The next section addresses the limits of critique and introduces the reconstructive move that authentic sustainability makes possible.

16.10 From Deconstruction to Reconstruction

Over the past several decades, sustainability discourse has been shaped as much by critique as by construction. It has challenged industrial excess, corporate irresponsibility, environmental exploitation, social inequality, institutional opacity, and technocratic overreach. These critiques have been necessary. Without them, many destructive practices would have remained invisible, unquestioned, and institutionally protected. However, critique alone cannot build the future.

A growing tension has emerged within sustainability discourse. The capacity to deconstruct has far outpaced the capacity to reconstruct.[17] Systems are criticized faster than they are redesigned. Institutions are exposed faster than they are renewed. Narratives are dismantled faster than new ones are stabilized. As a result, sustainability increasingly operates within a field of permanent contestation, without shared ground upon which durable reconstruction can occur.

This deconstructive saturation produces several unintended consequences. It generates chronic instability, as institutions become defensive and risk-averse rather than imaginative and regenerative. It intensifies ideological polarization, transforming sustainability from a systemic necessity into a contested identity position. It also produces reconstruction fatigue, where constant exposure of failure without credible pathways forward leads to disengagement rather than mobilization.

The paradox is that critique is indispensable for diagnosis, yet insufficient for renewal. Deconstruction identifies what no longer works. It does not specify what must replace it in a structurally coherent way.

What is missing is not effort, data, or intention, but reconstruction.

[17] **Habermas, J. (1987)** *The Theory of Communicative Action, Vol. 2*. Boston: Beacon Press.
Bhaskar, R. (2008) *A Realist Theory of Science*. London: Routledge.
Latour, B. (2004) 'Why has critique run out of steam?', *Critical Inquiry*, 30(2), pp. 225–248.

Introduced in *Sustainabilism*[18] as a key component of the Authentic Sustainability Framework, the Reconstructive Ontology of Sustainability (ROS) addresses this gap by shifting sustainability from a performative activity to a structural condition. It begins from a simple premise: sustainability cannot be stabilized at the level of action if it is fractured at the level of Being, sense-making, and systemic integrity. In other words, sustainability must be rebuilt from the inside out.

Rather than introducing another prescriptive framework, the reconstructive approach integrates the ontological dimensions that precede behavior. It brings together Being, sense-making, meaning-making, and systemic integrity into a coherent architecture through which sustainability can emerge as a lived and enduring property of systems.

An analogy helps clarify this distinction. Traditional sustainability models often operate like downstream engineering. They regulate flow, build barriers, and monitor outputs. Yet if the source of the river is polluted, no amount of downstream intervention can restore the system. Reconstruction begins upstream. It asks whether the source itself is coherent.

In this ontological framing, Being functions as the source. The way individuals and collectives relate to reality, responsibility, and truth shapes every downstream decision. Sense-making functions as the stream through which information is interpreted. Meaning-making gives direction and motivation to action. Systemic integrity forms the riverbed, holding action within coherent boundaries over time. When these elements are aligned, sustainability flows naturally. However, when they are fragmented, sustainability becomes theatrical.

This movement from awareness to action follows a simple but often overlooked progression: content to clarity to conduct. Contemporary systems are saturated with content. Data, policies, reports, and targets abound. Yet content alone does not produce sustainability. Clarity requires disciplined sense-making and shared meaning. Conduct emerges only when clarity is embodied. Without this progression, action either stalls at analysis or collapses into performance.

The reconstructive ontology therefore ensures that sustainability is not reduced to compliance or signaling. Instead, it restores sustainability as conduct that arises from coherence and clarity rather than enforcement.

This shift in how sustainability is understood and structured also explains a recurring pattern across sectors. Organizations publish impressive sustainability reports while internal trust erodes. Governments announce ambitious reforms while public confidence declines. In both cases, action exists without shared meaning and metrics operate without integrity. Reconstruction does not reject these efforts. It re-grounds them by restoring the conditions under which coherence can endure.

Seen through this lens, sustainability becomes less about controlling outcomes and more about designing systems capable of regenerating their own conditions of continuity. This reconstructive shift does not abandon critique. It completes it.

[18] Tashvir, A. (2025) *Sustainabilism: Exposing the Sustainability Illusion*. Engenesis Publications, Sydney.

Critique remains essential for revealing fracture. Reconstruction becomes essential for restoring coherence.

What follows is not a departure from the practical frameworks presented throughout this volume, but a deepening of them. The next section therefore completes this integration by explicitly linking systemic integrity and the reconstructive ontology to the Three-Pillar Model that grounds this book.

16.11 Ontological Integration with the Three-Pillar Model

The Three-Pillar Model[19] has functioned throughout this volume as a pragmatic and unifying framework for translating sustainability aspiration into organizational capability. Its strength lies in its operational clarity. It provides leaders and institutions with a shared structure for navigating complexity, uncertainty, and multistakeholder coordination.

What the reconstructive ontological perspective introduced in this chapter offers is not an alternative to this model, but a deeper foundation beneath it. It clarifies why the Three-Pillar Model functions when it does and why it falters when deeper conditions are absent.

Seen through the lens of systemic integrity, the Three-Pillar Model can be understood as the surface expression of deeper ontological conditions.

Sustainable Purpose is not sustained by strategic alignment alone. It requires existential and intrapersonal integrity. Existential integrity refers to alignment with objective reality, including ecological limits, systemic risks, and material constraints. Intrapersonal integrity refers to coherence between intention, values, and lived behavior within leaders and institutions. Where either form of integrity is absent, purpose fractures into rhetoric or idealism without structural durability.

Traveling Organization reflects the reality that sustainable systems cannot remain static. They must learn, adapt, and evolve across uncertain futures. This adaptive capacity depends not only on agile processes but on developmental coherence. Developmental coherence refers to the ability of a system to change without losing its identity. Without this coherence, adaptation becomes volatility rather than evolution. Systems drift rather than grow. They react rather than regenerate.

Connectivity expresses the relational dimension of sustainability. It acknowledges that no organization, institution, or sector sustains itself in isolation.

[19] The so-called Three-Pillar Thinking Model for organization design and leadership consists of the pillars **Sustainable Purpose** (the raison d'être of an organization, bringing orientation and certainty to the people for their joint endeavor and success), **Traveling Organization** (an organization accepting to be in a permanent state of flux and transformation), and **Connectivity** (the ability to interconnect on the journey with all needed). Details see: Wollmann, P.; Kühn, F.; Kempf, M. (Eds.): Three Pillars of Organization and Leadership in Disruptive Times—Navigating Your Company Successfully through the 21st Century Business World. Cham: Springer Nature, © 2020. See also Chap. 2 of this book or Chaps. 1 and 2 of the previous book: Peter Wollmann / Doris Pemler / Mersida Ndrevataj (eds.): The Sustainable Organization – How Organizations Address the 17 UN SDGs Using the 3-P-Model, Vol. 1, Cham: Springer Nature 2025.

Connectivity depends on interpersonal and structural integrity. Interpersonal integrity refers to trust, ethical reciprocity, and communicative honesty. Structural integrity refers to the coherence of institutional arrangements, governance mechanisms, and legal architectures that enable cooperation rather than competition to dominate system relationships. Connectivity collapses when trust erodes or when structures reward fragmentation rather than coordination.

When viewed ontologically, the Three-Pillar Model becomes more than a management framework. It becomes the practical interface through which deeper principles of integrity, coherence, and regeneration are enacted. The model does not need to be expanded conceptually. It needs to be grounded ontologically.

This grounding also clarifies a pattern observed across sustainability practice. Where Sustainable Purpose is pursued without intrapersonal integrity, it becomes performative. Where the Traveling Organization is pursued without developmental coherence, it becomes reactive. Where Connectivity is pursued without interpersonal and structural integrity, it becomes transactional rather than regenerative.

The reconstructive ontology offered in *Sustainabilism*[20] therefore does not sit alongside the Three-Pillar Model. It functions beneath it. It explains the conditions under which the model becomes transformational rather than merely instrumental.

This integration reinforces an important insight. Sustainability cannot be implemented solely through structural design. It must be enacted through ontological alignment. Frameworks operate on the nature of systems as they already exist. When that underlying nature is misaligned, frameworks cannot compensate indefinitely.

The Three-Pillar Model provides the operational language for sustainability. The ontological perspective provides the grammar of that language. Together, they offer a pathway that is both practically actionable and philosophically coherent. One without the other remains vulnerable. Together, they allow sustainability to function not only as a strategy but as a stable mode of systemic continuity.

16.12 Conclusion and Takeaways: A New Horizon for Sustainability Thinking

The sustainability challenges confronting humanity today are not occurring at the margins of civilization. They are occurring at its core. They touch how value is produced, how meaning is constructed, how power is exercised, how institutions are designed, how economies function, and how human beings understand their responsibility to one another and to the world they inhabit.

This concluding chapter has argued that sustainability cannot be secured through technical progress, regulatory instruments, or institutional innovation alone. Although these remain essential, they operate downstream of far deeper forces. Sustainability ultimately depends on the ontological conditions that shape how

[20] Tashvir, A. (2025) *Sustainabilism: Exposing the Sustainability Illusion.* Engenesis Publications, Sydney.

systems emerge, how human beings participate within those systems, and how continuity is either preserved or eroded across time.

When sustainability is treated primarily as an objective, it remains vulnerable to short-term pressure, ideological capture, and performative dilution. In contrast, when it is understood as a systemic property grounded in coherence, integrity, meaning, and regenerative logic, it acquires durability. It becomes capable of enduring complexity rather than merely responding to it.

The central proposition advanced here is that sustainability is not sustained by reaction. It is sustained by reconstruction. It requires the capacity to redesign systems around first principles that align human intention, institutional behavior, ecological reality, and temporal responsibility within a coherent whole.

This reconstructive task cannot be delegated solely to governments, corporations, or regulatory bodies. It implicates leadership, education, culture, economic design, and the deeper architectures of human sense-making. It requires a shift from asking what must be defended toward clarifying what must be structurally true for sustainable life to remain possible at all. This responsibility rests with all of us.

Within this perspective, sustainability is no longer experienced primarily as a restraint on freedom or a burden of regulation. It becomes the organizing logic through which genuine human continuity is made possible. Responsibility ceases to be experienced as an imposition and becomes a form of participation in the ongoing regeneration of shared systems.

The work developed in *Sustainabilism*[21] approaches sustainability from this reconstructive ontological position. It treats sustainability not as a policy agenda or moral campaign, but as a design condition for living systems. It seeks to restore coherence where modern systems have fragmented meaning, responsibility, integrity, and long-term orientation.

Seen through this lens, the practical frameworks, leadership models, institutional mechanisms, and case studies presented throughout this volume acquire a deeper significance. They are not isolated solutions but expressions of an underlying requirement. Their durability depends on the degree to which integrity, coherence, and meaning are structurally embedded within the systems that apply them.

The Three-Pillar Model provides a powerful operational language for sustainability in practice. What this chapter contributes is the ontological grounding that explains why that language either generates continuity or collapses into performance. Sustainable Purpose endures only when existential and intrapersonal integrity are present. The Traveling Organization remains adaptive only when change does not fracture identity. Connectivity becomes regenerative only when trust, reciprocity, and structural coherence replace fragmentation and transactional logic.

Sustainability, therefore, is not secured by models in isolation. The Three-Pillar Model operates through the underlying nature of the systems to which it is applied. When the orientation of Being, sense-making, and responsibility is misaligned, even the most robust models become vulnerable to dilution, reversal, or simulation.

[21] Tashvir, A. (2025) *Sustainabilism: Exposing the Sustainability Illusion.* Engenesis Publications, Sydney.

It is precisely at this level that the Authentic Sustainability Framework becomes critical—not as an alternative model, but as the ontological architecture that enables these models to function as intended.

The future of sustainability will not be determined by the sophistication of metrics, the volume of reporting, or the frequency of intervention. It will be determined by whether human beings and the systems they design are capable of holding coherence across time, responsibility across consequence, and meaning across uncertainty.

Where systemic integrity is present, sustainability emerges as a natural property of well-designed systems. Where it is absent, sustainability must be continuously imposed and continuously defended.

This is the deeper horizon toward which this volume points. Not sustainability as a response to crisis, but sustainability as the architectural logic of a civilization capable of enduring with minimal intervention for generations to come.

Further Reading

IPCC (2023) *AR6 Synthesis Report*. Geneva: Intergovernmental Panel on Climate Change.
Kahneman, D. (2011) *Thinking, Fast and Slow*. New York: Farrar, Straus and Giroux.
Meadows, D.H. (2008) *Thinking in Systems*. White River Junction: Chelsea Green.
Stiglitz, J.E., Sen, A. and Fitoussi, J.-P. (2009) *Report by the Commission on the Measurement of Economic Performance and Social Progress*. Paris

Ashkan Tashvir is a multidisciplinary author, philosopher, philanthropist, and entrepreneur whose work bridges ontology, civilizational architecture, organizational design, human development, technology, and systems thinking. His intellectual pursuit is not limited to observing the world but transforming it. Through a rare integration of philosophical inquiry and real-world execution, he has developed frameworks that directly advance human potential, leadership, and the sustainability of systems.

His body of work includes the **Being Framework™**, an ontological paradigm for leadership, performance, and effectiveness, adopted by leaders across more than 50 countries. Supported by applied tools such as the **Being Profile®**, his work enables measurable growth in individuals and organizations by illuminating the human qualities that shape—or compromise—performance, integrity, and systemic resilience.

In parallel with this book, Ashkan has also led the development of the **Sustainability Profile™**—an ontometric diagnostic and transformation tool designed to elevate how leaders and organizations assess and build sustainability at the human, cultural, and structural levels. Mirroring the transformative impact of the Being Profile, it offers a practical pathway for leaders to cultivate authentic sustainability in themselves, their teams, and the systems they shape.

Ashkan has published five books to date, including *Being*, *Human Being*, *Becoming*, *Metacontent*, and *Sustainabilism*. In *Sustainabilism*, he extends his inquiry from the inner structure of the individual to the architecture of institutions, economies, and civilizations, examining how human coherence, meaning, and ethical grounding shape whether systems endure or deteriorate.

Ashkan's thinking is informed by Eastern, Western, Islamic, and Persian philosophical traditions, as well as faith-based and scientific perspectives. Rather than taking sides, he synthesizes and transcends them, offering a practical and regenerative path forward for leaders, organizations, and societies. In recognition of his contributions to leadership and organizational development, he received an Honorary Doctorate in Business Administration from the Collège de Paris in 2024.

Beyond theory, he builds. Ashkan is the founder of **Engenesis**—a platform for human transformation—and **Engenesis Ventures,** a venture-building organization shaping ventures, technologies, and leadership cultures using his ontological and systemic frameworks. His work exists not only in books and philosophy, but in enterprises, economies, and futures being shaped today.

Recently, Ashkan has focused increasingly on philanthropic work through **GHEST**—a stewardship-driven community on the Engenesis platform dedicated to human empowerment and sustainable transformation. Recognizing that many individuals who would benefit most from his body of work may not pursue traditional commercial pathways, he created GHEST as a philanthropic channel to make these ideas, tools, and developmental structures accessible to broader communities. His commitment in this realm is grounded in expanding human potential, fostering conscious and ethical participation and enabling individuals and groups to navigate uncertainty with integrity, humility, and collective responsibility.

Outside his professional life, Ashkan enjoys time with his family and dogs (especially in nature), cooking, collecting, playing music, and exploring the beauty and complexity of the natural world.

Conclusions and Take-Aways 17

Peter Wollmann, Doris Pemler, and Mersida Ndrevataj

Abstract

It is important to stress again that the second and the third volume of *The Sustainable Organization* were created at the same time and in close coordination by a joint author community and with a joint general framing which means that the two books complement each other and refer to each other—this is especially valid for the conclusion chapters and the reason why we also show in this chapter also conclusions from volume 2 to create a comprehensive picture for our readers.

Thus, in this conclusions chapter, we attempt to summarize the many and varied messages, key findings, and most interesting recommendations from the individual chapters of this third volume at a meta-level complemented with content from the second volume. In doing so, we cover, among other things, the necessary classification of events and activities in historical dimensions, an understanding of the breadth of the sustainability movement in terms of topics and content, the necessary readjustment of perceptions of progress and levels already reached, and the need for an action-oriented mindset even under volatile, unclear, complex, or ambiguous external conditions. In this context, we present an integrated and commented overview of selected important conclusions from the various contributions in the book and interpret them within the book's overall

P. Wollmann (✉)
Bonn, Germany
e-mail: pw@peterwollmann.com

D. Pemler
Ciudad Condal GmbH, Zürich, Switzerland
e-mail: doris@ciudad-condal.ch

M. Ndrevataj
IUAV University of Venice, Venice, Italy
e-mail: mndrevataj@iuav.it

© The Author(s), under exclusive license to Springer Nature Switzerland AG 2026
P. Wollmann et al. (eds.), *The Sustainable Organization – Volume 3*, Future of Business and Finance, https://doi.org/10.1007/978-3-032-28949-0_17

framework, complemented by key outcome from the previous book. Thus, the aim of the chapter is to give the readers a set of fundamental messages and suggestions for further thinking and/or acting, and also animations which chapter especially they should study in detail to cover their individual key demand. Compared to Volume 1 and 2, the conclusion covers many additional new topics and facets, but the general direction remains unchanged.

17.1 A Fundamental Foreword[1]

"The Truth is Rarely Pure and Never Simple"
Oscar Wilde, The Importance of Being Earnest
Volume 2
"Experience is one thing you can't get for nothing"
Oscar Wilde
Volume 3

As the second and the third volume of *The Sustainable Organization* were created at the same time and in close coordination by a joint author community and with a joint general framing which means that the two books complement each other and refer to each other, we will introduce in this chapter also all relevant content from the conclusion chapter of the second volume in order to create a comprehensive picture for our readers.

This second and the third book begin both with a quote from Oscar Wilde (see above) in order to raise awareness of the urgent need to change the way we perceive the world in general, and tasks, events, progress, etc., in terms of the Sustainability Movement and to give guidance for our attitude toward learning from experience especially accepting the underlying efforts. Thus, our perception has to be challenged and recalibrated as our actions often follow the analysis of our spontaneous impulses (which might be a victim of thinking traps, perception biases, and neurological patterns[2]). The reality of the VUCA[3] world is confusing, unclear, ambiguous, and often an unknown territory which we have to explore and for which we have to tailor or even to reinvent[4] our thinking styles, mindsets, methods, and tools. And we must always question ourselves—not only in terms of our perceptions: ("*What if we were wrong*"—Barack Obama 2016.[5] This is a challenging question which is very useful to avoid excessive certainty about one's own worldview. The question also exactly refers to Oscar Wilde's quote "The Truth is Rarely Pure and

[1] Parts of the fundamental foreword quoted almost verbatim from the previous book The Sustainable Organization Volume 2, s. https://link.springer.com/book/9783032238016

[2] See Chap. 1 Sect. 1.2.4.

[3] **V**olatile, **U**nclear, **C**omplex, **A**mbiguous World, today often also described as BANI World (**B**rittle, **A**nxious, **N**on-linear, and **I**ncomprehensible).

[4] Using the Three-Pillar Thinking Style and Mindset Model, for detailed description, see: Chap. 2.

[5] See https://www.nytimes.com/2018/05/30/us/politics/obama-reaction-trump-election-benjamin-rhodes.html

Never Simple" and fits to the very limited human ability to fully and finally understand a complex situation and especially to predict the future or even scenarios for the future). And if it turns out that we were wrong then we have to accept this demanding experience because without efforts—and without failure—there is no real learning possible.

In these disruptive and turbulent times, attempts at long-term, differentiated thinking in analyzing the present and assessing the future are often overshadowed by hectic headlines and selective attention. In Chap. 1 of both books, volume 2 and 3, we therefore described this circumstance as a risk for a book like this one: between the completion and submission of the manuscript and its publication, will a new situation have arisen that could potentially devalue parts of the book? We countered this fear with a fundamental, long-term perspective, including from a historical point of view, but in the case of both books, volume 1 and 2, waited until shortly before the manuscript was submitted to write this last chapter. And interestingly, 1 week before the deadline of volume 2, an interesting column by Ullrich Fichtner has been published in the magazine "Der Spiegel"[6] supporting the book's outcome and direction and we already mentioned its quintessence in the second volume's conclusion chapter. Fichtner refers to a thought experiment to better understand the character of time and, in particular, the age of the universe. For this, the age of the universe is mapped to only 1 calendar year, which means that humanity would have existed for only 8 min—and 8 or 10 years at the helm of a country would be mapped to less than the blink of an eye. What does this have to do with our perception of time? We should not overestimate the moment. And as Fichtner stresses: "Thinking about space and time can help us stay calm."[7]

Fichtner continues: "Our planet is facing an exorbitant emergency: climate change, resource depletion, economic upheaval, and the extinction of more and more species. Hope through action shows us how to face this crisis and respond to it with unexpected resilience and creativity. This hope does not necessarily require optimism; we can even use this process where we see no hope left. The guiding impetus is our intention. Instead of weighing up the chances of success first and only taking action when we feel hopeful, we focus on our intention and let it guide us."[8]

Florence Gaub's recent considerations about the "character of future," we showed in the first chapter of this book, strongly support this view from Fichtner: "The future emerges something like this: humans—possibly the only species capable of vividly imagining a time that has never existed before—take cues from the past and present and construct the future as a space full of possibilities, both good and bad. This future can trigger two things: either the desire to make it a reality, or fear, if it appears undesirable. Where one lands on this emotional scale depends less on

[6] Der Spiegel on 18.02.2026, 14.14 Uhr, see: https://www.spiegel.de/politik/donald-trump-warum-der-us-praesident-nur-ein-staubkorn-im-universum-ist-a-04d1c084-ad19-4fb4-8005-f5831e3894e7

[7] Ibid, translated into American English using DeepL.

[8] Ibid, translated into American English using DeepL.

intellect or knowledge than on personality—and the feeling itself has only limited predictive power. Optimists tend to achieve better results, that is true; but a certain degree of fear—not too much, as that leads to paralysis, known as the Cassandra effect—can spur us to act. And this action then changes the course of the future."[9]

And so, the circle is complete: Fichtner, in the article we introduced, also mentions the American Buddhist and ecologist Joanna Macy[10] and praised her making many valid ideas powerful and popular, for example, with her spiritual classic "Hope Through Action."[11] Key ideas from it are consistent with the American political scientist and psychologist Sarah Stein Lubrano's research and her publication on the subject.[12] Her credo is: "Don't discuss. Act." Action is in (political or societal) communication always more powerful than pure talking and discussing. In relation to the subject matter of our books, Volume 1, 2 and 3, this means focusing on the acting. And here, a reference from Chap. 6 of this book is of high importance to be mentioned: the ME310 courses of Stanford University, a blueprint for innovation learning, giving students "in a real-world, engineering design experience in the spirit of a Silicon Valley start-up, managing the uncertainty inherent in entrepreneurial design"[13] which has to be more understood as a mentoring concept than a classic university course. This is training in starting to act even if not everything is clear.

It is important to know that this proceeding is explorative, starting without a detailed plan over the full project life cycle, agile proceeding, rapid prototyping, and learning on the journey (a classic "Travelling Organization Approach," see details in Chap. 2). To repeat: students are taught that they have to start acting in partly unknown territories and how to best cope with this. Many parts of the Sustainability Movement Activity Landscape have exactly these preconditions. The question of what to do under these circumstances is also answered in this book in the sense of an "Ethics of Responsibility" and a clear call for action and requirements for future education. In this context, a quote from Hannah Arendt from the 1950s might be interesting: "*Education is the point at which we decide whether we love the world enough to assume responsibility for it, and by the same token save it from that ruin which except for renewal, except for the coming of the new and the young, would be inevitable. And education, too, is where we decide whether we love our children enough not to expel them from our world and leave them to their own devices, nor to strike from their hands their chance of undertaking something new, something*

[9] Florence Gaub's contribution was published at the renowned German newspaper Süddeutsche Zeitung on April 3rd, 2026, under the title (translated into English): "The Future Has Always Sucked" (see: www.sz.de/li.3457426).

[10] https://en.wikipedia.org/wiki/Joanna_Macy

[11] https://openlibrary.org/books/OL25139864M/Active_hope

[12] https://www.sarahsteinlubrano.com

[13] https://explorecourses.stanford.edu/search?q=ME+310A%3A+Global+Engineering+Design+Thinking%2C+Innovation%2C+and+Entrepreneurship&view=catalog&filter-coursestatus-Active=on&academicYear=20252026

unforeseen by us, but to prepare them in advance for the task of renewing a common world."[14]

The good news is that there are overwhelmingly many similar (successful) initiatives on education—and on better communication—in place worldwide as the readers will see browsing the book. Contrary to all fears, neither multilateralism nor measures against climate change, etc., have been stopped across the board or even come to the end of their tether. As already mentioned in Volume 1 and 2[15] and stressed in the debate contribution of Florence Gaub cited in Chap. 1 of this book and also above, the human lack of capabilities to imagine and think the future and to design future aspirations has a massive impact on understanding change developments which are nearly never linear[16] and to neutrally evaluate the progress of (political and societal) transformations—a fact that on the other hand severely affects the prognosis skills of people, especially if the situation is a lot better than perceived.

These systemic and personal human incapabilities were also impressively shown by Hans Rosling in his book "Factfulness"[17] where he—only based on available facts—demonstrates the development of the state of the world in multiple perspectives. The interesting point here is that people, when asked about their perception of the state of the world in the most important perspectives covered by Rosling, are nearly always wrong (and too pessimistic), by the way, the more educated they were, the more pessimistic they were. It therefore seems as if brains like dramatic narratives, as can also be seen when looking at the success of attention seeking content in social media. So, Barack Obama's question might be further developed "What if we were wrong and still are wrong?". Is my thinking and perception styles effective and efficient—or does it produce wrong convictions and perception? To which extend do I have the capability to predict the future?

Considering the facts and results from the chapters and their authors carefully and critically, the message from Chap. 1 of this book and (also of volume 2) still remains very valid:

- There is a really broad movement of activities and actors in place covering a large number of important and significant areas. We call it the Sustainability Movement Activity Landscape covering Sustainability and the UN Sustainable Development Goals (SDG)[18]
- There is a lot more progress already achieved in many of the activities than perceived in the public perception

---

[14] Hannah Arendt: The Crises in Education, 1954. https://thi.ucsc.edu/wp-content/uploads/2016/09/Arendt-Crisis_In_Education-1954.pdf

[15] In the following partly quoted from Volume 1 and 2.

[16] See Chap. 1 in Wollmann, P.; Pemler, D.; Ndrevataj, M. (Eds.): The Sustainable Organization - How Organizations Address the 17 UN SDGs Using the 3-P-Model. Cham Springer Nature, © 2025; (https://link.springer.com/book/10.1007/978-3-031-89549-4).

[17] Rosling, H: Factfulness New York 2008, Flatiron Books.

[18] https://sdgs.un.org/goals

- In general, there is a perception bias in the judgement of large transformational movements in place as people tend to expect linear developments (also as a consequence of neurological imprints)
- The movement is to a very large degree well rooted in the middle of society and irreversible, also as it covers systemic logic
- The movement will force many organizations (and private persons) to journey through unknown territories which requires the mindset of a Traveling Organization. The respective societal transformation will need a historical time span
- The coordination and cooperation across activities, projects, and transformations is a Herculean task and can only be handled in an explorative and agile way. The expectation of a waterfall model with a detailed masterplan in place is totally unrealistic (all such transformations have to be agile as the future cannot be predicted as we stressed above and also in Chap. 1 in Sect. 1.2.1).

In this context, it might make sense to cite from Chap. 3 of volume 2: "In the long run, the intrinsic logic of (global and national) systems striving for optimized effectiveness and efficiency demands from different perspectives respective suitable global knowledge-based institutions with high legitimacy, rule-based interaction, cooperation and procedures and a reasonable degree of public support for the concept to achieve Sustainability and UN SDG Coverage and the respective Movement, which demands adequate public communication. However, reaching Sustainability and UN SDG Coverage including the institutions' interactions is clearly a highly complex, explorative and incremental step by step creation and optimization task with many variables to be taken into consideration. Expectations of politicians, experts and the public have to be respectively calibrated. In addition, the (interim and final) results of the transformation journey will for sure change over time under changed conditions, this has to be taken into account—and it is not a sign for bald planning or failure but for the success of an explorative, flexible proceeding."[19]

These considerations are complemented by our statement from Chap. 1: "fundamental system logics beats individual power and influence" and by our explanations that in the case of the lateral system "Global Economy"[20] in the end, sustainability will not be key from ethical reasons but from economic reasons: for example, energy generation follows economic principles like the sustainable (lower) price. This might be regarded as a "hidden law of evolution" and strongly connected to scientific principles and rules.

It is again very encouraging that the Three-Pillar (Mindset and Thinking) Model[21] perfectly fits the challenging situation and was applied in nearly all chapters or use cases, respectively.

[19] See the full text and its framing in Chap. 3 of volume 2, s. https://link.springer.com/book/9783032238016

[20] ibid.

[21] Described in detail in Chap. 2 of this book—and nearly identically in Chap. 2 of volume 2. The Three-Pillar Model is based on three key building blocks (pillars), defining the Three-Pillar (Mindset or Thinking) Model for organization design and leadership for the VUCA world. The

(1) The first Pillar: **Sustainable Purpose**
Sustainable Purpose is the *raison d'être* of an organization. Having a clear guiding objective provides orientation and certainty to people, fostering a sense of shared effort and success. The ***special challenge with Sustainable Purpose*** linked to a general idea of sustainability and the coverage of selected UN SDGs in detail lies in the perception bias, the gap between the public perception of status, success, reachability of sustainability targets including the judgement of concrete parts of the activity landscape, on the one hand, and the fact-based reality (of the comprehensive big picture) on the other hand. To fight doubt, discouragement, even despair (up to doomsday mood), it is important to act, not to talk, and to develop and show positive use cases and what an organization can reach with the help of good leadership.

(2) The second Pillar: **Traveling Organization**
Traveling Organization refers to an organization accepting to be in a permanent state of flux and transformation while interacting with the external requirements of its journey by means of rapid learning and adaptivity. Ideally, this is the mindset of loving the Unknown and being well able to manage it.[22] The ***special challenge with a Traveling Organization Mindset and Capabilities*** is today, on the one hand, that the North Star of Sustainable Purpose in the context of Sustainability and UN SDG Coverage is challenged by a pessimistic view and a public perception bias already described above which has an impact on motivation and momentum. Going on an explorative journey through the Unknown being confronted with change is always demanding and even more challenging under the current preconditions. But our use cases show that even though the challenges for leadership are obviously tremendous, the journeys are manageable and at the end a mind-changer.

(3) The third Pillar: **Connectivity**
Connectivity focuses on the ability to interconnect during the journey with all necessary parties and resources inside and outside the silos, thus creating high effectiveness and consistency and also building shared learning networks[23] during this process. The ***special challenge with Connectivity*** in the context of

---

author community's starting point in 2019 was to establish guiding principles for a world seemingly devoid of them. The model has since been frequently applied—and its applicability and usefulness proven—in various global use cases (so far >80) in the private and public sectors. The goal of the Three-Pillar model is to support understanding and navigating the complexities of today's corporate and institution territory in all sectors.

[22] One has an intuitive idea of a Traveling Organization from the metaphor of an expedition to Africa in former times. The participants of the Traveling Organization must be curious, open to learn, courageous, keen to experiment, and they deal well with uncertainty, stress, unforeseen incidents in a totally new world. This mindset of the Traveling Organization is the core of a Learning Organization. It is impossible in and for the Traveling Organization not to learn. Learning is a question of survival.

[23] This means that organization has to be thought of within a network perspective, including interfaces to adjacent organizations and across organizational borders. In many cases, it might be impossible to learn effectively without collaborating closely with people from outside your silo.

activities on Sustainability in general and the coverage of selected UN SDGs in detail is the increasingly strong fragmentation among stakeholders in terms of opinions, perceptions, and ethics (ethics of conviction versus ethics of responsibility), all of which are sometimes strongly linked to a certain emotionality. This makes it difficult to agree on a common starting point for cooperation or even neutral exchange. High, often unattainable expectations are met with defeatism, but also with total rejection. The situation often requires the utmost diplomatic skills and special integrative leadership qualities. But the principle: start and don't talk too much, also fits here very well and seems reasonable.

17.2 Proceeding

We selected from each chapter or part, respectively, those conclusions which are of overarching importance and sorted them. Some of them were already mentioned in the fundamental foreword of the chapter. We also added, respectively, selected conclusions from volume 2 and integrated them. Out of this completed selection, we chose the following top 10 general conclusions overarching the whole books volume 2 and 3. They share the character of recommendations for thinking, judging, and acting.

For the detailed framing of these conclusions, we recommend reading the chapter they are taken from. We did not want to produce too much redundancy by adding framing text from the original chapters as we believe that the conclusions are self-explanatory and do not require any additional explanations.

17.3 Our Top 10 General Conclusions and Recommendations from Volume 3 and 2[24]

1. **Be always aware of your intellectual, mental, neurological, and cultural limitations**
   The key question is: What if we were wrong and are still wrong? We must be aware that our intellectual, mental, neurological, and cultural conditioning (including our usual thinking styles and analytical methods) limits us in our ability to holistically understand the world in general—especially in historical dimensions—and especially the complex, multidimensional and, in sum, unmanageable landscape of activities on sustainability and coverage of the UN SDGs and to analyze their development. We expect familiar patterns, linear developments, and methodological clarity, but we are finding ourselves—at least in part—in *terra incognita*. It is necessary to accept this and to find ways to best cope with this, for example, actively look for different perspectives and thinking styles to approach a holistic understanding—and never be too con-

[24] Those of the Top 10 General Conclusions and Recommendations which are only basing on content from volume 2 are specially marked.

vinced to know now the full truth and all its facets, there could always be an additional surprise. And we should often take the "Gapminder test" of Rosling's foundation[25]

2. **Be aware that your perception bias is often coming from your appetite for drama**
   In particular, we must accept that our ability to recognize progress in the development of the world is very poorly developed. It is not for nothing that particularly educated people judge many issues (see development of child mortality, literacy and school attendance, spread of child labor, mortality, etc.).[26] This conclusion was also shown in Volume 1 and 2—and it is truer now than ever.
3. **Regard "Sustainability" together with "Coverage of UN SDGs" empirically as a (historically) powerful long-term mega-trend**
   Striving for "Sustainability" and "Coverage of UN SDGs" is obviously a long-term megatrend to be judged in historical time dimensions, not short-term. It is important to challenge the world's understanding and perception of time. We show in the book again (as in Volume 1 and 2) that there are well documented and significant efforts on a global scale (UN level) and both on the regional and national scale in place to push the connected topics "Realization of the UN SDGs" and "Sustainability." It is clear that the implementation of individual projects and measures does not proceed in a straight line but is subject to their own problems, deviations, and many counter-movements. This follows a well-known pattern and has also been the case with all former major developments and transformations. It will therefore only be possible to measure the success of this whole transformation effort for future generations from a historical perspective.
4. **Be aware that "Sustainability" and "Coverage of UN SDGs" have first and foremost a strong long-term economic and systemic logic and therefore a strong long-term systemic and operative underpinning**
   The general targets of the so-called Agenda 2030[27] are simply common sense—only the means of achieving them are controversial and must, of course, be discussed. In the long run, a green and circular economy will be more effective and efficient—and in parallel the costs of not striving for sustainability and coverage of the UN SDGs are overwhelming, so that even economic rationality requires following the megatrend. Even political changes in relevant countries will not be able to change its direction mid- or long-term as the main driver is economic efficiency, not ethical conviction. And this logic drives the operative underpinning of efforts to cover UN SDGs and reach sustainability for the world as well on a global scale (UN level) as also on regional and national level—even when some countries preliminarily withdraw support and efforts.

[25] Go to https://www.gapminder.org and select an icon under the headline "You are probably wrong about."

[26] Quote of Hans Rosling: "We have to learn to get a grip on our drama consumption. Without such control, our appetite for drama goes too far and prevents us from seeing the world as it is, leading us terribly astray" (Rosling, H: Factfulness, New York 2008, Flatiron Books, S. 27).

[27] https://sdgs.un.org/2030agenda

As we show in the book, there are many concrete measures in place with named responsible organizations and individuals (see exemplarily the wonderful use case from the UK in volume 2 in Chap. 8). All official activities are subject to regular controlling and reporting.

5. **From Volume 2 specifically: Be aware that the Movement for "Sustainability" and "Coverage of UN SDGs" requires in the long term strong international institutions following the logic of effective and efficient systems and processes**
As shown in the discussion in Chap. 3 in volume 2 (and also there in Chaps. 5, 6, and 7), the meta-level systemic logic driving progress of humanity in history, namely improving effectiveness and efficiency, strives for strong international institutions in the context of the Sustainability Movement. This is proven from different perspectives, firstly from a mathematical or project/project portfolio driven logic, secondly from Rule of Law as well from a country as also from a multilateral UN perspective and thirdly, from the Global Governance view of system theory, applied in a knowledge-driven world. All perspectives are mainly based on the progress driving principles of gaining effectiveness and efficiency as system logic. A detailed deep dive use case regarding WIPO[28] is shown in Chap. 7; it is a role model for other institutions with a strong logical argumentation. It is obvious that the broad-based Movement as we might regard the nearly infinite Activity and Project Landscape to reach Sustainability and UN Sustainable Development Goals Coverage with billions of entries, which was reintroduced in Chap. 1, needs some institutional and rule-based fundaments to reach the needed effectiveness and efficiency—but this also valid for many other aspects of the world's development (same principle applied to other contexts).
6. **From Volume 2 specifically: Be aware that there are successful models to sustainably develop whole countries beyond the pattern of the Western world but follow long-term the same goals and principles**
As shown in Chap. 5 of volume 2, countries might choose another journey to Sustainability and UN SDG Coverage as simply to follow the example given by the Western world. It would be a perception and judgement bias to underestimate these alternative journeys or respectively their development options as not each country has the exact same preconditions. It is important to prevent superficial or even arrogant perspectives but to take a more ethnological approach and check ambitions and progress. At the end, there might be in each of the use cases parts to regard as very positive as well as parts which need significant improvement. And coming back to the historical dimensions: even the West needed significant time to reach its current level of maturity.
7. **Be aware that academia and academic transformational education are a special cornerstone for the success of the Movement for "Sustainability" and "Coverage of UN SDGs"**
As shown in Chaps. 4–8 and 15 of this book, universities worldwide are very active in transformational research and education; we show use cases from

[28] World Intellectual Property Organization.

Germany, Switzerland, USA, and Australia. Even though the few chapters only give a first limited idea, it becomes obvious that a large set of topics are touched upon, we show topics around the data value chain in organizations (Chap. 4), transformative education and the Earth Charter (Chap. 5), purpose-driven innovation as well as Stanford University's ME320 model (Chap. 6), improvement of Corporate Environmental Responsibility (Chap. 7), and leadership concepts as conscious and archetypal way of being (Chap. 8). Chapter 15 (Sect. 15.4) additionally shows some already existing activities of selected universities. The number of similar use cases and examples is infinite.

8. **From volume 2 specifically: Be aware that the Sustainable Development in urban and rural areas is a special cornerstone for the success of the "Movement for Sustainability and Coverage of UN SDGs"**
   The book shows some few but impressive use cases from different countries in Chap. 4 of volume 2 and there also in Chaps. 10–16, covering different topics including: the challenges of sustainable management of the food value chain, given the heterogeneity of actors it involves (Chap. 10); the rural development and the institutional learning in the Schleiden-Eifel region in Germany, following the 2021 flood event (Chap. 11); the rural development in India through the case of Bunker Roy's Barefoot College (Chap. 12); social innovation in tourism in Venice, Italy, through the local cultural heritage initiatives (Chap. 13); proximity welfare centers and the horizontal networks activated through and for such initiatives (Chap. 14); the World Heritage policies and management for sustainable development at national, regional, and local level in Italy (Chap. 15); co-working spaces and their contribution to Sustainability, examined through a comparative study across France, Italy, and Czechia (Chap. 16)—while Chap. 4 provides the conceptual foundation for all of these contributions, situating them within a common argument: that urban and rural areas form a system of interdependence, and that sustainable development depends on governance frameworks capable of managing it as such.
9. **From Volume 2 specifically: Be aware that the right and significantly improved communication is a special cornerstone for the success of the "Movement for Sustainability and Coverage of UN SDGs"**
   One of the most important points of analysis and findings in Volume 1 was that communication is absolutely essential, not only within the Movement but especially with the public, and that it must be greatly improved. This is particularly true because of the perception and thinking traps mentioned above. The topic of communication is therefore addressed in detail at various points in this book: in a separate Chap. 9 of volume 2, showing animating use cases in different contexts, implicitly also in Chap. 1 of both books, volume 2 and 3, including in the interview with Ovais, former deputy head of UNFCCC and responsible for several COP events, further implicitly in the second volume's Chaps. 3, 6–8, and also the third volume's Chap. 15 (with a context to the urgent need of investment in transformational education and consultancy). It remains a hot topic that will require a great deal of attention in the coming years and demands a different perspective: it is not only important that experts and stakeholders within the

system understand each other, but also that the system is understood and accepted by people outside it. There are already various initiatives around the world for a new, more advanced form of journalism (in times of seemingly overpowering social media).

10. **Be aware that the Sustainable Development in enterprises finds different and very individual concepts and solutions to support the success of the "Movement for Sustainability and Coverage of UN SDGs"**
We only are able to show some few use cases in the book to give a taste of the diversity of concepts, approaches, and results, starting with a use case from Swiss Post (Chap. 9), touching then concepts on integrated governance, risk management, and compliance (GRC) and internal control systems (ICS) forming the operational backbone of sustainable enterprises (applicable in all industries, Chap. 10), a use case from MedTech Industry (Chap. 11), concepts of circular economy in different contexts and companies (Chap. 12), the collaborative sourcing concept with three high-level application examples (Chap. 13), and a use case from fashion (Chap. 14). It makes sense for the reader to go into detail in the mentioned chapter and collect those conclusions which might fit best the readers' demands.

17.4 Some Selected Additional Specific Individual Conclusions from the Chapters of Volume 3[29]

- **The use case "Data Value Chain"** (Chap. 4)—**Data as a Force for Sustainable Transformation**
Transforming data management from an environmental liability into a force for sustainable development is crucial. As global data volumes double roughly every 2 years, today's choices will profoundly impact our collective contribution to the UN SDGs. This path requires courage to challenge data hoarding, wisdom to balance performance and sustainability, and persistence. Organizations embedding sustainability into their data practices will not only contribute to global goals but also lead the sustainable digital economy. Every data choice has environmental consequences. Through strategic purpose, organizational agility, and collaborative connectivity, we can transform data management into a catalyst for achieving the UN SDGs and a sustainable future.
- **The use case "Transformational Education—Earth Charter"** (Chap. 5)
The implementation of the idea of transformative learning begins with the shift from what we know to how we know[30] (Baumgartner, 2001). Transformative learning in higher education with the Earth Charter is not about adding content to the curriculum; it is not about including a sustainability checklist to module plans. Rather, it helps educators and students to place the entire learning process

[29] Mostly taken as a 1:1 excerpt from the conclusion section of the chapter.

[30] Baumgartner, L. M. (2001). An update on transformational learning. New directions for adult and continuing education, 2001(89), 15–24.

within an ethical and transformative framework. This foundation helps to implement the concept of emergent teaching which means to address the "whole person" through a sustainable teacher or lecturer mindset, a culture of sustainability and sustainable structures in the institutions.

By embedding the principles of the Earth Charter, learning becomes a reflective and action-oriented experience. It promotes dialogic learning, in which students and teachers engage in open, critical discussions, and fosters a mindset characterized by responsibility, empathy, and global citizenship. This approach transforms education into a space where knowledge, values, and action are interconnected, preparing students to address complex social and environmental challenges with insight and ethical commitment.

The Earth Charter helps learners to hold the tension of trade-offs between different Sustainable Development Goals. For instance, advancing industrial growth (SDG 8) may conflict with climate action (SDG 13), or promoting economic development can challenge responsible consumption (SDG 12). These conflicts often appear as visible "problems at the tip of the iceberg" but addressing them effectively requires looking deeper—into the underlying structures, values, and mental models that shape decision-making. By providing an ethical and systemic framework, the Earth Charter supports students and educators in developing a both/and mindset, enabling them to navigate complexity, uncertainty, and ambiguity without reducing challenges to simple solutions.

In the AI era, we are increasingly confronted with the question of how to fundamentally shape the feasible and desired boundaries of human–machine interaction. According to widespread opinion, only the concept of "human-in-the-loop" (HITL) should be applied, according to which humans are ultimately responsible or actively involved in data preparation, model training, and in a supervisory role in automated decision-making.

Nevertheless, we are increasingly seeing a shift in the concept of epistemic injustice,[31] in which AI systems are achieving positions of power that are leading to a change in their role in the process of knowledge sovereignty. AI systems are increasingly taking control of knowledge generation processes and their representation in the education sector. We run the risk here of creating, as humanity as a whole, a **generative epistemic injustice**[32] that will not only fundamentally change future economic, social, and ecological structures, but may also systematically reinforce existing inequalities. For these reasons, the concept of transformative learning should be applied and institutionally anchored, especially in the context of academic education.

Additionally, we can also emphasize again that it is important in transformative learning journeys to include different forms of knowing, because otherwise

[31] Fricker, M. (2017). *Epistemic Injustice. Power and Ethics of Knowing*, Oxford University Press.

[32] Anneborg, Raymond. (2020). Computing Machinery and Knowledge. 10.48550/arXiv.2012.06686.; Origgi, G., Ciranna, S. (2017). Epistemic Injustice, The Case of Digital Environments. Kidd, I.J., Medina, J., Pohlhaus, G. (Eds). *The Routledge Handbook of Epistemic Injustice*, Routledge, 303–312.

"when those shared epistemic resources are biased, this transformation prevents students from retaining aspects of their personal epistemic resources which it is strongly in their interests (as well as in the interests of the broader epistemic community) to retain".[33]

We see transformative education and learning as a fundamental prerequisite for counteracting these developments and contributing to the achievement of the SDGs: There is an emphasis on capacity building, empowerment, and action competence, stressing the ability to engage creatively, to manage successfully in conditions of uncertainty, complexity, and ambiguity, to reflect critically and learn iteratively over time from engaging with real-world experience. This may involve "third order learning" where a change of mindset occurs toward one which is more holistic, connected, agile, and open in outlook (epistemic change).

- **The use case "Purpose-Driven Innovation"** (Chap. 6)

  Chapter 20 set out to examine how purpose-driven innovation can be learned and enacted in the life sciences to enable circular outcomes. Drawing on three interdisciplinary projects implemented by Stanford University, ZHAW, and industry partners, the chapter demonstrates that the primary constraints to circular transformation are coordinating heterogeneous actors, governing shared resources, and learning to navigate uncertainty, trade-offs, and competing disciplinary and institutional logics. Across the cases, the project purpose emerged as a practical coordination and legitimization mechanism that enabled systemic problem framing and ecosystem-level collaboration.

  The chapter makes three related theoretical contributions. First, it advances a learning-centered perspective on purpose-driven system innovation by showing how a shared purpose operates in practice as a stabilizing reference that legitimizes exploration and compromise. Second, it extends research on circular economy by demonstrating that circular innovation materializes only through ecosystem-level coordination anchored in physical and digital artifacts, orchestrated governance arrangements, and economically viable business models. Third, it contributes to system innovation theory by positioning education as a micro-foundation: a protected enactment space in which actors develop ambidextrous capabilities to move between exploration and exploitation, design and analysis, creativity and compliance.

  For practitioners, the findings underscore that purpose-driven circular innovation cannot be achieved through firm-centric optimization or isolated sustainability initiatives. Managers in the life sciences must invest in ecosystem orchestration capabilities, shared data infrastructures, and business models that align incentives across manufacturers, healthcare providers, regulators, and downstream actors. Purpose, when jointly articulated and continuously revisited, can be operationalized as a governance mechanism that legitimizes data sharing, negotiated trade-offs, and experimentation within regulatory constraints. For policymakers,

[33] Fairbairn, F. (2023). Epistemic injustice through transformative learning, *Journal of Philosophy of Education*, Volume 57, Issue 4-5, August-October 2023, Pages 964–982, https://doi.org/10.1093/jopedu/qhad082

the results highlight the importance of enabling conditions for ecosystem collaboration, including neutral coordinating bodies, interoperable data standards, and regulatory spaces that allow learning without compromising safety and accountability.
The chapter also has clear implications for teaching and educational design. Preparing future leaders for sustainability-critical domains requires learning environments that mirror the ambiguity, interdisciplinarity, and coordination challenges of real-world system innovation. Open-ended, design-oriented, and purpose-anchored educational architectures such as those building on the Stanford ME310 tradition and extended through international business perspectives, provide protected spaces in which students can rehearse ecosystem coordination and develop ambidextrous thinking. Education thus emerges as a central lever for accelerating purpose-driven system transformation.
Finally, the chapter points to several avenues for future research. Longitudinal studies are needed to examine how purpose-driven innovation capabilities developed in educational settings translate into organizational practice over time. Comparative research across industries, regulatory regimes, and national contexts could further test and refine the proposed framework. Future work should also examine power asymmetries, exclusion dynamics, and the conditions under which purpose loses—or regains—its coordinating force in complex ecosystems. Together, these directions reinforce the chapter's central insight: advancing circular transformation in the life sciences requires sustained attention to the intertwined dynamics of learning, innovation, and governance dynamics that are likely to become increasingly salient across sectors facing systemic sustainability challenges.
- **The use case "Leadership"** (Chap. 8)—**The Leader as Re-weaver of the World**
  We are not simply experiencing a period of disruption; we are undergoing collective initiation into a whole new way of living. Established models and paradigms are disintegrating, making way for an emergent reality that has yet to fully take form. What is required now is not an improved iteration of the past, but a fundamentally different mode of innovation and way of being that is based on the full acknowledgment on what is serving humanity and their integration into the environment. By shifting from scarcity to interdependence, from separation to sacredness, from mechanistic to life-affirming systems, we reclaim the essence of leadership, not as power over, but as service, not as conquest, but as contribution.
  When leaders take this responsibility to heart, they bring humanity back into the systems they guide. They shift from simple transactions to meaningful, transformational exchanges. They create room for pause, curiosity, creativity, and care. And they recognize that the greatest resource in any organization isn't money or technology, but people who are aware, authentic, and courageous.
  Leading in this way means becoming a re-weaver of the world, someone who recalls what has been forgotten, helps repair what is broken, and dares to imagine what could be. It's a path that is neither easy nor conventional, but absolutely

necessary. And it's not a journey to be undertaken alone, but one joined with the wider flow of life and community.

- **The use case "Swiss Post"** (Chap. 9)
  The case of Swiss Post shows that sustainability transformation requires a shift in how an organization understands its purpose, governs itself, and organizes its operations. Credible change begins with a recognition of the institutional foundations that shape present-day choices—and with the willingness to redirect established trajectories when they are no longer aligned with long-term relevance. This perspective is reflected in the Three-Pillar Model, which highlights three essential capabilities for navigating transformation: a clearly articulated Sustainable Purpose, the adaptability of a Traveling Organization, and the cross-boundary collaboration embodied in Connectivity. Swiss Post's experience illustrates how these capabilities enable an organization to operate under uncertainty while contributing to broader sustainability priorities, including the UN SDGs and the Swiss Sustainable Development Strategy.
  A further insight from the case is the central role of integrated thinking as the connective tissue of sustainability transformation. By aligning financial and non-financial steering, creating shared information systems, and linking resource allocation to long-term value creation, integrated thinking ensures that sustainability becomes embedded in governance and managerial routines rather than remaining a parallel narrative.
  The Climate & Energy case study demonstrates how ambition can be translated into tangible outcomes. Science-based targets, a net-zero 2040 trajectory and operational levers—from fleet electrification to fossil-free heat and Scope3 engagement—show what it takes to move from purpose to execution at scale. A recurring lesson is that progress becomes real only when top management commitment is explicit, continuous, and reflected in governance, decision-making, and resource allocation. This leadership dimension underpins a structured five step model of sustainability transformation: from executive commitment to materiality integration, to science-based goal setting, to integrated steering, and finally to cultural change and partnerships.
  Despite the transformation success illustrated by Swiss Post, the case also implicitly reveals inherent constraints that shape sustainability transformation. Climate & energy benefit from comparatively mature metrics and established standards, whereas social and biodiversity dimensions lack similar methodological clarity. This creates a bias: organizations tend to focus on what can easily be measured and audited while under-addressing areas that are societally important but harder to quantify. In addition, a universal service organization operates within dependencies it cannot fully control. Large-scale decarbonization hinges on the availability of technologies, the readiness of national infrastructure, regulatory developments, and political alignment—factors that determine what is feasible and at what pace. These external conditions make resilience and adaptive sequencing essential components of any credible transition pathway. Sustainability transformation also relies on stable financing and enduring legitimacy. Investments in electrification, buildings, or value chain measures require

long-term commitment, yet their acceptance depends on leadership support, customer acceptance, and political confidence—all of which are sensitive to economic cycles and shifting public sentiment. This creates an inherent fragility even in well-designed transition programs. These critical reflections bring the central tensions in sustainability transformation to the forefront: organizations must pursue bold, long-term change while remaining constrained by what can currently be measured, controlled, and institutionally executed.

Looking ahead, the Swiss Post's case opens two deeper questions that extend far beyond this single organization. First, how can institutions design transformation architectures that remain robust when essential sustainability impacts lie outside their direct control—in value chains, infrastructures, regulatory environments, and societal behavior? Second, what forms of leadership, collaboration, and incentive structures are needed to ensure that sustainability becomes a shared endeavor across ecosystems rather than an isolated corporate journey? Addressing these questions will determine whether organizations can move from transforming themselves to shaping the wider systems on which sustainability ultimately depends.

- The use case **"Circular Economy"** (Chap. 12)

  This chapter examines how circular economy implementation can move beyond pilots and compliance toward durable competitive advantage. Across the conceptual framing on the Three-Pillar Model, the comparison with digital transformation, and the case evidence, one conclusion stands out: circularity becomes economically compelling when it is designed into the core logic of value creation and managed as a transformation with long horizon perspectives—not added as a sustainability layer.

  The comparison with digital transformation highlights four recurring challenges that also shape Circular Economy implementation: business-model uncertainty, delayed investment returns, unclear performance metrics, and cross-boundary coordination needs. Digital transformation addressed these constraints not through technology alone, but through business model redesign, structured experimentation, and governance and finance aligned with long-term value creation. A similar approach might be required for circular economy strategies.

  The case of the German water-meter manufacturer shown in detail provides empirical support for these parallels. Over two decades, the firm embedded circularity into a digitally enabled product-service system by combining design-for-remanufacturing, lifecycle control via take-back, and upgradeable digital architectures. The result was not only ecological benefit but also clear economic outcomes: lower long-term material costs, more stable margins, and greater resilience under supply disruptions. The central mechanism was multi-cycle value capture—enabled by product architecture and reinforced by organizational routines and financing patience.

  Overall, the main barriers to circular economy implementation are less technological than organizational and temporal: misaligned incentives, short-term evaluation logics, and insufficient coordination across ecosystems. When these constraints are addressed—through coherent business model design, experimentation that matches multi-cycle realities, and financing structures that absorb long-term risk—circularity can function as a competitive operating model.

- The use case **"Collaborative Sourcing"** (Chap. 13)
- Sourcing, a key function in both private and public organizations, plays a crucial role in meeting the respective environmental, social, and economic targets of any for-profit and non-profit enterprise. By applying the Three-Pillar approach of organizational development, the need for a new operational model and capability shift in procurement and sourcing has surfaced.

  (1) **Sustainable purpose**: By sourcing intangibles (besides tangibles) for sustainability, the nature of the procurement objects is reshaping the procurement function itself. A more service-dominant portfolio requires a transition from simply buying goods to curating proof (evidence) and crafting a story (narrative) in support of ESG claims. Co-creating and sharing sustainable value propositions in the upstream markets has become a source of competitive advantage in the downstream markets.[34]
  (2) **Traveling organization**: Complex internal and external collaboration requires the blurring of organizational boundaries, proactive end-to-end value chain alignment, smart pooling of resources, and sharing of knowledge.[35] Individual navigation and organizational orchestration are essential skills; only successful intra- and inter-organizational teamwork will make achieving the SDGs dream work.[36]
  (3) **Connected resources**: Working together for environmental, social, and economic impact is dependent on the interconnection and interplay of private, public, and (non) governmental organizations in multi-stakeholder partnerships and extended ecosystems. Proprietary interests and incentives are to be replaced by "creating shared value."[37] Achievements will be measured by independent benchmarking indices, like the Dow Jones Sustainability Index, ultimately pushing for "broader societal interests over narrow corporate interests."[38]

As "two-thirds of the average company's environment, social, and governance footprint lies with suppliers, procurement leaders (…) can make a decisive difference in sustainability."[39] Capitalizing on this opportunity is based on the

[34] Scherer, Jürgen; Marell, Susanne; Wenzl, Jutta: Building a Sustainable Brand in Specialty Chemicals (Carve Out from a Big Corporate), in: Wollmann, Peter; Kühn, Frank; Kempf, Michael; Püringer, Reto (Eds.): Organization and Leadership in Disruptive Times, Cham 2021: 253–265.

[35] Zarei et al., 2023, op cit.

[36] Chauhan et al., 2021, op cit.

[37] Strand, Robert; Freeman, R. Edward: Scandinavian Cooperative Advantage: The theory and practice of stakeholder engagement in Scandinavia, in: Journal of Business Ethics 127 (2015): 65–85.

[38] Strand, Robert: Global Sustainability Frontrunners: Lessons from the Nordics, in: California Management Review, 66/3 (2024): 5–26.

[39] Cherel-Bonnemaison, Celine; Erlandsson, Gustav; Ibach, Ben; Spiller, Peter: Buying into a more sustainable value chain, www.mckinsey.com, 22.09.2021.

successful implementation of a collaborative sourcing approach, as illustrated in Fig. 13.2 of the chapter. A first possible structure for executing collaborative sourcing to meet the 17 SDGs is outlined in Table 13.1 of the chapter. In many of the showcased multi-stakeholder alliances and initiatives, the sourcing and procurement function is the driving force for sustainable business development[40] and is "holding the key to sustainable supply chains."[41]

- The use case "**A Philosophical Deepening of Sustainability Beyond Perception and Performance**" (Chap. 16)

  The sustainability challenges confronting humanity today are not occurring at the margins of civilization. They are occurring at its core. They touch how value is produced, how meaning is constructed, how power is exercised, how institutions are designed, how economies function, and how human beings understand their responsibility to one another and to the world they inhabit.

  This concluding chapter has argued that sustainability cannot be secured through technical progress, regulatory instruments, or institutional innovation alone. Although these remain essential, they operate downstream of far deeper forces. Sustainability ultimately depends on the ontological conditions that shape how systems emerge, how human beings participate within those systems, and how continuity is either preserved or eroded across time.

  When sustainability is treated primarily as an objective, it remains vulnerable to short-term pressure, ideological capture, and performative dilution. In contrast, when it is understood as a systemic property grounded in coherence, integrity, meaning, and regenerative logic, it acquires durability. It becomes capable of enduring complexity rather than merely responding to it.

  The central proposition advanced here is that sustainability is not sustained by reaction. It is sustained by reconstruction. It requires the capacity to redesign systems around first principles that align human intention, institutional behavior, ecological reality, and temporal responsibility within a coherent whole.

  This reconstructive task cannot be delegated solely to governments, corporations, or regulatory bodies. It implicates leadership, education, culture, economic design, and the deeper architectures of human sense-making. It requires a shift from asking what must be defended toward clarifying what must be structurally true for sustainable life to remain possible at all. This responsibility rests with all of us.

  Within this perspective, sustainability is no longer experienced primarily as a restraint on freedom or a burden of regulation. It becomes the organizing logic through which genuine human continuity is made possible. Responsibility ceases

[40] van Hoeven, 2009, op cit.

[41] Hargreaves, Libby: WEF: CPOs hold the key to sustainable supply chains, in: Procurement Magazine, 07.01.2025.

to be experienced as an imposition and becomes a form of participation in the ongoing regeneration of shared systems.

The work developed in *Sustainabilism*[42] approaches sustainability from this reconstructive ontological position. It treats sustainability not as a policy agenda or moral campaign, but as a design condition for living systems. It seeks to restore coherence where modern systems have fragmented meaning, responsibility, integrity, and long-term orientation.

Seen through this lens, the practical frameworks, leadership models, institutional mechanisms, and case studies presented throughout this volume acquire a deeper significance. They are not isolated solutions but expressions of an underlying requirement. Their durability depends on the degree to which integrity, coherence, and meaning are structurally embedded within the systems that apply them.

The Three-Pillar Model provides a powerful operational language for sustainability in practice. What this chapter contributes is the ontological grounding that explains why that language either generates continuity or collapses into performance. Sustainable Purpose endures only when existential and intrapersonal integrity are present. The Traveling Organization remains adaptive only when change does not fracture identity. Connectivity becomes regenerative only when trust, reciprocity, and structural coherence replace fragmentation and transactional logic.

Sustainability, therefore, is not secured by models in isolation. The Three-Pillar Model operates through the underlying nature of the systems to which it is applied. When the orientation of Being, sense-making, and responsibility is misaligned, even the most robust models become vulnerable to dilution, reversal, or simulation. It is precisely at this level that the Authentic Sustainability Framework becomes critical—not as an alternative model, but as the ontological architecture that enables these models to function as intended.

The future of sustainability will not be determined by the sophistication of metrics, the volume of reporting, or the frequency of intervention. It will be determined by whether human beings and the systems they design are capable of holding coherence across time, responsibility across consequence, and meaning across uncertainty.

Where systemic integrity is present, sustainability emerges as a natural property of well-designed systems. Where it is absent, sustainability must be continuously imposed and continuously defended.

This is the deeper horizon toward which this volume points. Not sustainability as a response to crisis, but sustainability as the architectural logic of a civilization capable of enduring with minimal intervention for generations to come.

---

[42] Tashvir, A. (2025) *Sustainabilism: Exposing the Sustainability Illusion*. Engenesis Publications, Sydney.

Urban and rural areas are interdependent and must be governed as a relational system. Treating them as distinct entities generates policy silos that actively hinder sustainable development.

17.5 Some Closing Thoughts[43]

To start with some fundamental statements:

- Since Sustainability is increasingly held hostage in the process of worldwide political polarization,[44] we have to stress that in a mindset of equality of all people, it is in our shared interest to ensure sustainable solutions for the future, regardless of political views.
- We can hope and trust that action and our shared endeavors will expose the politics adverse to a sustainable future for all as a fight analogous to the one of Don Quixote, who is fighting a lost cause when the times had already changed.

The conclusions above cover a broad range of different focus topics and show different facets of the potential of intellectual capital in the global transformation endeavor to cover the UN SDGs and reach a sustainable development of the world. It is evident that even though some impressive concepts and methods are already in place, there exists no overarching recipe to solve the individual problems and challenges. There can be no "one size fits all"—also because we are breaking completely new ground in many areas. It turned out that the Three-Pillar Model as a meta-level thinking style and mental setting is in all cases helpful as it does not provide recipes but rather an attitude on how to approach challenges regarding the ever-unknown territories of change and the future.

Peter Wollmann has been a responsible manager, initiator, mentor, and facilitator in large, predominantly global transformations and strategic developments for almost 40 years in large organizations and since 2017 as independent consultant. He is also a member of an advisory board of a R&D oriented company. Peter holds a degree in mathematics and physics from the University of Bonn. He is the author and publisher of numerous books and articles on strategy, leadership, organization design and project and project portfolio management.

[43] Identical to the respective section in volume 2.

[44] We kept this statement unchanged in volume 3 even though it seems as if the conflicts in Middle East make many things appear in a different light and surprisingly give sustainability topics a new momentum additional to the long-term momentum it has anyway from the often-mentioned system logic reasons.

Doris Pemler is an entrepreneur, artist, and researcher based in Zurich. She received her M.A. in Egyptology, Sudan Archaeology, and Archaeology of the Ancient Near East from Humboldt University zu Berlin and FU Berlin. Her research focuses on iconography, gender, and intercultural communication in Ancient Egypt and Sudan. As a photographer, she contributed to several exhibitions in Germany and Switzerland. As CEO of Ciudad Condal GmbH, she leads an aspiring company focused on import and distribution of high-end sustainable products in Switzerland.

Mersida Ndrevataj is an urban scholar with a background in architecture, urban planning, and environmental psychology. She holds a PhD in Urban Planning and Public Policy from the IUAV University of Venice, where her research focused on the implementation of public policies in cultural heritage management. She is currently a lecturer at the IUAV University of Venice and collaborates with several COST Action research networks focused on urban studies. Her work brings together academic research and field-based inquiry by adopting audiovisual and sensory approaches to investigate urban spaces, everyday socio-spatial practices, and lived experiences.

FSC
www.fsc.org
MIX
Papier | Fördert gute Waldnutzung
FSC® C083411

Kolibri 360 GmbH
Ferdinand-Jühlke-Straße 7
99095 Erfurt, Deutschland
produktsicherheit@kolibri360.de